P9-DVY-698

Studies
in the History of Mathematics and
Physical Sciences

4

Editors
M. J. Klein G. J. Toomer

C. TRUESDELL

THE TRAGICOMICAL HISTORY OF THERMODYNAMICS 1822–1854

With 8 Figures

Springer-Verlag
New York Heidelberg Berlin

Clifford Ambrose Truesdell, III
Latrobe Hall 119, The Johns Hopkins University, Baltimore,
Maryland 21218, USA

AMS Subject Classification: 01A55, 80–03

Library of Congress Cataloging in Publication Data. Truesdell, Clifford Ambrose, 1919– . The tragicomical history of thermodynamics, 1822–1854. (Studies in the history of mathematics and physical sciences; 4.) Bibliography: p. . Includes indexes. 1. Thermodynamics—History. I. Title. II. Series. QC311.T83 536′.709 79–11925

Printed in the United States of America

9 8 7 6 5 4 3 2 1

ISBN 0–387–90403–4 Springer-Verlag New York Heidelberg Berlin
ISBN 3–540–90403–4 Springer-Verlag Berlin Heidelberg New York

To

Bernard D. Coleman,

principal architect of rational thermodynamics

Acknowledgment

The first draught of this book, a single lecture, was written early in 1970. Many men have helped me in many ways during the years of perplexity and frustration the work has cost me since then—so many that I fear my memory cannot muster them. Specific acknowledgments in regard to details will be found at certain points in the text. I thank Professor E. DAUB for some helpful observations in 1970/1971. Mr. STEPHEN WINTERS discussed and criticized much of the text as it evolved in 1973/1977; he checked most of my translations and references and brought to my attention some sources I might otherwise have overlooked. Collaboration with Mr. BHARATHA in 1974/1975, which resulted in the book often cited below as *Concepts and Logic*, itself an outgrowth of the studies leading to this one, provided essential understanding without which I could never have released for publication any work with the title under which I present my results now. He has also criticized some passages in the text of this book. Conversations and correspondence with Mr. SERRIN, mainly in 1978/1979, have added much to my understanding of the essential issues of thermodynamics itself, of some of the ideas of KELVIN, CLAUSIUS, and GIBBS, and of the specific defects in much of the later work, especially CARATHÉODORY's. I owe most to Mr. CHI-SING MAN, whose untiring perseverance in searching out, probing, and perpending the sources major and minor is matched by his gentle refusal to let pass any vagueness that can be cleared, any gap in the record that can be closed. Word by word and equation by equation, he scrutinized the final manuscript and the galley proofs with a devotion seldom spent on the work of another. At several points in the text specific attributions to him will be found, and he is the author of the appendix to § 9D. Finally I thank the other members of my seminar in the spring of 1978: Messrs. C. DAVINI, RICHARD JAMES, and M. PITTERI, for their critical analysis of some central aspects of the work of the most important and most difficult authors: FOURIER, CARNOT, KELVIN, RANKINE, CLAUSIUS.

As for all my work since 1960, I gratefully acknowledge the support of the U.S. National Science Foundation. A grant from its program in the History and Philosophy of Science supported directly the research presented here. Grants from the programs in Applied Mathematics and Solid Mechanics enabled me to pursue in the context of current research such old ideas as seemed to deserve new development, the results of which may be seen in my recent papers on the efficiency of irreversible engines, absolute temperatures, and the teaching of thermodynamics to undergraduate engineers. Thus past and present have again, as often before, illuminated each other.

"Il Palazzetto" C. TRUESDELL
Baltimore
March 1, 1980

Contents

Mon but n'a jamais été de m'occuper de
ces matières comme physicien, mais seulement
comme logicien . . . .

F. Reech [1856, pp. 65–66, footnote].

The historical development of thermodynamics
has been . . . particularly susceptible to
logical insecurity, . . . and there has been
no adequate reexamination of the fundamentals
since.

Bridgman [1953, p. 226 of the edition of 1961].

Buio d'inferno e di notte privata
d'ogne pianeto, sotto pover cielo,
quant' esser può di nuvol tenebrata
non fece al viso mio sì grosso velo
come quel fummo ch'ivi ci coperse,
né a sentir di così aspro pelo,
che l'occhio stare aperto non sofferse . . . .

Dante, *Purgatorio* XVI, 1–7.

1. The Producer's Apology to the Spectators

We are not such optimists as were our teachers and parents. We do not have to equate "progress" with every $\delta f(t)$ if $\delta t > 0$, t being the time. In discussing the interplay between mathematics and physics[1] I feel myself permitted, therefore, to select instead of fields of brilliant success like hydrodynamics, elasticity, and electromagnetism, one accursed by misunderstanding, irrelevance, retreat, and failure. Thus I write of thermodynamics in the nineteenth century. No-one will be surprised, consequently, by my use of a delta to define progress, since thermodynamics is the kingdom of deltas. However, the single δ just used will suffice. In return, I bring the time back into its rightful, central place—a place it occupied at the start but from which it was wrongly driven by late authors who confused dynamics with statics.

Thermodynamics is the kingdom also of running current history as well as polemics, not to mention verbosity. In no other discipline have the same equations been published over and over again so many times by different authors in different ill-defined notations and therefore claimed as his own by each; in no other has a single author seen fit to publish essentially the same ideas over and over again within a period of twenty years; and nowhere else is the ratio of talk and excuse to reason and result so high. In no other part of mathematical physics have so many claims and counterclaims of priority been issued by the leading creators of the subject, and in no other have these same men turned aside from research to write historical papers or long historical notes within a decade or two of their first attacks on the theory itself. Small wonder then that histories and historical papers by secondary authors and historians abound, yet the field seems ever fresh to the newcomer.

Only now could a real history of thermodynamics be written, since only in the last twenty years have the expressed aims of the creators of thermo-

[1] This essay began as an hour's lecture for the symposium on "The Interplay between Mathematics and Physics in the Nineteenth Century" held at Aarhus in August, 1970. I am grateful to Professor OLAF PEDERSEN for having invited me to take part in that symposium and for having released me from my obligation to publish my lecture in its proceedings. I was unable to do so because my text was not then ready for the press. The comments upon the lecture had shown me that only an exhaustive, fully documented treatise might make today's reader, whether scientist or historian, come to see that logic and clean mathematics had a place—indeed, a place mainly left vacant—in classical thermodynamics.

dynamics been achieved. In blunt terms, only now do we know a decent theory of the scope the creators sought, so only now can we see just where the old authors stopped short or even went wrong. While this remark of mine may shock some by its quaintness, it ought not. However much it is the vogue nowadays to pretend that always everybody was just as right as everybody else, or that truth in science is no more than the vote of some time-dependent minority, even the staunchest proponents of the "new" history when protesting adherence to ancient innocence adjoin footnotes, expressed in very modern English, in which they compare old science with that currently received (for otherwise their subject of study might not be recognized), and if they do refer to the correct answers, they put "correct" in quotation marks. In what follows, such quotation marks may be imagined set around such few "corrects" as may be found, despite my intent to banish them along with the useless δs. Nevertheless, much of what I write now about the classical papers on thermodynamics I could not have written twenty years ago, because I did not then have the grasp of rational thermodynamics that today we may and do teach our beginning students. This knowledge does not change the historical record one whit; rather, it teaches us to read it better.

This essay is a conceptual analysis: I aim not only to recount but also to marshal. I will outline the assumptions and logic, pointing out the abundant vagueness of the former and the scarcity of the latter, in the major early works of thermodynamics. The contents of the minor works will be mentioned where they belong: in interludes and—of course—footnotes, a writer's most dull and deadly proofs that he is not an author but a scholar. The text presents the tragicomedy entire, as produced. The spectators will not see the footnotes, which a scholiast has provided, as scholiasts will do, for the edification of other scholiasts.

Among the reasons for which my first and short draught[2] was criticized was the unhistorical character it showed in those passages where I applied my own reason to certain early equations. The blemish of thinking has been largely removed by my subsequent discovery (no surprise to me) that most of my logical observations had been made already by one or another early writer. Thus the very same observations are now become (I trust) respectable history instead of "present-mindedness".

---

[2] That draught, somewhat revised and extended, was delivered in three lectures at Udine in June, 1971, and has been published as *The Tragicomedy of Classical Thermodynamics* (1971), International Centre of Mechanical Sciences, Udine, Courses and Lectures, No. 70, Wien and New York, Springer-Verlag [1973], 41 pp. I take this occasion to remark that the text of that pamphlet is no more than a preliminary sketch toward a part of this essay. I complied with the International Centre's requirement that I hand over the manuscript of my lectures to be duplicated and sent gratis to a small list of interested persons. Its publication two years later as a separate work for sale by a commercial publisher was without my consent or even knowledge beforehand.

Meanwhile, moreover, I have found it possible to organize CARNOT's general ideas axiomatically and develop them through a mathematical analysis both rigorous and exhaustive, to the point that CARNOT's particular theory and CLAUSIUS' appear as special, mutually exclusive cases within a general scheme, and the possibilities and limitations of each and every statement in either theory are laid bare. This analysis, pro-historical in character, appears in the tractate by Mr. BHARATHA and me, *The Concepts and Logic of Classical Thermodynamics as a Theory of Heat Engines, Rigorously constructed upon the Foundation laid by S. Carnot and F. Reech*, New York, Springer-Verlag, 1977. Although in that tractate we scrupulously limit our mathematics to *what was widely available in the 1820s*, so our analysis calls upon nothing but what the pioneers themselves might have used, had they mastered the mathematics of their own day, in this essay I will not repeat it but will rest content to direct the reader to appropriate passages of that tractate, which I shall cite as "*Concepts and Logic*".

I shall refer only to the published sources, and roughly in the order they appeared. Such callousness not only to the ever-widening alluvium of secondary literature[3] on minutiae but also to the infinite subtleties of the withheld and the rejected, may seem equally quaint in the modern hives of scholarship. My reason, again, is a blunt one: While chaste and laconic if not secretive private intercourse was a major channel of creative science in the seventeenth century, the mass of notebooks and letters of the abundantly public Victorian era should remain the province of biographers and doctorands.

This essay was written for students of science and for the creative and critical young thermodynamicists of our day. Should any Historian of Science chance upon it, he would do well to omit all sections labelled "critique" and all words confined between square brackets, for in that way he will save himself such pain as my "ahistorical" approach might otherwise inflict. A reader whose interest lies in applicable analysis of scientific method, on the other hand, may find my ahistorical moralizing of greater worth than the often tedious involutions which the strictly narrative parts trace and abstract. A serious student who makes the effort needed to follow the analysis, line by line and proof by proof, will need no previous acquaintance with thermodynamics. The historical method is not the easiest way to learn a science; neither is it the worst.

[3] Secondary literature, whether old or new, raises a difficult question. While making no attempt to search it, I confess to having consulted some of it. I wish I had not, for in most cases it led me into sociology and protophysics and historiography and away from history of science: the analysis of specific concepts in their historical origins and settings. In order to do justice to the secondary literature, I should have to read more of it; if I cited it, I should have to do so largely in contest rather than credence; therefore, in regard to the central theme of this essay I have decided not to cite at all what little secondary matter I have scanned. For neighboring domains and periods earlier or later I cite with gratitude a number of studies by others.

Finally, I confess to a heartfelt hope—very slender but tough—that even some thermodynamicists of the old tribe will study this book, master the contents, and so share in my discovery: *Thermodynamics need* never *have been the Dismal Swamp of Obscurity* that from the first it was and that today in common instruction it is; in consequence, it need not so remain.

> Ben puoi veder che la mala condotta
> è la cagion che 'l mondo ha fatto reo,
> e non natura che 'n voi sia corrotta....
> "Drizza," disse, "ver' me l'agute luci
> de lo 'ntelletto, e fieti manifesto
> l'error de' ciechi che si fanno duci...."
> Ma quello ingrato popolo maligno...
> ti si farà, per tuo ben far, nimico;
> ed è ragion: ché tra li lazzi sorbi
> si disconvien fruttare al dolce fico.
> Vecchia fama nel mondo li chiama orbi,
> gent'è avara, invidiosa e superba;
> dai lor costumi fa che tu ti forbi.
>
> DANTE, *Purgatorio* XVI, 103–105; XVIII, 16–18; *Inferno* XV, 61, 64–69.

Notation

I. *Letters for quantities.* The letters chosen by the early authors to stand for temperature, heat, *etc.* differed from one to the next. So as not to lay a pointless burden on the reader who would follow the analysis (for I desire no other), I adopt a single set of letters once and for all. These, although mainly ones used by some or another early author, are selected so as to conform pretty nearly with those current in rational thermomechanics today, which also refers explicitly to the time t.

Even in quoted passages I shall for the most part silently reduce the original notation to that of this essay.

II. *Relations.* The symbol $\equiv$ is to be read "is defined as".

III. *Functions and derivatives.* Classical thermodynamics considers many different functional relations among triples of variables. The physical interpretations of the values of these functions need to be kept in mind continually. For this reason the same letter serves well to denote both a function and its value, or two different functions having the same value at corresponding arguments, whenever such can be done without danger of confusion. For example, if

$$p = \varpi(V, \theta) = \varpi^*(\rho, \theta) \ ,$$

we shall usually write

$$\frac{\partial p}{\partial V} \quad \text{and} \quad \frac{\partial p}{\partial \rho} \ ,$$

respectively, for the functions

$$\frac{\partial \varpi}{\partial V} \quad \text{and} \quad \frac{\partial \varpi^*}{\partial \rho} \ ,$$

the advantage being that the letter p recalls "pressure". Moreover, usually we shall use p to denote not only the pressure itself but also that function of time whose value at the time t is the pressure, but when confusion might otherwise result, we shall write $p(t)$ for that value.

Although differentials occur frequently in the early literature, I prefer the explicitness gained by employing derivatives. Use of the ordinary notations of calculus should suffice to set aside the strange superstition that thermodynamics has a mathematics all of its own[1]—a prime example to show that physicists are not exempt from the madness of crowds[2].

Symbols Frequently Used

Symbol	Name	Page on which introduced
c	speed of sound	13
C, C^+, C^-	heat added, absorbed, emitted	15, 25
F	function in CARNOT's Special Axiom	102
G	function in CARNOT's General Axiom	101
J	mechanical equivalent of a unit of heat in various circumstances	128, 150, 157, 159, 189
L	work done	24
p	pressure	9
P	power	192
Q	heating	15
R	gas constant	9
V	volume	9
γ	ratio of specific heats	24
E	internal energy	71, 192
H	entropy	214, 223
$\mathrm{H}_{\mathrm{L}}, \mathrm{H}_{\mathrm{C}}$	LAPLACE's and CARNOT's heat functions	35, 85
θ	ideal-gas temperature	9
$\mathrm{K}_V, \mathrm{K}_p$	specific heats at constant volume and constant pressure	16, 22
Λ_V, Λ_p	latent heats with respect to volume and pressure	16, 22
μ	"CARNOT's function"	111
π	pressure function	12
ρ	mass-density	12
τ	KELVIN's first abolute temperature	171, 308
T	KELVIN's second absolute temperature	309

Citations in square brackets refer to the list of sources printed at the end of the book.

---

[1] *Cf.* BRIDGMAN [1941, p. 4 of the 1961 ed.]: "an unfamiliar brand of mathematics". For a specimen note the common "Second Law" $TdS \geqq \delta Q$, which would have us believe not only that one differential can be bigger than another but also that a multiple of a differential can be bigger than something that is not a differential.

[2] *Cf.* CHARLES MACKAY, *Memoirs of Extraordinary Popular Delusions and the Madness of Crowds*, London, 1841, revised 1852, many times reprinted. See especially "The Witch Mania" and "Relics" in Volume II.

2. The Common Inheritance

la fama nostra il tuo animo pieghi
a dirne chi tu se', che i vivi piedi
così sicuro per lo 'nferno freghi.
DANTE, *Inferno* XVI, 31–33.

2A. The Thermal Equation of State

On the basis of data from experiments regarding the compressibility of air at constant temperature collected by BOYLE and interpreted by TOWNELEY and POWER, and of data from experiments at constant volume obtained and interpreted by AMONTONS and many later experimenters, especially GAY-LUSSAC and DALTON, by 1820 it was generally agreed that the *pressure* p, the *volume* V, and the *temperature* θ of a *body* of aeriform fluid at rest obeyed the relation

$$pV = R\theta, \qquad R = \text{const} \,. \tag{2A.1}$$

Here the zero of the temperature θ is suitably selected[1].

---

[1] In taking advantage of the convenience of what is now called a temperature measured from "absolute zero", I do not violate historical truth. The early authors, selecting θ_0 as some particular temperature, usually wrote $\theta_0 + \theta$ or $\theta_0(1 + \theta/\theta_0)$ for what I here call θ; EULER left θ_0 arbitrary. I do not mean, of course, that all early authors used the letter θ to denote the temperature. That letter was used by POISSON, by FOURIER in his last work [1833], and by MAXWELL in his papers on the kinetic theory. I have followed this usage of theirs in my own research since 1948, and it is nowadays standard in the literature of rational thermodynamics.

Anyone who looks at (1), no matter what the notation in which it be written, sees that it implies the existence of an "absolute cold", at which the product pV vanishes. Since AMONTONS (1703) was the first to be able to see a relation equivalent to (1), we should not be surprised that it was he who first suggested that there was an "absolute cold". Some early authors regarded the existence of such a temperature as thereby proved, while others regarded the conclusion as ridiculous and hence interpreted the relation (1) as valid only for sufficiently high temperatures. The story is recounted by W. E. KNOWLES MIDDLETON in §§6–7 of Chapter IV of his *A History of the Thermometer and Its Use in Meteorology*, Baltimore, The Johns Hopkins Press, 1966. Converting

Throughout the period of this history, partly on the basis of more accurate experiments and partly in consequence of the evolution of theoretical beliefs, people's conception of the "gas laws" will change, and (1) will be interpreted differently by different authors. In general, all this is of slight moment for the development of thermodynamics, but some knowledge of it helps us understand what various early authors presume of their readers. A brief history of the "gas laws" in the eighteenth century has been written by FOX[2]; the tables on his pp. 324–326 show that experiments long failed in effect to yield a consistent value for the pressure coefficient $\alpha = (p_s - p_i)/100p_i$ (V = const.) or for the volume coefficient $\beta = (V_s - V_i)/100V_i$ (p = const.), where the subscripts s and i denote the steam point and the ice point, respectively.

For the following further facts I am indebted to Mr. C.-S. MAN. The experiments of DALTON and even more those of GAY-LUSSAC, some of them[3] published in 1802 and others done probably before 1805 but first described in print by BIOT[4] in 1816, are epoch-making in the sense that they eliminated the source of inconsistency in the work of their predecessors, namely[5] "the presence of water in the apparatus", and produced results consistent enough to let GAY-LUSSAC conclude[6], "All gases, whatever their density or the quantity of water which they hold in solution, and all vapors expand to the same extent for the same degree of heat." The experiments of DULONG & PETIT, which in part dealt with thermometry, were published in 1816 and 1817. One of their conclusions was taken as confirming GAY-LUSSAC's rather than DALTON's form of the law of dilatation, so GAY-LUSSAC's value for the

---

the data to the modern centigrade scale, he reports the following values of "absolute cold":

AMONTONS (1699): $-248°$
LAMBERT (1779): $-270°$
REGNAULT (1847): $-272.75°$
RANKINE (1853): $-274.6°$

The equation of state used by CARNOT corresponds to absolute cold at $-267°$; that used by CLAUSIUS, to $-273°$. MIDDLETON reports also other values ranging from $-1250°$ to $-853°$, some of them attached to great names in physics and chemistry.

The "absolute cold" remained a concept for philosophers, chemists, *etc.*, until the kinetic theory afforded a mechanical model which gave it conceptual concreteness in a major special case. The early history of "absolute cold" on this basis may be read in my "Early kinetic theories of gases", *Archive for History of Exact Sciences* **15**, 1–66 (1975).

For the general theory, I repeat, all this makes no difference.

[2] R. FOX, *The Caloric Theory of Gases from Lavoisier to Regnault*, Oxford, Clarendon Press, 1971; see especially pp. 61–67. *Cf.* also GAY–LUSSAC [1802, §II].

[3] GAY-LUSSAC [1802].

[4] BIOT [1816, *1*].

[5] GAY-LUSSAC [1802, p. 141].

[6] GAY-LUSSAC [1802, pp. 174–175].

volume coefficient of expansion at constant pressure became standard for some time: $\beta = 0.00375$. Thus, if we write $\theta_0 + \theta$ or $\theta_0(1 + \theta/\theta_0)$ instead of θ in (1), LAPLACE and POISSON took $1/\theta_0$ as 0.00375, CARNOT took θ_0 as 267°C; the θ here in °C was to be measured by an air thermometer. LAPLACE used his Caloric Theory to justify (1) and the use of the air thermometer for measuring θ.

In experiments published in 1842 both MAGNUS and REGNAULT found that α varied from one gas to another. REGNAULT found the same to be true of β as well; also for a given gas he observed that α varied with the density, β with the pressure. However, the results of his experiments led REGNAULT[7] to conjecture at the end of his paper that the law (1) and "all those which have been discovered for gases, such as the law of volumes, etc., are true *at the limit*, that is, that they come nearer to conforming with the results of observation in proportion as we use the gas in a more expanded condition. These laws hold good for a perfect gaseous state, which the gases that nature places before us more or less approach according to their chemical characteristics, according to the temperature at which we study them..., finally and above all, according to their condition of less or greater compression."

From that time until the end of the period treated in this history, different authors are to adopt different attitudes towards (1). For those like RANKINE and CLAUSIUS, who will have molecular models in mind, (1) will hold for a "perfect gas" and θ in (1) will be some kind of "absolute temperature" defined in terms of the molecular motions. For others, (1) will hold only approximately for most gases in a certain range of pressure and temperature, and θ in (1) may be defined differently for different gases. For example, in the law (1) written in the form $pV = R\theta_0(1 + \theta/\theta_0)$ HOLTZMANN will use MAGNUS's value of α as $1/\theta_0$ for air, and for various vapors values such as to make his theory fit experimental data with θ counted from the boiling point. In his early papers KELVIN will take REGNAULT's α for his standard air thermometer as $1/\theta_0$ and will regard θ in $\theta_0(1 + \theta/\theta_0)$ as to be measured by that thermometer; it is in this sense that we shall have to understand KELVIN's interpretation of θ in (5N.7), below.

The several creators of thermodynamics generally will name (1) or its special cases, variously interpreted as we have just explained, after BOYLE, MARIOTTE, DALTON, and GAY-LUSSAC.

In theoretical studies EULER had used (1) from 1757 onwards and had explained it in detail in his treatise on fluid mechanics, the relevant part of which was published in 1777. This work became more widely known through an annotated German translation[8] published in 1806. Much earlier, EULER and DANIEL BERNOULLI had projected kinetic theories which delivered (1) only as an approximation for high densities; the latter reported experimental deviations from (1); and other geometers of the eighteenth century

---

[7] REGNAULT [1842, p. 83].

[8] BRANDES [1806].

followed in regarding (1) as only roughly valid. EULER preferred to use a more general functional relation

$$p = \varpi(V, \theta) > 0 \ , \tag{2A.2}$$

which he thought appropriate to any condition of motion of any substance that could be regarded *fluid*. In this essay I follow the usage of calling (2) the *thermal equation of state* of a fluid body, and ϖ the *pressure function* of that body. The symbols $\partial p/\partial V$ and $\partial p/\partial \theta$ shall stand for the partial derivatives of ϖ, assumed to exist and to be continuous functions of (V, θ). Sometimes I will follow the confusing custom of books on physics and write p for the function ϖ.

The special equation of state (1) is to bulk large in early writings on thermodynamics. Often the pioneers' appeals to it were unnecessary. To distinguish general ideas and reasoning from essentially irrelevant uses of (1), in this essay I will mainly use the general equation of state (2). The particular body of fluid defined by (1) I shall call the body of *ideal gas* having the *constitutive constant R*. The pressure function ϖ of the body of ideal gas is $R\theta/V$.

If the mass of a homogeneous fluid body at rest is M, we can define the density ρ as usual, $\rho \equiv M/V$, and express p as the value of a function ϖ^* of density and temperature:

$$p = \varpi^*(\rho, \theta) \ . \tag{2A.3}$$

Then, since M is constant,

$$\rho \frac{\partial p(\rho, \theta)}{\partial \rho} = -V \frac{\partial p(V, \theta)}{\partial V} \ . \tag{2A.4}$$

While (2) makes no sense in a field theory like hydrodynamics, (3) does, and the researches on aeriform fluids by EULER, LAGRANGE, LAPLACE, and POISSON adopt (3) as an *a priori* relation between the fields of pressure, density, and temperature. The function ϖ^* is appropriate to a material, while ϖ is appropriate to a body. For an ideal gas $\varpi^* = r\rho\theta$. All the early students knew AVOGADRO's hypothesis, published in 1811, which makes r inversely proportional to the combining mass of the substance: $r = k/m$, and if m is chosen correctly for each substance, k is a universal constant. However, this fact will play little or no part in the development of thermodynamics, because AVOGADRO's hypothesis was not generally accepted during the period before 1860.

Early studies of heat always presumed, tacitly if not expressly, the following *constitutive inequalities*:

$$\frac{\partial p}{\partial V} < 0, \qquad \frac{\partial p}{\partial \theta} > 0 \ . \tag{2A.5}$$

The former asserts that the volume of a body must decrease if pressure is applied isothermally, and that isothermal increase in volume requires a decrease of pressure. It is equivalent to

$$\frac{\partial p}{\partial \rho} > 0 , \tag{2A.6}$$

which implies that sound may propagate in a body of fluid, no matter what be its density and temperature. The latter, taken together with the former, implies that a fluid contracts as the temperature is decreased at constant pressure. Such is not always the case for some fluids, the most familiar being water, which at atmospheric pressure expands when cooled below 4°C. Although this "anomalous behavior" of water was well known, it is not mentioned in any early work on thermodynamics[9]. Accordingly, we shall presume that $(5)_2$ as well as $(5)_1$ holds until there is reason to consider the contrary possibility, namely in our terminal year, 1854, for only in that year, as we shall see in §9F, will the "anomalous" behavior of water find a place in thermodynamics.

The two inequalities (5) together imply that (2) may be inverted locally for V or θ, as was commonly assumed in the early studies.

2B. The Theory of Sound in Aeriform Fluids

NEWTON's imaginative and semirational theory of sound had presumed the sonorous vibrations subservient to the TOWNELEY-POWER-"BOYLE" law $p = N\rho$ and had concluded that the speed of sound c must satisfy the relation $c^2 = N = p/\rho$, a result abundantly contradicted by experiment for the next 100 years. NEWTON's successors had considered the more general possibility $p = p(\rho)$ and had obtained the famous formula

$$c^2 = \frac{dp}{d\rho} . \tag{2B.1}$$

Adopting a general equation of state (2A.3), they had assumed further that in sonorous motion the temperature was everywhere and always the same. They had found that

$$c^2 = \frac{\partial p}{\partial \rho} . \tag{2B.2}$$

---

[9] Although most modern textbooks mention the "anomalous behavior" of water, the formulae they use to discuss it are derived from considerations based tacitly upon use of $(5)_2$, which excludes it. To derive classical thermodynamics from classical ideas but without prejudice of the sign of $\partial p/\partial \theta$ is no easy matter. It provided a major barrier to the program of *Concepts and Logic*.

Further remarks on the "anomalous behavior" of water will be found below in §4D and §9F.

Thus for an ideal gas they fell back inescapably upon NEWTON's result. The painstaking researches of EULER[1] and others had set aside, one by one, various suggested causes of the discrepancy: loose mathematics, impurities in the fluid, the shape of the wave front, the amplitude of the disturbance. While the analysis does not always convince a modern reader, particularly in the last regard, the main conclusion is correct, as was to be shown with finality by HUGONIOT[2] a hundred years later: The only way to square theory with experiment is to admit that although the TOWNELEY-POWER-"BOYLE" law $p = N\rho$ is confirmed at least roughly for a gas in equilibrium, it cannot be valid in sonorous oscillation. As LAGRANGE had remarked, we may simply assume that $p = C\rho^{1+k}$ and then determine the positive constant k so as to make (1) fit the measured speed, but, in accord with his strictly algebraic approach to mechanics, he could give no conceptual reason for this cheap if prophetic trick.

Various researches in the eighteenth century made it abundantly plain that (2A.1) and (2A.2) could not apply to solids. For them, neither is there in general any natural concept of a single, scalar pressure, nor does change of volume furnish an adequate description of change of shape. Since all early work on thermodynamics presumes (2A.1) or (2A.2), it applies mainly to fluid bodies.

---

[1] Most of EULER's discoveries in the theory of vibrations are traditionally attributed to LAGRANGE, LAPLACE, or RAYLEIGH. I have written the history of the matter: "The theory of aerial sound, 1687–1788," L. EULERI *Opera Omnia* (II) **13**, Zürich, Füssli, 1956, pp. XIX–LXXII, and "The rational mechanics of flexible or elastic bodies, 1638–1788", L. EULERI *Opera Omnia* (II) **11**$_2$, Zürich, Füssli, 1960.

[2] For a modern treatment one may refer to p. 712, especially footnote 4, of FLÜGGE's *Encyclopedia of Physics* III/1, Berlin *etc.*, Springer, 1960. In any motion such that, for whatever reason, the pressure at the typical fluid-point X is given by $p = f(\rho, X)$, then $\dot{p} = (\partial f/\partial \rho)\dot{\rho}$, the dot denoting the material time derivative, so according to HUGONIOT's theorem

$$c^2 = \frac{\dot{p}}{\dot{\rho}} .$$

This form, which follows trivially from (1) if f does not depend upon X, will be used below repeatedly and without further comment.

The "material time derivative" is the derivative with respect to t when X is held constant. A celebrated formula of EULER, which was known by every serious student of mechanics and mathematical physics in the period with which this essay deals, expresses $\dot{f}$ in terms of field derivatives:

$$\dot{f} = \frac{\partial f}{\partial t} + \mathbf{v}\cdot \operatorname{grad} f \ ;$$

$\mathbf{v}$ is the spatial velocity field, and the f on the right-hand side stands for that function of place $\mathbf{x}$ and time t whose value is $f(X, t)$, that f being the one that appears on the left-hand side. Any reader not familiar with hydrodynamics should consult some standard treatment of the subject, *e.g.* §5 of H. LAMB's *Hydrodynamics*, Cambridge University Press, 2nd ed., 1895, or any later edition or reprint.

2C. The Doctrine of Latent and Specific Heats

Toward the end of the eighteenth century calorimetry became a regular branch of experimental physics. Units of heat were introduced, and the amounts of heat needed to produce specified increases of temperature, volume, or pressure under various conditions were measured. Since in all mathematical calculations, however elementary, the increments were assumed to be related by the rules of differential calculus, and since all natural changes occur as time elapses, these increments make sense only if they are referred to time rates. So as to avoid the needless and trivially obviable obscurity of the early work, as such I will express them.

A "process"[1] is the assignment of V and θ as positive functions of time:

$$V = V(t) > 0, \qquad \theta = \theta(t) > 0 \ . \tag{2C.1}$$

By (2A.2), p becomes the value of a function of time, namely, $p(t) = \varpi(V(t), \theta(t))$; thus, when the process is smooth,

$$\dot{p} = \frac{\partial p}{\partial V}\dot{V} + \frac{\partial p}{\partial \theta}\dot{\theta} = \frac{\partial p}{\partial \rho}\dot{\rho} + \frac{\partial p}{\partial \theta}\dot{\theta} \ . \tag{2C.2}$$

The special case in which $\theta(t) = \text{const.}$, which today we call an *isothermal* process, in the early literature was again and again mentioned or assumed to hold.

Let Q denote[2] the *heating*, namely, the function of time whose value is the time-rate at which heat is put into a given body. Then the *heat added* C between the times t_1 and t_2 is given by

$$C \equiv \int_{t_1}^{t_2} Q(t)dt \ . \tag{2C.3}$$

Q is assumed only to be a function integrable in the sense of EULER and CAUCHY. Thus it may fail to exist at a finite number of times in the interval $[t_1, t_2]$.

Early students used the term "heat" vaguely. What they meant is not always clear. Sometimes they referred to "the total heat" of a body or in a body, which in some cases may be regarded as something like internal energy or total energy or even equivalent to one of those. The concepts of "heating" and "heat added" as related by (3) suffice to make sense of the way the term "heat" was used in early works on thermodynamics proper. Accordingly we

---

[1] CARNOT [1824, pp. 10, 19, *et passim*], une opération.

[2] The notation Q for the heating is used in papers on rational thermodynamics today; in older works it would have been written as dQ/dt or $\delta Q/dt$ or $\Delta Q/dt$, had the authors chosen to indicate that an increment of heat is always associated with an increment of time, as of course it must be in nature.

shall stay with (3) throughout this work except in one or two passages where it will be necessary to depart from it briefly.

All the pioneers of thermodynamics assumed that in every process the heating Q would equal a linear function of the rates of increase of volume and temperature, with coefficients which were functions of V and θ only and hence independent of the process. That is, at all times when $\dot{V}$ and $\dot{\theta}$ exist[3],

$$Q = \Lambda_V(V, \theta)\dot{V} + \mathrm{K}_V(V, \theta)\dot{\theta} \ , \tag{2C.4}$$

Λ_V being called the *latent heat with respect to volume*[4] and K_V the *specific heat*[5] *at constant volume*. The coefficients Λ_V and K_V were assumed to be positive functions[6]:

$$\Lambda_V > 0, \qquad \mathrm{K}_V > 0 \ , \tag{2C.5}$$

---

[3] This basic assumption is implicit in *all* of the works we shall review in this essay except FOURIER's and DUHAMEL's. Both CARNOT and CLAPEYRON (see Chapter 5 and §6A below) assumed that there was a heat function $\mathrm{H_C}(V, \theta)$, so for them

$$\Lambda_V = \frac{\partial \mathrm{H_C}}{\partial V}, \qquad K_V = \frac{\partial \mathrm{H_C}}{\partial \theta} \ .$$

CLAUSIUS [1850, §1], who rejected $\mathrm{H_C}$, wrote dQ/dV and $dQ/d\theta$ for what we here denote by Λ_V and K_V; he called $\int Q dt$ "the quantity of heat which must be communicated to a gas . . ." and explained that it was not a function of volume and temperature (v and t in his notation). THOMSON [1849, footnote to §26] introduced the symbols M and N for Λ_V and K_V, respectively, but he then regarded them as being $\partial \mathrm{H_C}/\partial V$ and $\partial \mathrm{H_C}/\partial \theta$. When, later [1851, *1*, §20], he discarded this unnecessary restriction, he continued to use M and N, thus abstaining from CLAUSIUS' double talk with notations. Coming at last to an analysis sufficient to compare the proposals of different theorists in terms of a common framework, THOMSON [1852, *1*, §63] described reasoning based on (2A.2) and (4) alone as being "without any assumption admitting of doubt" and "without hypothesis".

An extensive and valuable secondary source is the book of R. Fox, cited above in Footnote 2 to §2A. Had I seen this book early enough, it would have saved me a good deal of tiresome study of vague early writings about heat before the first steps toward thermodynamics were taken. Fox discusses mainly experiment and physical speculation; he describes also some rather rudimentary efforts at mathematical theory of the physical kind; and I am not sure that either he or I have seen all the early works on mathematical theory. Fox nowhere writes the basic equation (4) except subject to the unnecessary and largely irrelevant assumption that there is a heat function, though of course he includes many statements that express special cases of it. The following quotation from his p. 31 suggests that he may ascribe to LAVOISIER & LAPLACE [1784, p. 388] the idea that (4) succinctly embodies:

> But in one important respect they went beyond Black, for they suggested that the absorption of heat was necessary in order to effect not only melting and vaporization but also expansion. Thus, when a body was heated, some heat would go to raise its temperature and some to increase its volume. The idea, skilfully developed by Laplace, became a most important one during the first quarter of the nineteenth century, . . . but it provoked little immediate reaction . . . .

[4] In lectures deriving from about 1757 BLACK [1803, Vol. I, p. 157] wrote, "the heat absorbed does not warm surrounding bodies . . . . [C]onsidered as the cause of warmth,

and to have continuous partial derivatives. These inequalities assert that heat must be added to the fluid body in order to effect either isothermal expansion or isochoric rise of temperature, and conversely, that heat is given off by the fluid body in isothermal contraction or isochoric fall of temperature. We

---

we do not perceive its presence: it is concealed, or latent, and I gave it the name of LATENT HEAT." The *Oxford English Dictionary* quotes three earlier instances of the term in print; the earliest, of 1765, attributes the "doctrine of latent heats" to BLACK.

Most of BLACK's work concerns changes of phase, and the term "latent heat" in common modern use seems to be restricted to "latent heat of fusion" and "latent heat of vaporisation". Neither of these, obviously, is what we denote by Λ_V in this book. The sense that concerns us here was specified by IVORY [1827, *1*] in the context of the Caloric Theory: "the absolute heat which causes a given rise of temperature, or a given dilatation, is resolvable into two distinct parts; of which one is capable of producing the given rise of temperature, when the volume of the air remains constant; and the other enters into the air, and somehow unites with it while it is expanding.... The first may be called the *heat of temperature*; and the second might very properly be named the *heat of expansion*; but I shall use the well known term, *latent heat*, understanding by it the heat that accumulates in a mass of air when the volume increases, and is again extricated from it when the volume decreases."

So far as I can learn, IVORY and MEIKLE make an exception among early British authors, the rest of whom did not use the concept of latent heat except in reference to changes of phase. *Cf.* the booklet of KELLAND [1837, *3*, §§17–19]. *A fortiori*, there is no early British contribution to thermodynamics. The first comes in 1848 with KELVIN, who had trained himself in French mathematical physics and hence accepted the Doctrine of Latent and Specific Heats as a matter of course. See §7H, below.

RANKINE [1853, *3*, §47], writing after Λ_V had been used fluently by KELVIN and CLAUSIUS, explained "latent heat" much more clearly: "... when divested of ideas connected with the hypothesis of a subtle fluid of caloric, and regarded simply as the expression of a fact, this term denotes heat which has disappeared during the appearance of expansive power in a mass of matter, and which may be made to reappear by the expenditure of an equal amount of compressive power." RANKINE's reference to "expansive power" reflects the assumption, common to all early authors on thermodynamics and usually but by no means always true of real bodies, that $\Lambda_V > 0$, as is explained in Footnote 6.

In FOX's *Caloric Theory*, cited in Footnote 2 to §2A, I have found only three uses of the term "latent heat" in the sense symbolized by Λ_V: (p. 131) "latent heat that was necessary simply to bring about expansion"; (p. 174) in a quotation from LAPLACE (see also Footnote 11 to §3C, below); and (p. 174) "the presence of latent as well as sensible heat in gases did have the additional support of Delaroche and Bérard's paper of 1812, as Laplace pointed out." FOX's extensive discussion of "expansion by heat" (pp. 60–67, 69–79) indicates effects governed by Λ_V but does not name it.

THOMSON [1878, §2] in explaining "latent heat" to the intelligent layman wrote as follows:

> It has become of late years somewhat the fashion to decry the designation of latent heat, because it had been very often stated in language involving the assumption of the materiality of heat. Now that we know heat to be a mode of motion, and not a material substance, the old "impressive, clear, and wrong" statements regarding latent heat, evolution and absorption of heat by compression, specific heats of bodies and quantities of heat possessed by them, are summarily discarded. But they have not yet been generally enough followed by

shall find below in §9F that in 1854, the terminal year of this history, KELVIN will see that in some cases $\Lambda_V < 0$, and we shall discuss the matter further in §11Hβ.

---

equally clear and concise statements of what we now know to be the truth. A combination of impressions surviving from the old erroneous notions regarding the nature of heat with imperfectly developed apprehension of the new theory has somewhat liberally perplexed the modern student of thermodynamics with questions unanswerable by theory or experiment, and propositions which escape the merit of being false by having no assignable meaning. There is no occasion to give up either "sensible heat" or "latent heat"; and there is a positive need to retain the term latent heat, because if it were given up a term would be needed to replace it, and it seems impossible to invent a better. Heat given to a substance and warming it is said to be sensible in the substance. Heat given to a substance and not warming it is said to become latent. These designations express with perfect clearness the relation of certain material phenomena to our sensory perception of them.

A footnote to "materiality of heat" reads

...We shall not now be in danger of any error if we use latent heat as an expression meaning neither more nor less than this:—

"DEFINITION.—*Latent heat is the quantity of heat which must be communicated to a body in a given state in order to convert it into another state without changing its temperature.*"—Maxwell's *Theory of Heat* [MAXWELL [1871, p. 73] [1891, p. 73]].

Among the modern authors who understand the matter is J. R. PARTINGTON; he writes as follows in §2 of Chapter II of his *An Advanced Treatise on Physical Chemistry*, Volume 1, London *etc.*, Longmans Green, 1949: "Heat absorbed by a body *at constant temperature*... is called *latent heat*." In his §3 he presents (4) and (8) and writes that those equations "are definite and their legitimacy follows from the physical justification of the concept of 'quantity of heat', which is based on experimental calorimetry...." He mentions that "they rarely appear in the later books on Thermodynamics...."

[5] LAVOISIER & LAPLACE [1784, p. 289 of *Œuvres de Lavoisier* **2**] introduced the terms "capacité de chaleur" and "chaleur spécifique" as interchangeable. They refer both to bodies of unit mass. They present a table of ratios of specific heats of various substances to the specific heat of water. They allude also to latent heats (p. 301 of *Œuvres de Lavoisier* **2**) when they state that it would be interesting to augment the table to include "the specific weights of bodies, the variations that heat induces in these weights, or, what amounts to the same thing, both the dilatabilities and the specific heats of the bodies...."

[6] *Cf.*, for example, the following remarks of CARNOT [1824, pp. 29–32], after each of which I bracket an interpretation in terms of the notations of this essay, small increments being replaced by time rates.

"When a gaseous fluid is rapidly [adiabatically as defined in §3C below] compressed, its temperature rises; on the contrary, its temperature falls when it is rapidly expanded. This is one of the facts best confirmed by experiment. We shall take it as the basis of our proof." [If $Q = 0$, then $\Lambda_V \dot{V} + K_V \dot{\theta} = 0$; $K_V \neq 0$, and $\Lambda_V / K_V > 0$.]

"If, when a gas has been brought to a higher temperature as an effect of compression, we wish to bring it back to its original temperature without causing its volume to change further, we must withdraw some caloric from it." [If $\dot{V} = 0$, then $Q = K_V \dot{\theta}$, and $K_V > 0$.] "This caloric could be drawn off also in proportion as pressure such as to

The ***Doctrine of Latent and Specific Heats*** as laid down by its promulgers is expressed entirely by (4) and (5). Those who used it appealed freely to EULER's axiom as well:

$$p = \varpi(V, \theta) > 0 \ , \qquad (2A.2)_r$$

with the adscititious inequalities

$$\frac{\partial p}{\partial V} < 0, \qquad \frac{\partial p}{\partial \theta} > 0 \ . \qquad (2A.5)_r$$

We shall regard (2A.2), (2A.5), (4), and (5) as defining the ***theory of calorimetry***, and the consequences of that theory alone we shall call *calorimetric*.

Any theory of the passions of bodies rests upon certain *generic principles* or *laws*. These laws express the features *common* to all bodies the theory intends to describe. The *diversity* of these bodies is represented by the constants or functions that are left unspecified by the generic principles. Relations that restrict or specify these constants or functions are called *constitutive*. The generic principles of the theory of calorimetry are (2A.2) and (4), along with their adscititious inequalities (2A.5) and (5). They assert the existence of the *constitutive functions* of the body of fluid: ϖ, Λ_V, and K_V, and they specify the roles of those functions in determining p and Q. The three functions are assumed to be defined over a common set of pairs (V, θ); this set is the *constitutive domain* of the particular body. The theory allows no more diversity in the behavior of different bodies than can be represented by choice of that domain and of the three constitutive functions over it. Early authors never specified the constitutive domain, and all their analysis was local. They tacitly presumed in the constitutive functions whatever smoothness was necessary in the simple formal manipulations to which they subjected those functions. In reporting their work in this history we shall always presume that the constitutive domain is non-empty and open, and that the constitutive functions ϖ, Λ_V, and K_V are continuously differentiable. Like the pioneers, we

---

maintain the temperature of the gas constant were applied." [If $\dot{\theta} = 0$, then $Q = \Lambda_p \dot{p}$, and $\Lambda_p < 0$, the quantity Λ_p being defined by (8), below.]

"Likewise, if the gas is rarefied, we can prevent its temperature from falling by giving it a certain amount of caloric." [If $\dot{\theta} = 0$, then $Q = \Lambda_V \dot{V}$, and $\Lambda_V > 0$.] "The caloric used in the circumstances when its temperature does not change we shall call caloric due to change of volume. This term does not mean that the caloric belongs to the volume; it belongs to the volume no more than it does to the pressure, and it could just as well be called the caloric due to change of pressure." [If $\dot{\theta} = 0$, then $Q = \Lambda_V \dot{V} = \Lambda_p \dot{p}$.] "We do not know what laws it follows in regard to changes of volume. Possibly its quantity varies with the nature of the gas, with its density, or with its temperature. Experiment has taught as nothing on this subject ..." [Both Λ_V and Λ_p are functions of V and θ; what functions they are has not been determined; they may be constitutive functions.] The assumption that $K_V > 0$ DUHEM was to call "Helmholtz's postulate"; see pp. 164–165 of P. DUHEM, §2 of Chapter X of *Traité Elémentaire de Mécanique Chimique*, t. 1, Paris, Hermann, 1897.

In the later thermodynamics it is not necessary that $\Lambda_V > 0$ but only that $\Lambda_V \partial p/\partial \theta \geqq 0$. See Footnote 9 to §2A and §9F.

shall pass over in silence such difficulties as may arise when results proved only locally are applied in the large.

The early theorists recognized that the effects of heat were proportional to the mass M of the body in which they occurred. We can express this idea by writing $\upsilon \equiv 1/\rho = V/M$, $\lambda_\upsilon \equiv \Lambda_V/M$, $\kappa_\upsilon \equiv \mathrm{K}_V/M$ and regarding λ_υ and κ_υ as functions of υ and θ; the functions ϖ^*, κ_υ, and λ_υ are constitutive functions of a *material*, while ϖ, Λ_V, and K_V are constitutive functions of a *body*. When only one body at a time is being considered, it is all the same, especially since the early theorists did not always select bodies of unit mass[7].

In (4) the symbols $\dot{V}$ and $\dot{\theta}$ denote the derivatives of functions of time. These functions are known as soon as a particular process (1) is specified. The same specification makes Λ_V and K_V the values of certain functions of time, namely, $\Lambda_V(V(t), \theta(t))$, $\mathrm{K}_V(V(t), \theta(t))$. Thus by (4) the specification of a process also specifies Q uniquely as a function of time, for a given fluid.

We may represent the pairs (V, θ) as points in a quadrant[8] (Figure 1).

If we suppose that $V(t_1) = V_1$, $V(t_2) = V_2$, $\theta(t_1) = \theta_1$, $\theta(t_2) = \theta_2$, then a process having these two points as endpoints is represented by a curve $\mathscr{P}$ connecting them, and by putting (4) into (3) we may calculate C as a line integral along the curve $\mathscr{P}$:

$$C = C(\mathscr{P}) = \int_{\mathscr{P}} [\Lambda_V(V, \theta)dV + \mathrm{K}_V(V, \theta)d\theta] \,. \tag{2C.6}$$

---

[7] CARNOT [1824, p. 74] refers to "a given quantity of air"; his constitutive constant is what we call R, not r; what he calls "specific heat" in his Equation (5) on p. 77 refers to a given volume V, not a given specific volume υ.

CLAUSIUS [1850, just before his Equation (I.)] also used R but referred to "a certain quantity, say a unit of weight". He stated that R was inversely proportional to the specific gravity. Later, after his Equation (10a), he again referred to "a unit of weight of the gas considered", so his specific heats are taken with respect to weight. In his Equation (11) he converts them to unit volume.

[8] Diagrams in the p–V quadrant were invented by WATT and regarded by him as a great and profitable secret. Perhaps their debut in the literature of thermodynamics is in the paper of CLAPEYRON [1834]. Although some commentators upon the history of thermodynamics make much of them, they do no more than facilitate the discourse. As we shall see below, some of the pioneers used diagrams so as to infer this or that, but only from their own unhandiness or insecurity in the common integral calculus of their times, not from any need.

It is true that line integrals were not commonly familiar in the early years of the nineteenth century. The transformation of an integral around a simple closed path into an integral over the included region of the plane is traditionally attributed to AMPÈRE, GAUSS, and GREEN. Nevertheless, line integrals are not to be found in KELVIN's first paper on thermodynamics (THOMSON [1849]), although he was soon to become expert in use of them. It is tempting to attach the theory of line integrals to CAUCHY's theory of integration of functions of a complex variable; although the two theories are intimately connected, CAUCHY's path of discovery was obscure as well as tortuous. As may be seen from H. FREUDENTHAL's fine analysis of CAUCHY's work in Volume 3 of the *Dictionary of Scientific Biography*, Scribners, New York, 1971, CAUCHY did not publish a clear account of his theory of integration until 1846.

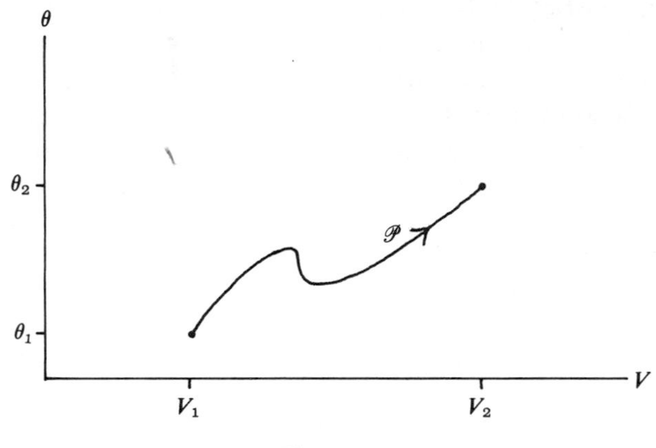

Figure 1

As the notation indicates, $C(\mathscr{P})$ depends in general upon both the fluid body and the choice of the path $\mathscr{P}$ which connects (V_1, θ_1) to (V_2, θ_2), but it does not depend upon the rate at which the path is traversed. That is, for given functions Λ_V and K_V, *all processes that correspond to the same path* $\mathscr{P}$ *from* (V_1, θ_1) *to* (V_2, θ_2) *give rise to the same value of* $C(\mathscr{P})$. In this sense the time becomes irrelevant: Any function of t with positive derivative would do as well to parametrize $\mathscr{P}$ and hence make it possible to evaluate $C(\mathscr{P})$. Of course the differentials imply the use of some parameter to describe the curve, but that parameter need not be specified, and the notation conceals it. In this legitimate and apparently innocuous way the differentials—those accursed differentials famous as vehicles of thermodynamic obscurity—enter the subject.

If we let $-\mathscr{P}$ denote the path $\mathscr{P}$ traversed in the opposite sense, or, as we shall say, the *reverse* path, then from (6) we see that[9]

$$C(-\mathscr{P}) = -C(\mathscr{P}) : \qquad (2C.7)$$

If a body receives a certain quantity of heat as it traverses a certain path, it will lose an equal quantity upon the reverse path.

---

[9] LAVOISIER & LAPLACE [1784, pp. 287–288 of *Œuvres de Lavoisier* **2**]: "All variations of heat, be they real or be they apparent, that a system of bodies experiences in changing its state are reproduced in an inverse order when the system returns to its original state." LAVOISIER & LAPLACE seem to introduce this statement as a "principle" compatible both with the hypothesis that heat is a substance and also with the hypothesis that heat is a mere manifestation of the kinetic energy of the tiny parts of the body in which it appears. Although the paper reports the results of experiments, the "principle" does not seem to be derived from them. It is far more general and far more vague than the specific and demonstrable theorem (7) in the text above. Nevertheless LAVOISIER & LAPLACE state that their general hypothesis is "confirmed by experience" and even given "a sensible proof" by an experiment of their own on the detonation of nitre.

Cf. the statement of CARNOT [1824, p. 35]. REECH [1853, p. 359] stated (7) explicitly in words. The fact it expresses he did not imply to be anything but well known.

Perhaps because of this fact, time and its effects are rarely mentioned in the early thermodynamics or in standard textbooks today. So long as Q be given by (4), the time plays no essential part. Although most of the work we shall analyse in this essay takes (4) for granted, nevertheless I prefer to keep the time ever in mind. First of all, it does occur in early studies of the speed of sound and the conduction of heat; it cannot be eliminated from them, and, accordingly, we cannot form a unified picture of early work on all aspects of heat and temperature except in terms of changes in time. In the second place, when apparently timeless variables like V and θ change, they do so in the course of time; time is the basic descriptor of natural changes. Finally, it is only the special relation (4) that makes it possible to dispense with t in (3). Rational thermodynamics today does not rely upon anything so special as (4), and the roots of the ideas upon which rational thermodynamics is constructed go back to the pioneer studies, to analyse which is the purpose of this essay. I think that much of the confusion in some early work on thermodynamics and some textbooks today, and in particular the use of various peculiar *d*s and δs, reflects overconfidence in (6), forgetting that it is only a special case of the primary definition (3). Although heating is a primitive concept of thermodynamics, heat need not be.

By use of the equation of state (2A.2) we may express (4) alternatively in terms of $\dot{p}$ and $\dot{\theta}$:

$$Q = \Lambda_p(V, \theta)\dot{p} + \mathrm{K}_p(V, \theta)\dot{\theta} \; ; \tag{2C.8}$$

the new coefficients Λ_p and K_p are expressed as follows in terms of Λ_V and K_V:

$$\Lambda_p = \Lambda_V \bigg/ \frac{\partial p}{\partial V}\,, \qquad \mathrm{K}_p - \mathrm{K}_V = -\Lambda_V \frac{\partial p}{\partial \theta} \bigg/ \frac{\partial p}{\partial V}\,. \tag{2C.9}$$

The function Λ_p is the *latent heat with respect to pressure*[10], and K_p is the *specific heat at constant pressure*[11]. These relations[12] show that for a fluid with a given thermal equation of state (2A.2), *the specific heats* K_p *and* K_V

[10] It is CARNOT's "caloric due to change of pressure", *cf.* Footnote 6, above. RANKINE [1859, §212] was to call it "the latent heat of expansion".

[11] The early literature often refers to "the" specific heat, leaving the reader to infer or guess which be meant. On pp. 37–38 of his *Caloric Theory*, cited above in Footnote 2 to §2A, Fox ascribes to CRAWFORD (1788) the distinction between K_p and K_V; CRAWFORD interpreted his experiments on air as showing that $\mathrm{K}_p > \mathrm{K}_V$ but $\mathrm{K}_p/\mathrm{K}_V = 113/110$, so K_p and K_V were long thereafter regarded as virtually interchangeable although distinct.

[12] The earliest statement of this kind I have found is that of CARNOT [1824, pp. 58–60, *cf.* also pp. 43–46]. CARNOT's argument, purely verbal, refers to ideal gases and is set within the Caloric Theory of Heat. The result to which it leads if rendered formal is that which is stated below as (5Q.6). If the Caloric Theory is abandoned, the same argument leads to (5Q.7), which rests upon CARNOT's General Axiom. The Doctrine of Latent and Specific Heats by itself, without use of any of CARNOT's further assumptions, if applied to an ideal gas leads directly to (14), which CARNOT did not state. I take this evidence as sufficient to show that CARNOT did not see the full power of the theory of calorimetry, at least in this context, and did not arrive at (9) in generality. The

may be any functions of V and θ we please, but the two latent heats Λ_V and Λ_p are determined uniquely by the difference $K_p - K_V$. Thus we may choose to specify bodies by specifying their two specific heats, which seem to be somewhat more accessible to direct experiment than is either Λ_V or Λ_p.

By appeal to (2A.5) and $(5)_1$ we see from (9) that[13]

$$\Lambda_p < 0, \qquad K_p > K_V \,. \tag{2C.10}$$

If $(2A.2)_1$ is invertible for V, as is true in many cases, we can express Λ_p and K_p as functions of p and θ, but for this history we do not need to.

If we use (2) to eliminate $\dot{\theta}$ from (8), we find that

$$Q = \frac{1}{\dfrac{\partial p}{\partial \theta}} \left[K_V \dot{p} - K_p \frac{\partial p}{\partial V} \dot{V} \right] . \tag{2C.11}$$

In terms of the equation of state (2A.3) appropriate to fields, we can write (11) in the form

$$Q = \frac{1}{\dfrac{\partial p}{\partial \theta}} \left(K_V \dot{p} - K_p \frac{\partial p}{\partial \rho} \dot{\rho} \right) . \tag{2C.12}$$

For future reference we note also the alternative form

$$\begin{aligned} Q &= -(K_p - K_V) \frac{\partial p/\partial \rho}{\partial p/\partial \theta} \dot{\rho} + K_V \dot{\theta} \,, \\ &= \frac{K_p - K_V}{\partial \rho/\partial \theta} \dot{\rho} + K_V \dot{\theta} \,; \end{aligned} \tag{2C.13}$$

---

relation $(9)_2$ is due to W. WEBER [1830, §10] in the context of solids; he wrote it in the form $\beta' = \beta - (3k')(\alpha/r)$, in which $\beta = K_p/M$, $\beta' = K_V/M$, $\alpha/r = V\Lambda_V/M$ and $3k'$ is the fractional increase of volume with respect to temperature at constant pressure, that is, $-(\partial p/\partial \theta)/(\partial p/\partial V)$.

In the literature of thermodynamics the earliest explicit statement and satisfactory derivation of $(9)_2$ I have found is that of THOMSON [1851, Eq. (15)]; he does not state $(9)_1$. PARTINGTON, in §6 of Chapter II of the work cited in Footnote 4, is one of the few modern authors who presents (9) before he states the laws of thermodynamics and hence implicitly recognizes that $(9)_1$ and $(9)_2$ follow from the theory of calorimetry alone and do not presume any relation between heat and work. However, he does not emphasize this centrally important fact. Contrary to his custom, for these relations he does not cite any source.

[13] The later thermodynamics does not adopt the constitutive inequalities $(2A.5)_2$ and $(5)_1$, although it usually does adopt $(2A.5)_1$ and $(5)_2$. Using the assumption $(2A.5)_1$ alone, we need only glance at (9) to see that

$$K_p > K_V \quad \Leftrightarrow \quad \Lambda_V \frac{\partial p}{\partial \theta} > 0 \,,$$

$$K_p = K_V \quad \Leftrightarrow \quad \Lambda_V \frac{\partial p}{\partial \theta} = 0 \,.$$

POISSON [1833, §634] regarded it as "evident *a priori*" that $K_p > K_V$, presumably in reference to gases.

The possibility that $\Lambda_V \leqq 0$ was first noticed in 1854, as we shall see below in §9F.

in the denominator we consider ρ as being a function of p and θ, obtained by inverting (2A.3) for ρ, which of course is not always possible. To derive (13), we need only substitute (9) into (4).

In the special case of an ideal gas, defined by (2A.1), the relations (9) reduce to

$$\Lambda_p = -\frac{V}{p}\Lambda_V , \qquad \mathrm{K}_p - \mathrm{K}_V = \frac{R\Lambda_V}{p} = \frac{V\Lambda_V}{\theta} , \tag{2C.14}$$

while (11), (12), and (13) reduce to

$$Q = \theta\left(\mathrm{K}_V\frac{\dot{p}}{p} + \mathrm{K}_p\frac{\dot{V}}{V}\right) = \theta\left(\mathrm{K}_V\frac{\dot{p}}{p} - \mathrm{K}_p\frac{\dot{\rho}}{\rho}\right) = \theta\left[-(\mathrm{K}_p - \mathrm{K}_V)\frac{\dot{\rho}}{\rho} + \mathrm{K}_V\frac{\dot{\theta}}{\theta}\right] . \tag{2C.15}$$

Of frequent reference will be the *ratio of specific heats*, which we denote by γ:

$$\gamma \equiv \frac{\mathrm{K}_p}{\mathrm{K}_V} . \tag{2C.16}$$

Because of $(5)_2$ we conclude from (10) that

$$\gamma > 1 , \tag{2C.17}$$

Cf. the remarks in Footnote 13, above.

For an ideal gas (14) shows that

$$V\Lambda_V = (\gamma - 1)\theta\mathrm{K}_V . \tag{2C.18}$$

In studies of fluid mechanics in the eighteenth century the concept of the *work* L done in $[t_1, t_2]$ by a fluid body subject to the pressure p had been introduced and studied:

$$L \equiv \int_{t_1}^{t_2} p(t)\dot{V}(t)dt . \tag{2C.19}$$

If (2A.2) holds, then L is determined uniquely by a process (V, θ):

$$\begin{aligned} L = L(\mathscr{P}) &= \int_{t_1}^{t_2} \varpi(V(t), \theta(t))\dot{V}(t)dt , \\ &= \int_{\mathscr{P}} \varpi(V, \theta)dV . \end{aligned} \tag{2C.20}$$

In the second line θ stands for an assigned function of V on the path $\mathscr{P}$ connecting $(V(t_1), \theta(t_1))$ to $(V(t_2), \theta(t_2))$. Thus L, like C, is determined by the path $\mathscr{P}$ traversed in the V–θ quadrant and is independent of the rate at which that path is traversed. From (20) we see that

$$L(-\mathscr{P}) = -L(\mathscr{P}) . \tag{2C.21}$$

Adjoining (21) to (7), we conclude that *both the heat added and the work done on the reverse* $-\mathscr{P}$ *of a path* $\mathscr{P}$ *are the negatives of their counterparts on* $\mathscr{P}$. These two *reversal theorems* arise in consequence of the *generic principles* (4) and (2A.2), respectively; they hold for *all choices of the constitutive functions* ϖ, Λ_V, and K_V. In this sense, and *in this sense only*, they express properties common to many fluids—as thermodynamicists were soon to claim, "to all bodies".

The reversals of sign just noticed have been associated traditionally with "reversible processes". This term has sown confusion from the day it was born. So long as both (2A.2) and (4) hold, *all processes are "reversible"* in this sense. To get free of this restriction and consider processes that need *not* be reversible, it would be necessary to *replace at least one of the generic assumptions* (4) *and* $(2A.2)_1$ by something else, preferably something more general.

Except, perhaps, for FOURIER and DUHAMEL, *all the writers whose work we shall follow in the first four acts of this tragicomical history will assume that both* (4) *and* (2A.2) *hold.* Some early obscure claims and verbal inferences have been thought by historians to refer to irreversible processes and have been so explained by them, but they are wrong, misled by the most dangerous because most unconscious form of "present-mindedness", that which imposes today's divisions of science into compartments upon the science of times when no such compartments existed. *In thermodynamics down to 1852*, again with the exception of works on the conduction of heat, *all processes are "reversible"*. Thus there is no need to use the term "reversible" before it appears literally, as indeed it does in Act V of our drama.

For later use we remark here that (4) enables us to define the *heat absorbed* and the *heat emitted*[14] by a given fluid body in the interval of time $[t_1, t_2]$. Denoting the former by C^+ and the latter by C^-, we have

$$\begin{aligned} C^+ &\equiv \tfrac{1}{2}\int_{t_1}^{t_2} (Q + |Q|)dt \geqq 0 \ , \\ C^- &\equiv \tfrac{1}{2}\int_{t_1}^{t_2} (|Q| - Q)dt \geqq 0 \ . \end{aligned} \tag{2C.22}$$

Thus

$$C = C^+ - C^- \ , \tag{2C.23}$$

and because of (6) both C^+ and C^-, for a given fluid body, are functions of $\mathscr{P}$ alone and satisfy the following reversal theorems:

$$\begin{aligned} C^+(-\mathscr{P}) &= C^-(\mathscr{P}) \ , \\ C^-(-\mathscr{P}) &= C^+(\mathscr{P}) \ . \end{aligned} \tag{2C.24}$$

---

[14] CARNOT [1824, p. 37, footnote; p. 55; *et passim*]: "les quantités de chaleur absorbées ou dégagées"; p. 42, "la quantité de calorique absorbée ou abandonnée".

From $(5)_1$ we see that in an isothermal process[15]

$$C = \begin{cases} C^+ & \text{if } V \text{ increases} \\ -C^- & \text{if } V \text{ decreases} \end{cases} \text{monotonically} . \qquad (2C.25)$$

The modern student must be reminded that all this follows from nothing but calorimetry, the thermal equation of state, and the constitutive inequalities (2A.5) and (5). *Nothing* is presumed regarding the nature of heat, its conservation or its dissipation, or its power to do work—*nothing*. Moreover, the mathematics used to obtain the various consequences such as (9), (11), (12), (13), and (15) is merely standard for the eighteenth century.

Such was the inheritance of everyone who approached the basic problems of heat and temperature in 1800. We, the spectators, who take our seats before the curtain with old plays already played still fresh in memory, know all this, but we must not expect that each character who steps onto the stage shall know it. From some their inheritance, or at least a part of it, is to have been obscured or withheld. As the examples of *Œdipus*, *The Gondoliers*, and *Tom Jones* show, a man's ignorance of his ancestry can lead him into the tragic, the ludicrous, and the tragicomic.

All of us know the fable. That we are here nevertheless, proves the drama to be a good one, for only a cheap show relies on the unexpected at first hearing. A good play grows better the oftener seen, and sometimes a new production, profiting from old failings, can clear the text and heighten the action. While we know the dénouement, we expect drama in the contrasts and ironies of the working out, the balance of known against not known, fate's final conquest of the avoidable.

Had all the speakers we are soon to hear mastered the whole little budget of simple equations we have just written down, thermodynamics might have had a shorter and clearer history; it might have matured, like mechanics and electromagnetism, into an adult science long before 1963. While there is in those equations nothing any geometer or physicist of the early 1800s would have denied, there was no one place then where all could be found clearly stated[16]. Writings on calorimetry abound in tiny increments and differentials

---

[15] This result does not always hold in the later thermodynamics, because in it Λ_V need not be positive.

[16] The *Annales de Chimie et de Physique* in the 1820s published a number of notes which in proposing this or that relation among total heats, specific heats, and temperatures bear witness to at best a very limited understanding of the logical connections of these quantities. The same may be said of most of the researches Fox discusses in his *Caloric Theory*, cited above in Footnote 2 to §2A, which faithfully reproduces the general level of theoretical physics in the periods it describes. The booklet of KELLAND [1837], which is perhaps the first monograph on heat in English, is particularly bad. It does not present anything at all about what is called here "latent heat" and is denoted by Λ_V. *A fortiori*, it contains not a word on the motive power of heat, although the memoir of CLAPEYRON had been published three years earlier. KELLAND rejects also all the work of LAPLACE and POISSON but offers no substitute for it. Of the French geometers

of functions of unspecified variables, leading to the fogginess such usage always fosters. Thermodynamics inherited from calorimetry the handicap of lacking *the clarity which explicit mathematical statement gives to physical assumptions*, be they right or be they wrong.

---

only FOURIER draws anything but criticism from this insular author, who gives his readers pages of formal series expansions and physical beliefs.

The earliest systematic and fairly complete exposition of the theory of calorimetry that I have seen is that of REECH [1868, §22–44]. It obtains most of the results given above in the text and a good many more, and it obtains them in essentially the same way, though not quite so simply. It appeared far too late to be of use to any of the creators of thermodynamics; it was published obscurely and has never been cited by anyone before now, so far as I know. In irony typical of thermodynamics, I did not see it until the summer of 1978, years after I had written the treatment above.

3. Prologue: LAPLACE, BIOT, and POISSON

Dinanzi parea gente; . . .
. . . a' due mie' sensi
faceva dir l'un "No," l'altro "Sì, canta."
DANTE, *Purgatorio* X, 58–60.

3A. BIOT, and POISSON's First Attempt

Urged by Citizen LAPLACE, Citizen BIOT[1] undertook "to examine the influence that the variations of temperature which accompany the dilatations and condensations of air might have on the speed of sound. . . . It is a fact known to the physicists that atmospheric air, when it is condensed, loses a part of its latent heat, which goes into the state of sensible heat, and on the contrary when it is rarefied, it takes back a portion of sensible heat, which it converts into latent heat." The sonorous condensations must therefore be accompanied by changes of temperature. Since both of these are very small, "we shall regard them as proportional. . . ." Thus BIOT assumes that[2]

$$\dot{\theta} = \beta \frac{\dot{\rho}}{\rho}, \tag{3A.1}$$

β being a coefficient to which he attributes no particular functional dependence. By use of $(2C.2)_2$ we conclude from (1) that[3]

$$\dot{p} = \frac{\partial p}{\partial \rho} \dot{\rho} + kp \frac{\dot{\rho}}{\rho}, \tag{3A.2}$$

[1] BIOT [1802, *1*], extract in BIOT [1802, *2*]. *Cf.* the critical paraphrase by BRANDES [1804], who also included an account of BIOT's work in his annotations to EULER's treatise on hydrodynamics [1806, §§429–434].

[2] As has been stated in Footnote 2 to §2B, the superimposed dot denotes the "substantial" or "material" derivative. For infinitesimal changes, which are the only ones allowed in the particular hydrodynamical researches we shall consider, this time derivative is approximated by the partial derivative at a fixed place, so the distinction is blurred. I make it partly for the convenience of those accustomed to hydrodynamics and partly so as to indicate that the reasoning here, to the extent it is valid at all, is valid also for finite motion of a gas.

[3] BIOT considers only an ideal gas, but to clarify his reasoning I apply it a general equation of state. Also in his definition of k he omits the factor $1/p$, which is required to render his later formulae correct. He makes other slips as well.

where

$$k \equiv \frac{\beta}{p}\frac{\partial p}{\partial \theta} = \frac{\beta}{\theta}, \tag{3A.3}$$

the latter expression being appropriate to an ideal gas. By (2B.1) we obtain for the speed of sound the relation

$$c^2 = \frac{\dot{p}}{\dot{\rho}} = \left(1 + \frac{kp}{\rho \frac{\partial p}{\partial \rho}}\right)\frac{\partial p}{\partial \rho}, \tag{3A.4}$$

which for an ideal gas reduces to

$$c^2 = (1 + k)\frac{p}{\rho}. \tag{3A.5}$$

Taking up the subject a few years later, in the course of a discursive memoir on sound in general POISSON[4] went through essentially the same steps; his otherwise thorough historical preface does not mention BIOT. While BIOT had not said anything about the nature of k, POISSON states in §3 that it is a constant, in §21 that it varies with the temperature in an unknown way.

[Like BIOT before him,] POISSON claims to reconcile theory with experiment but suggests that the experimentally measured value of c should be used to determine k. Thus he concludes that $k = 0.4254$ at 6°C. It follows from (1) that in a sonorous vibration air rises in temperature by 1° when its volume is reduced by the 116th part (§22).

3B. Critique of BIOT's Theory

At first sight there is little difference between BIOT's result (3A.5) and LAGRANGE's old comment that to square theory with experiment one need only suppose that $p = C\rho^{1+k}$ and then give the number k the right value. However, the assumption (3A.1) and the references to perceptible oscillations of temperature suggest that more could be done than was done. If we start from (3A.5) and regard it as a fact of experiment that the gas is ideal and that $k = \text{const.} > 0$, we can work backward from (2B.1) and conclude that (3A.1) does hold and that $\beta = k\theta$.

The modern reader who is familiar with the later work of LAPLACE may think that BIOT here assumes the sonorous vibrations to be *adiabatic*: $Q = 0$. Indeed, that assumption put into $(2C.15)_3$ yield's BIOT's starting point (3A.1) at once and shows that $\beta = (\gamma - 1)\theta$ and $k = \gamma - 1$. BIOT's words about heat certainly refer to the terms $\Lambda_V \dot{V}$ and $K_V \dot{\theta}$ in (2C.4), but he states only that in sonorous condensations neither term is null, not that their sum is null. On the contrary, he chooses to "regard" $\dot{\rho}$ as proportional to $\dot{\theta}$ because both

[4] POISSON [1808, §§3 and 21].

are small, an assumption quite unnecessary if the motion is adiabatic. The concept of adiabatic heating and cooling was at this time far from clear[1]. If we may trust the published record, the first man to formulate it was LAPLACE, as we shall see presently.

BIOT does not make his debt to LAPLACE clear. Later authors were always to call the theory LAPLACE's, and LAPLACE in his own publications on the theory of sound was never to mention BIOT[2]. If in 1802 LAPLACE had ideas more definite than BIOT's, he did not then reveal them.

3C. LAPLACE's Theory of Sound and Heat

In a short note[1] published fourteen years after BIOT's, LAPLACE wrote that modern discoveries on the nature of atmospheric air "offer us a phenomenon which seemed to me the true cause of the excess of the observed speed of sound over the calculated one, as most mathematical physicists have since agreed. This phenomenon is the heat which the air develops by its compression... One may suppose without sensible error that during the time of a vibration the quantity of heat remains the same between two neighboring molecules. Thus these molecules in approaching one another repel each other more; first because, their temperature being supposed constant, their mutual repulsion increases in reciprocal ratio to their distance, and then because the latent caloric so developed raises their temperature. Newton took account only of the first of these causes of repulsion, but it is plain that the second cause must increase the speed of sound since it increases the spring of air. By introducing it into the calculation I arrive at the following theorem:

"The real speed of sound equals the product of the speed according to the Newtonian formula by the square root of the ratio of the specific heat of the air subject to the constant pressure of the atmosphere at various temperatures, to its specific heat when its volume remains constant."

That, according to LAPLACE, if

$$\gamma \equiv \frac{K_p}{K_V}, \tag{2C.16)$_r$ }$$

then

$$c^2 = \gamma \frac{\partial p}{\partial \rho}, \qquad (3C.1)$$

[1] Effects we now recognize to be associated with adiabatic flow of air were known but not understood. The turbid early physics of the concept is ably traced by FOX, pp. 39–60 and 79–99 of his *Caloric Theory*, cited above in Footnote 2 to §2A.

[2] *Cf.* LAPLACE [1823, §1]: "I was the first to remark, ...," *etc.*, and "Mr. POISSON has developed my remark in a learned memoir...."

[1] LAPLACE [1816].

at least for ideal gases. If (1) holds, comparison with (3A.4) shows that BIOT's coefficient k is in fact $(\gamma - 1)\, \partial p/\partial \rho \div p/\rho$. For an ideal gas (1) reduces to

$$c^2 = \gamma p/\rho = \gamma r\theta, \qquad r \equiv R/M, \tag{3C.2}$$

M being the mass of the body of ideal gas whose constitutive constant is R.

LAPLACE discussed experimental results pertinent to his conclusion but let another five years pass before even beginning[2] to explain his ideas on the nature of heat. He developed them fully the next year[3] and immediately thereafter included them in his celebrated and widely read *Celestial Mechanics*[4]. As was his custom, he constructed an elaborate semi-quantitative explanation in terms of attractions and repulsions between the infinitesimally small parts of a static continuous medium, or, rather, a mixture of such media. This is the kind of theory hodiernal Historians of Science are wont to call "Newtonian". LAPLACE (p. 101) supposes that the molecules of a gas "retain their caloric by their attraction, and that their mutual repulsion is due to the repulsion of the molecules of caloric, a repulsion plainly indicated by the increase of the spring of gases when their temperature increases. I suppose finally that this repulsion is sensible only at imperceptible distances." Thus, for him, caloric is corpuscular. He writes also of "caloric rays", which are easily pictured as streams of corpuscles. Finally (p. 104), "each molecule of a body is subject to the action of these three forces: 1°, the attraction of the molecules all around it; 2°, the attraction of the caloric of these same molecules, plus their attraction upon its own caloric; 3°, the repulsion of its caloric by the caloric of these molecules." He assumes a law of central force for these attractions and repulsions, and he claims to calculate their resultants by integration.

Amplifying his remarks of 1816, LAPLACE writes that (p. 109) "the time of a vibration of a molecule of air is less than a sexagesimal tierce [*i.e.* 1/60 second]. In this short interval the absolute caloric of the molecule can be supposed constant, for it can be lost only by the radiation of the molecule or by its communication with neighboring molecules, and to render this loss sensible, a time much longer than a tierce is needed."

Finally, LAPLACE concludes from experiments of GAY-LUSSAC & WELTER that the ratio of the specific heats of air is very nearly constant[5].

[I find LAPLACE's calculations based on inverse-square attractions and repulsions altogether incomprehensible. Since, beyond what he claims to derive from them, he has to make phenomenological assumptions in order to get his conclusions on sound and heat, and since his results follow from

---

[2] LAPLACE [1821].

[3] LAPLACE [1822, *1–3*]. FOURIER [1822, end of preface] refers to these works and to the one cited in the preceding footnote as having already appeared in print.

[4] LAPLACE [1823]. The passages translated above are from this final exposition, which differs only in inessential details from the papers cited in the preceding footnote.

[5] All the foregoing quotations and paraphrases are from LAPLACE [1823, §1]. Page numbers refer to the reprint in LAPLACE's *Œuvres*.

those assumptions alone, without the apparatus of attractions, the phenomenological aspects are all we need to consider here[6].]

First, LAPLACE assumes outright[7] that

$$\frac{\dot{p}}{p} = 2(1 - \beta)\frac{\dot{\rho}}{\rho}, \tag{3C.3}$$

so that

$$c^2 = \frac{\dot{p}}{\dot{\rho}} = 2(1 - \beta)\frac{p}{\rho}, \tag{3C.4}$$

as in the papers by BIOT and POISSON (above, §3A).

To evaluate the coefficient $2(1 - \beta)$, LAPLACE now brings to bear his assumption, stated above, that there is no gain or loss of heat, or, as we should say now, the sonic vibrations are *adiabatic*[8]:

$$Q = 0 . \tag{3C.5}$$

---

[6] Nevertheless I will point out some key passages in the pseudomolecular theory.

P. 119: For gases the entire pressure is due to the repulsion of the molecules of caloric. "Let c be the caloric contained in each molecule of the gas; the repulsion of two molecules will then plainly be proportional to c^2." Main conclusion (pp. 120–121):

$$p = 2\pi HK\rho^2 c^2 ;$$

the law of intermolecular repulsion is $Hc^2\phi(r)$, and K is a constant determined from ϕ by a triple integration. Pp. 121–122: The extinction of caloric rays on a surface is $q\Pi(\theta)$; here Π denotes a function of temperature which is independent of the nature of the gas, q is a constant depending on the gas, and $\rho c^2 = q'\Pi(\theta)$, q' being another such constant. Hence $p = i\ \Pi(\theta)$, and $i = 2\pi KHq'$.

While LAPLACE uses these results again and again, a reader of sufficiently dogged will to calculate may verify that they cancel out of all his formulae that concern thermodynamics. That is, LAPLACE has not derived his conclusions from a molecular theory but rather has exhibited a molecular model that is consistent, sufficient license for mathematical manhandling being granted, with a few plausible phenomenological statements.

Cf. also Fox's discussion on pp. 165–174 of his *Caloric Theory*, cited above in Footnote 2 to §2A.

[7] LAPLACE [1823, §7].

The assumption is well buried. After 21 pages of horrid and useless calculations, LAPLACE writes on p. 134 in connection with a one-dimensional motion of a fluid in the direction of the co-ordinate x, "We suppose that

$$\frac{1}{\rho c}\frac{\partial(\rho c)}{\partial x} = (1 - \beta)\frac{1}{\rho}\frac{\partial \rho}{\partial x} \ \cdots,"$$

and this is the first occurrence of β. In view of the equation displayed in the preceding footnote, the assumption is

$$\frac{1}{2p}\frac{\partial p}{\partial x} = (1 - \beta)\frac{1}{\rho}\frac{\partial \rho}{\partial x} .$$

As p is a function of ρ for each fluid-point, this last result is equivalent to (3). LAPLACE's argument here, as usual, runs through a string of manipulations.

[8] The word "adiabatic" was coined by RANKINE [1859, §239].

For LAPLACE, heat is never created nor destroyed. We shall refer to this assumption by the traditional term *Caloric Theory of heat*[9]. LAPLACE renders the idea of the Caloric Theory definite by the specializing assumption that the heat[10] in a fluid body is the value of a function of pressure, density, and ambient temperature u, which later he takes to be constant[11].

---

[9] In this book the term "Caloric Theory" refers to the statement here annotated—nothing more.

As Fox shows in his *Caloric Theory*, cited above in Footnote 2 to §2A, the term "Caloric Theory" when it was current meant different things to different schools of thought. Some of these are now strange even to historians of physics; still familiar are some particular models such as subtle fluids or atoms of caloric. These may be seen as parallel to the atomic models which WATERSTON, RANKINE, CLAUSIUS, and others proposed for the later thermodynamics. Models do not fall within the scope of this book.

THOMSON [1851, §§19, 44] used "permanence of heat" to describe the Caloric Theory; in annotating his collected papers in 1881 he referred to it as "the assumption of the materiality of heat" (Volume 1, p. 127). THOMSON [1851, §§3–4] interpreted "[t]he recent discoveries made by MAYER and JOULE" as demonstrating "the immateriality of heat", in conformity with DAVY's "dynamical theory", again a sort of speculative model. Later he came to use the term "dynamical theory" for purely phenomenological thermodynamics, based on the uniform and universal Interconvertibility of Heat and Work.

The Caloric Theory was already known to have some shortcomings, but these were not considered fatal to it. RUMFORD's experiments were considered as having established three properties of heat:

1. No bound could be determined for the quantity of heat that could be extracted from a body by doing work upon it. Thus, presumably, the quantity of caloric in a body was so great as to be practically limitless. [The reader who thinks this objection ought have been fatal should recall that experimental possibilities are always limited. More than twelve decades were to pass before anyone could solidify helium, and even today it is unknown whether the space of physical experience be finite or infinite.]

2. At atmospheric pressure, water at 41°F is denser than water at 32°F. Thus thermal expansion cannot be explained as the effect of stuffing a body with some "caloric" substance or by any universal law of repulsion between particles of bodies and particles of a "caloric" substance. [This objection could not destroy LAPLACE's pseudomolecular theory, since the law of caloric-corporeal interaction, being constitutive, could be different and untypical for water, that "anomalous" substance.]

3. After large amounts of heat had been taken from or added to it, the body's weight remained the same, to within the limits of measurement then available. Thus the caloric substance would have to be very light, even "subtle".

RUMFORD regarded these and other facts of experiment as supporting the *vis viva* theory of heat. *Cf.* §7A, below.

The book of S. C. BROWN, *Benjamin Thompson, Count Rumford*, Oxford *etc.*, Pergamon Press, 1967, reproduces and analyses RUMFORD's major papers.

RUMFORD's importance in the history of the theory of heat is greatly exaggerated in popular accounts. As Fox remarks on p. 99 of his *Caloric Theory*, cited above in Footnote 2 to §2A, "a history of the theory could be written with scarcely any reference to Rumford".

[10] LAPLACE's notation for the heat is $c + i$, where c is the "free heat" and i is the latent heat". For the analysis, see pp. 136–137.

[11] P. 136: "The temperature u of the space or the density of the discrete fluid which represents it can thus be supposed constant during the time of an aerial vibration". LAPLACE

Thus in effect LAPLACE adopts the following *axiom* (p. 136): *the heat in a [fluid] body is the value of a heat-function* H_L, so the heat C added in a process from t_1 to t_2 is given by [12]

$$C/M = H_L(p_2, \rho_2) - H_L(p_1, \rho_1) ; \tag{3C.6}$$

here M is the mass of the body; subscript 1 and 2 denote evaluations at the times t_1 and t_2; and $p_2 = p(t_2)$, $p_1 = p(t_1)$, $\rho_2 = \rho(t_2)$, $\rho_1 = \rho(t_1)$. [If such a thing as "total heat" exists, its value may be identified with $H_L(p, \rho)$. However, only differences of such values enter the mathematical theory. LAPLACE seems not to have used (6) directly; an equivalent formula is to play a great part in the researches of CARNOT and others, as we shall see in §5B and many later passages.] Thus the heating Q/M per unit mass is the time-rate of change of H_L:

$$Q/M = \dot{H}_L = \frac{\partial H_L}{\partial p}\dot{p} + \frac{\partial H_L}{\partial \rho}\dot{\rho} . \tag{3C.7}$$

In adiabatic motion, then (p. 136),

$$0 = \frac{\partial H_L}{\partial p}\dot{p} + \frac{\partial H_L}{\partial \rho}\dot{\rho} . \tag{3C.8}$$

Therefore the assumption (4) is proved to hold if and only if the coefficient $2(1 - \beta)$ has the special value

$$2(1 - \beta) = \frac{-\rho\dfrac{\partial H_L}{\partial \rho}}{p\dfrac{\partial H_L}{\partial p}} . \tag{3C.9}$$

"It is easy to ascertain (p. 137) that the specific heat at constant pressure is $-\mu\rho\partial H_L/\partial\rho$, while the specific heat at constant volume is $\mu p\partial H_L/\partial p$; the factor μ is "a coefficient which, according to the experiments of Mr. Gay-Lussac, is 0.00375 at the temperature of melting ice." Such is LAPLACE's proof of (2), specialized to an ideal gas, and of the fact that in adiabatic motion of an ideal gas

$$\frac{\dot{p}}{p} = \gamma\frac{\dot{\rho}}{\rho} . \tag{3C.10}$$

Equivalently, for each fluid-point p is a function of ρ alone, and $dp/d\rho = \gamma p/\rho$.

LAPLACE remarks (p. 142) that since γ can be determined by experiment, no assumption about the heat function H_L need be made. "Nevertheless it would be very interesting to know it for the theory of the phenomena of

---

writes ν for "the temperature of the molecule", which seems to represent the temperature field of the gas. Since, in his notation, $k\rho c^2 = q\nu$, k being a constant, this temperature is a function of the density and pressure of the gas and so is taken into account by use of the general function H_L in (6).

[12] LAPLACE's notation for what we denote by H_L is V.

pressure and heat in atmospheric air." The experiments of GAY-LUSSAC & WELTER show that γ for air is constant "from the pressure represented by 144 mm to the pressure 1460 mm and from the temperature $-20°$ to the temperature 40°." If it is rigorously so, (9) may be integrated to yield

$$\mathrm{H_L} = \psi\left(\frac{p^{1/\gamma}}{\rho}\right) , \tag{3C.11}$$

ψ being an arbitrary function. "The simplest value of $\mathrm{H_L}$ included in this expression" is (p. 143)

$$\begin{aligned} \mathrm{H_L} &= F + K\frac{p^{1/\gamma}}{\rho} , \\ &= F + Kr\theta p^{1/\gamma - 1} , \qquad F = \text{const.}, K = \text{const} . \end{aligned} \tag{3C.12}$$

"On this supposition the absolute heat of a molecule of air at constant pressure increases with the temperature, which squares with the phenomena." From (12) we obtain the following expressions for the specific heats:

$$\mathrm{K}_p = MKrp^{1/\gamma - 1} , \qquad \mathrm{K}_V = \frac{1}{\gamma}\mathrm{K}_p . \tag{3C.13}$$

[I reserve my critique of LAPLACE's theory until after we shall have considered POISSON's last treatment of these matters.]

3D. POISSON's Second Treatment

Coming back to the subject in 1823, POISSON[1] states that at the "already remote" period of his first memoir on sound (above, §3A), "physicists had not yet done any experiment" which could determine "the increase of temperature corresponding to the condensation" in a sonorous vibration. For that reason he had then "reversed the question" and so determined the value of the condensation that would square with the observed speed of sound. Using the experimental results CLÉMENT & DESORMES had published in 1819, POISSON now derives a value of the adiabatic condensation not much different from that he had inferred in 1808. He then engages to derive "the new formula of Mr. Laplace" from the properties of gases "regarded as data of experiment", without use of LAPLACE's "hypotheses made so as to explain the laws of Mariotte and of the dilatation of gases". In this way the formulae would be rendered "independent of any particular explanation."

To this end POISSON considers a triangular cycle of infinitesimal changes for a fluid-point: heating so as to increase the temperature and volume at

[1] POISSON [1823, *1*] = [1823, *3*, §I]. We note that POISSON [1823, *1*, p. 14] [1823, *3*, p. 263] [1823, *2*, p. 339] cites LAPLACE [1823].

constant pressure, then suddenly compressing the air back to its original volume [*i.e.*, adiabatically], then at constant volume reducing the temperature and pressure to their original values. If the changes of volume and temperature in the first part are dV' and $d\theta'$, then $Rd\theta' = pdV'$; the heat added is $K_p d\theta'$. In the second part, let the increase of temperature be $d\theta$. Then because the third part completes the cycle, the temperature falls by the amount $d\theta + d\theta'$, so the heat subtracted is $K_V(d\theta + d\theta')$. "The volume, pressure, and temperature being become again...the same as they were before the expansion of the mass of air, the quantity of heat communicated to it is necessarily equal to what it has lost...." That is, $K_V(d\theta + d\theta') = K_p d\theta'$, or $d\theta = (\gamma - 1)d\theta'$. Hence in the adiabatic part

$$\frac{d\theta}{dV} = -\frac{d\theta'}{dV'}\frac{d\theta}{d\theta'} = -\frac{p}{R}(\gamma - 1) . \tag{3D.1}$$

Appealing to (3A.1) and $(3A.3)_2$, we see that

$$k = -\frac{V}{\theta}\frac{d\theta}{dV} = \gamma - 1 . \tag{3D.2}$$

Substituting this evaluation of k into [BIOT's] formula (3A.5), POISSON confirms LAPLACE's formula:

$$c^2 = \gamma p/\rho . \tag{3C.2$_{1r}$}$$

In another paper published in 1823 POISSON[2] shows in a few lines that LAPLACE's results concerning the heat function and adiabatic processes are likewise independent of the pseudomolecular trappings which LAPLACE laid upon them. POISSON assumes (3C.6) and hence writes down (his Equation (3))

$$\rho\frac{\partial H_L}{\partial \rho} + \gamma p\frac{\partial H_L}{\partial p} = 0 . \tag{3D.3}$$

[He has tacitly assumed that H_L = const. in the process considered; hence that process is adiabatic; but he does not at once say so.] "It is evident *a priori*" that $\gamma > 1$ because "necessarily more heat is required to raise the temperature of a gas when it expands than when its density remains constant, but experiment alone can let us know the value of γ for different gases and how that value depends upon pressure and density. According to the experiments of MM. Gay-Lassac and Welter, cited in the *Mécanique Céleste*, this quantity is sensibly constant...." Integration yields LAPLACE's determination of his heat function:

$$H_L = \psi\left(\frac{p^{1/\gamma}}{\rho}\right) . \tag{3C.11$_r$}$$

[If ψ is invertible,] it follows that

$$p\rho^{-\gamma} = f(C/M) , \qquad \theta\rho^{1-\gamma} = f(C/M)/r . \tag{3D.4}$$

[2] POISSON [1823, *2*, §I].

POISSON now mentions (p. 339) that C/M remains constant, so *in an adiabatic process of a body of ideal gas having constant ratio* γ *of specific heats*

$$p\rho^{-\gamma} = \text{const.}\ , \qquad p\theta^{\gamma/(1-\gamma)} = \text{const.}\ , \qquad \theta\rho^{1-\gamma} = \text{const}\ . \qquad (3D.5)$$

[These relations are traditionally attributed to POISSON. However, as they are obvious consequences of LAPLACE's formula (3C.11), and as the basic idea behind all of this was LAPLACE's, we will in this work refer to them as the ***LAPLACE–POISSON law of adiabatic change.***]

3E. Meikle's Claim

The work of LAPLACE and POISSON was accepted quickly but, as usually happens, bit by bit and with much discussion[1]. Their assumption that $\lambda = \text{const.}$ was regarded as confirmed by experiment, but only MEIKLE[2] subjected it to analysis. Working within the framework of the Caloric Theory of ideal gases, MEIKLE claimed to prove that if $\gamma = \text{const.}$, then[3] "neither the *magnitude* of a constant volume, nor the *intensity* of a constant pressure, have anything to do with the specific heat of a given mass of air." His first proof[4] starts from POISSON's equation (3D.3), but without any stated reason

---

[1] IVORY [1825] somehow extracted the LAPLACE-POISSON law of adiabatic change from a jumble of heat with temperature and some shady fluxional calculus. Then IVORY [1827, *1*] claimed to prove "a priori from the theory here laid down" that $\gamma = \text{const.}$ Thereupon IVORY [1827, *2*] by a still more incomprehensible argument derived an equation different from one published by POISSON [1823, *3*, §I] [=[1823, *1*]] and hence proclaimed POISSON's work erroneous! He generously added that his own treatment of 1825 was "liable to the same objection". POISSON [1827, postscript] in reply asked IVORY to show where his error lay. In the abusive kind of polemic then frequent in Britain MEIKLE [1828, *1*] attacked IVORY's work. IVORY [1828] replied in the same style. The whole eruption is insecure in mathematics, confused in physics, and careless in expression.

[2] MEIKLE [1826, *1*] and later papers, cited below.

[3] MEIKLE [1829, p. 67].

[4] MEIKLE [1826, *2*, p. 335] does not state that the quantity he denotes by B is a constant rather than an arbitrary function, but otherwise a very specific statement of his [1829, p. 62] is not true: "when the variations in the quantity of heat are uniform, those of its volume under a constant pressure form a geometrical progression; as do likewise the variations of pressure under a constant volume." *Cf.* also the similar earlier statements of MEIKLE [1826, *2*, p. 336] [1828, *2*, p. 319]. First, if $p = \text{const.}$, by (2C.4) and (2C.8) we conclude that

$$Q = \mathrm{K}_p\dot{\theta}\ , \qquad \Lambda_V\dot{V} = \left(1 - \frac{1}{\gamma}\right)Q\ . \qquad \text{(A)}$$

If $\gamma = \text{const.}$, we may use (3) and so conclude from (2C.18) that

$$\Lambda_V = M\frac{\gamma - 1}{\gamma V}\left[\frac{p^{1/\gamma}}{\rho}\psi'\left(\frac{p^{1/\gamma}}{\rho}\right)\right]\ .$$

Hence $(\mathrm{A})_2$ becomes

he chooses for ψ in its general integral (3C.11) the particular function B log and so concludes that

$$\mathrm{K}_V = \frac{B}{\theta}, \qquad \mathrm{K}_p = \gamma \frac{B}{\theta}, \qquad B = \text{const.} > 0 . \tag{3E.1}$$

In a later paper MEIKLE[5] provides a geometrical proof [which seems to me to start by assuming both K_p and K_V to be functions of θ alone.] He reproaches LAPLACE, POISSON, and IVORY for having failed[6] to draw "the necessary and unavoidable consequences" of their assumption that $\gamma =$ const., and he reproaches[7] "the eminent French philosophers" because they "built upon these errors an immense fabric of complex formulae, and have drawn from them a multitude of conclusions!"

3F. Critique of LAPLACE's and POISSON's Theories. Correction of MEIKLE's Claim

LAPLACE is the first author to present any concrete mathematical theory concerning heat. His work provides the first clear and formal concept of adiabatic process; the first calculation of the properties of such a process; the first explicit, assessable Caloric Theory of heat; discovery of a basis which makes it easy to see that in an adiabatic process of an ideal gas with constant ratio γ of specific heats, $pV^\gamma =$ const.; and unquestionable proof that the data on the speed of sound then available could be reconciled with gas dynamics if the sonorous motion of air were supposed to be an adiabatic process in a gas of that kind, provided only that γ when measured by other experiments should turn out to have a value close to 1.4.

LAPLACE's work falls into two parts:

1. His theory of adiabatic change in ideal gases.
2. His theory of specific heats, based upon his partial specification of the heat function $\mathrm{H_L}$ for an ideal gas with constant ratio of specific heats.

$$M \frac{\gamma - 1}{\gamma} \left[\frac{p^{1/\gamma}}{\rho} \psi'\left(\frac{p^{1/\gamma}}{\rho}\right)\right] \frac{\dot{V}}{V} = \left(1 - \frac{1}{\gamma}\right) Q .$$

Thus MEIKLE's first assertion about progressions is correct if and only if the quantity in square brackets is constant in processes at constant pressure. A parallel argument shows that his second assertion is true if and only if that same quantity is constant under processes at constant density. The two assertions hold if and only if (3E.1) holds.

[5] MEIKLE [1829, p. 67].

[6] MEIKLE [1829, p. 65].

[7] MEIKLE [1829, p. 67].

POISSON's last analysis shows, albeit crudely, that in both parts LAPLACE's semimolecular trimmings are unnecessary[1]. While to this end POISSON felt it necessary to do everything more or less afresh, our presentation in §3C shows that a careful and critical reader could have seen as much directly from LAPLACE's own exposition.

LAPLACE's work regarding specific heats is pretty clear and efficient, granted his starting point. Of course it rests essentially upon the Caloric Theory of heat. LAPLACE omits only the derivation of his formulae for the specific heats, which we can easily supply. Indeed, beginning from a general expression for the heating:

$$Q = \frac{1}{\dfrac{\partial p}{\partial \theta}} \left(\mathrm{K}_V \dot{p} - \mathrm{K}_p \frac{\partial p}{\partial \rho} \dot{\rho} \right) , \qquad (2\mathrm{C}.12)_\mathrm{r}$$

we see at once that if $Q/M = \dot{\mathrm{H}}_\mathrm{L}$ then

$$\mathrm{K}_p = -M \left(\frac{\partial p}{\partial \theta} \Big/ \frac{\partial p}{\partial \rho} \right) \frac{\partial \mathrm{H}_\mathrm{L}}{\partial \rho} , \qquad \mathrm{K}_V = M \frac{\partial p}{\partial \theta} \frac{\partial \mathrm{H}_\mathrm{L}}{\partial p} , \qquad (3\mathrm{F}.1)$$

formulae which for ideal gases reduce to

$$\mathrm{K}_p = -M \frac{\rho}{\theta} \frac{\partial \mathrm{H}_\mathrm{L}}{\partial \rho} , \qquad \mathrm{K}_V = M \frac{p}{\theta} \frac{\partial \mathrm{H}_\mathrm{L}}{\partial p} , \qquad (3\mathrm{F}.2)$$

so LAPLACE's factor μ is in fact $1/\theta$. Of course LAPLACE's formula for H_L:

$$\mathrm{H}_\mathrm{L} = \psi \left(\frac{p^{1/\gamma}}{\rho} \right) , \qquad (3\mathrm{C}.11)_\mathrm{r}$$

follows easily from these results if we assume that $\gamma = \text{const}$. Moreover, if we have (3C.11), we conclude from $(2)_2$ that

$$\mathrm{K}_V = M \frac{1}{\gamma \theta} \left[\frac{p^{1/\gamma}}{\rho} \psi' \left(\frac{p^{1/\gamma}}{\rho} \right) \right] . \qquad (3\mathrm{F}.3)$$

By inspecting (3) we see that for K_V to be a function of p alone it is necessary and sufficient that H_L have the special form (3C.12), which LAPLACE

---

[1] On p. 177 of his *Caloric Theory*, cited in Footnote 2 to §2A, Fox writes

> Poisson's greatest contribution, then, was rather to free Laplace's work of its more suspect elements, merely by picking up the argument at the stage $q = f(P, \rho)$. In doing so he was showing in a most effective manner just how irrelevant much of Laplace's theory was, so that even to a reader convinced of the physical reality of caloric Poisson's must have seemed undeniably the more fruitful approach.

We shall see below that in the most important parts of LAPLACE's and POISSON's work the heat-function H_L, which Fox denotes by f, is just as superfluous as LAPLACE's particles of caloric.

recommended as being its "simplest value". As early as 1805 LAPLACE[2] had suggested that a constant-pressure air thermometer would measure the "true" temperature; in BOWDITCH's translation,

> But, in the theory of heat, it is necessary to estimate the real degrees of heat indicated by those of a mercurial thermometer; and this will be given with great accuracy by the experiments just mentioned, if the increment of heat of a mass of air submitted to a constant pressure, be proportional to the increase of its volume. Now this hypothesis is at least very probable; for, if we imagine the volume of the mass of air to remain the same whilst its temperature increases, it is natural to suppose that the elastic force, of which heat is the cause, will increase in the same ratio. By submitting it in this last state to the pressure it suffered in the former case, its volume will increase as its elastic force, and therefore as its temperature. Hence it appears, that an air thermometer indicates accurately the variations of heat; but, its construction being difficult, it is sufficient to have compared its march with that of a mercurial thermometer by very exact experiments.

For an ideal gas at constant pressure $Q = \mathrm{K}_p\dot{\theta} = (p\mathrm{K}_p/R)\dot{V}$, so LAPLACE's desideratum is $\mathrm{K}_p = f(p)$. Since $\gamma =$ const., K_V is likewise a function of p alone, so LAPLACE's "simplest value" conforms with his earlier requirement.

Indeed, (3) serves also to show us at a glance that MEIKLE's claim $(3\mathrm{E}.1)_1$ is generally false[3] except when we add the assumption that K_V shall be a function of θ alone. Nevertheless there is a germ of truth in it! The LAPLACE-POISSON law (3D.5) shows that *in an adiabatic process* (3) *reduces to* MEIKLE's (3E.1) with the value of the constant B differing in general from one adiabat to another. As the adiabats of an ideal gas with constant γ are distinct from the isotherms, (3) show that K_V *cannot* be constant. *Hence the Caloric Theory does not allow both specific heats of an ideal gas to be constant.* This trivial, immediate, and essential consequence of the equations of the Caloric Theory was first published[4], so far as I can learn, in 1973. To some, this very lateness will abscind the fact itself from the history of thermodynamics. To others, it will serve as the prime example among many which show that thermodynamics, from its beginnings, has been a sick science, its sores unprobed by conceptual analysis and uncleansed by logical criticism. To still others, it will merely reveal ignorance of MEIKLE's claim (3E.1), even though

---

[2] LAPLACE [1805, p. xx in the reprint in his *Œuvres*]; *cf.* also POISSON [1833, §639]: "at constant pressure a gas dilates uniformly for equal increments of heat...." I am indebted to Mr. C.-S. MAN for pointing out these passages and their importance for the understanding of LAPLACE's later work.

[3] FOX on p. 193 of his Caloric Theory, cited above in Footnote 2 to §2A, pronounces MEIKLE's argument "perfectly sound".

[4] TRUESDELL [1973, *I*, §2].

that claim is generally false. There is also a certain prophetic quality in MEIKLE's assertion, and for two reasons: In §5T we shall see that (3E.1) is the only possibility consistent with CARNOT's thermodynamics, and Property 7 in Chapter 11 of *Concepts and Logic* asserts that MEIKLE's less specific statement in words holds in the entire class of theories compatible with CARNOT's General Axiom, which we shall state and analyse in Act II, below.

As we shall see in Acts II and III of this tragicomedy, the fact that the Caloric Theory forbids an ideal gas to have constant specific heats will prove fatal to it. Thus it is worthwhile to study the matter in another setting, without use of LAPLACE's explicit formula (3C.11) for H_L. While the Doctrine of Latent and Specific Heats allows K_p and K_V to be any functions of p and ρ we might desire, the existence of a heat function H_L leads to (1), which restricts them by the severe requirement

$$\frac{\partial}{\partial \rho}\left(\frac{K_V}{\dfrac{\partial p}{\partial \theta}}\right) + \frac{\partial}{\partial p}\left(\frac{K_p}{\dfrac{\partial p}{\partial \theta}\Big/\dfrac{\partial p}{\partial \rho}}\right) = 0 \ ; \tag{3F.4}$$

here K_V and K_p are regarded locally as functions of ρ and p, and $\partial p/\partial \theta$ are understood to be those functions of p and ρ that have the same values as do $\partial \varpi^*/\partial \theta$ and $\partial \varpi^*/\partial \rho$, calculated from the thermal equation of state $p = \varpi^*(\rho, \theta)$. Conversely, if K_p and K_V satisfy this condition of integrability, a heat function H_L such as to deliver them by (1) exists locally. When the fluid is an ideal gas, (4) assumes the form

$$K_p - K_V + p\frac{\partial K_p}{\partial p} + \rho\frac{\partial K_V}{\partial \rho} = 0 \ . \tag{3F.5}$$

Except for his somewhat imperfect statement of (2), LAPLACE does not give any of the relations (1)–(5), but from equations he does record they follow at once by simple mathematics which he himself mastered and used ordinarily. If we suppose both K_p and K_V to be constant, (5) yields $K_p = K_V$, which contradicts (2C.17): Thus, again, *the Caloric Theory forbids an ideal gas to have constant specific heats.*

So much for LAPLACE's work on specific heats, POISSON's purification of it, and MEIKLE's incorrect but suggestive claims. LAPLACE's theory of adiabatic change and the propagation of sound in ideal gases is another matter. While POISSON in his crude way could clear away the pseudomolecular claptrap, he missed the main point: *LAPLACE's theory of adiabatic change and the speed of sound is independent of the Caloric Theory of heat and all relations between heat and work. It derives from the theory of calorimetry alone.* Although LAPLACE uses his heat function H_L to derive his main results, namely

$$\dot{p}/p = \gamma\dot{\rho}/\rho \ , \qquad c^2 = \gamma p/\rho = \gamma r\theta \ , \qquad (3C.10)_r, (3C.2)_{1,2r}$$

he need not do so, and his pages of calculations serve only to obscure them. LAPLACE's crucial step is his statement that *the sonorous motion is adiabatic.*

Once this step has been taken, anyone who has at his disposal the Doctrine of Latent and Specific Heats as we have presented it in §2C can read off all of LAPLACE's conclusions at a glance[5]. Indeed, if we consider a general expression for Q for an ideal gas:

$$Q = \theta\left(\mathrm{K}_V\frac{\dot{p}}{p} - \mathrm{K}_p\frac{\dot{\rho}}{\rho}\right), \qquad (2C.15)_{2r}$$

and set $Q = 0$, then (3C.10) follows at once, showing that *in adiabatic motion of a body of ideal gas*

$$\frac{dp}{d\rho} = \gamma\frac{p}{\rho}, \qquad (3F.6)$$

which is neither more or less than POISSON's statement (3D.1). POISSON invoked the Caloric Theory needlessly at this point. As his argument in all its crudity refers only to infinitesimal changes, his conclusion follows directly from the calorimetric relation (2C.15). To obtain the corresponding formula when a general equation of state is assumed, we need only use in the same way the parent statement whence (2C.15) descended by specialization to an ideal gas:

$$Q = \frac{1}{\dfrac{\partial p}{\partial\theta}}\left(\mathrm{K}_V\dot{p} - \mathrm{K}_p\frac{\partial p}{\partial\rho}\dot{\rho}\right). \qquad (2C.12)_r$$

The result is

$$\frac{dp}{d\rho} = \gamma\frac{\partial p}{\partial\rho}, \qquad (3F.7)$$

so that by use of (2B.1) LAPLACE's famous correction to the Newtonian speed of sound, namely

$$c^2 = \gamma\frac{\partial p}{\partial\rho}, \qquad (3C.1)_r$$

follows in full generality[6].

---

[5] The theory of sound is a field theory (*cf.* Footnote 2 to §2B), while the Doctrine of Latent and Specific Heats as usually conceived takes Λ_V and K_V as functions of time associated with a whole body. None of the pioneers seems to have been disturbed by this fact. In effect, I believe, they considered an infinitely small sonorous particle. The modern reader does the same thing more precisely by taking (4H.3) as his axiom and supposing that along the path of each fluid-point in sonorous motion $\lambda_v\dot{v} + \kappa_v\dot{\theta} = 0$.

[6] The result (7) implies as a special case a famous formula of the later thermodynamics. Namely, if $p = \hat{p}(\rho, \eta)$, where η is the specific entropy, then in adiabatic motion $\dot{p}/\dot{\rho} = \partial\hat{p}/\partial\rho$ so (7) reduces to

$$\frac{\partial\hat{p}}{\partial\rho} = \gamma\frac{\partial p}{\partial\rho}.$$

This statement is called "REECH's theorem" in the French literature. The corresponding result first obtained by REECH (1853), which is in fact speciously general, is presented below in §10D. Later REECH [1868, §27] obtained (7) essentially as we have done in the

While modern books usually assume that $\gamma =$ const. in order to obtain the LAPLACE-POISSON law (3D.5) of adiabatic change as a basis upon which to prove LAPLACE's correction to the speed of sound, neither the assumption nor the detour is necessary. *Therefore, all of LAPLACE's results concerning the speed of sound are immediate consequences of the theory of calorimetry, and they hold whether or not γ be constant*[7]. They are not restricted to infinitesimal motion, and they do not require the existence of LAPLACE's heat function[8], let alone his semimolecular concepts of particles of air and caloric which attract and repel each other[9].

For later use we emphasize the fact just proved: *the LAPLACE-POISSON theory of the speed of sound allows the specific heats of gases to be arbitrary functions of ρ and θ.*

It is a different matter with the LAPLACE-POISSON law as expressed by (3C.11)–(3C.13) and (3D.4). To obtain those results, LAPLACE and POISSON assumed that $\gamma =$ const. From LAPLACE's own determination (3C.11) for his heat function[10] we have obtained (3). We know that the adiabats extend from

---

text above, making it entirely clear that LAPLACE's correction is independent of thermodynamics. See the Note Added in Proof on p. 301.

We may notice incidentally that setting $Q = 0$ in $(2C.13)_1$ yields BIOT's assumption (3A.1) and shows that BIOT's (and POISSON's) $\beta = (\gamma - 1)\rho[(\partial p/\partial\rho)/(\partial p/\partial\theta)]$. Likewise, putting $Q = 0$ in (2C.12) yields LAPLACE's assumption (3C.3) and shows that LAPLACE's $2(1 - \beta) = \gamma(\rho/p)(\partial p/\partial\rho)$. Although BIOT, POISSON, and LAPLACE considered only ideal gases, these conclusions from the theory of calorimetry are valid for any thermal equation of state.

[7] RANKINE [1852, §13] regarded it as a noteworthy achievement, based on his theory of molecular vortices, to have shown that LAPLACE's law applied "not only to a perfect gas, but to all fluids whatsoever." Of course, all that he did amounts to illustration of the consistency of his molecular model with the theory of calorimetry. *Cf.* §§8G and 9A, below.

PARTINGTON in Ch. II, §4 of the work cited above in Footnote 4 to §2C, is one of the few modern authors who prove (7) without any appeal to thermodynamics.

[8] LAPLACE's own argument, the one he claimed to be "easy to verify", doubtless was equivalent to that we have indicated by (1) and (2).

[9] *Cf.* the sarcastic criticism by WATERSTON in his article "On the theory of sound", *Philosophical Magazine* **16** Supplement, 481–495 (1858) (*Collected Papers*, p. 354):

> ... it is a question whether the reciprocal action between heat-atmospheres and molecules, which he expresses in mathematical symbols, can be realised by the mind.... [I]ndeed there seems to be no limit to this artificial and barren system of procedure, which is as far removed from the simplicity of nature as the hideous epicycles of Ptolemy.

[10] LAPLACE's (3C.11) follows from (3D.1) with no assumption beyond $\gamma =$ const. Nevertheless, it is appropriate only to ideal gases. For a general equation of state a correct differential equation for H_L is

$$\frac{\partial H_L}{\partial \rho} + \gamma \frac{\partial p}{\partial \rho}\frac{\partial H_L}{\partial p} = 0 ,$$

as we may see at once from (1). This equation reduces to (3D.1) if and only if $\partial p/\partial\rho = p/\rho$.

$\theta = 0+$ to $\theta = \infty$. Thus by applying (3) in an adiabatic process we see that low temperatures give rise to enormous specific heats, high temperatures to very small ones. Equivalently, a dense gas has a small specific heat, while a rare gas has a large one. Today we regard such behavior as most implausible. How can LAPLACE have endorsed it? We have seen that in 1805 he expected K_p to be a function of p alone, and that in 1822 he was able to claim that his formula (3C.12) squared with the phenomena. How so? In 1812 DELAROCHE & BÉRARD had published measurements from which they concluded that K_p for gases was a slowly increasing function of p! This claim of theirs was to have a disastrous effect upon the development of thermodynamics for thirty-eight years, as we shall see below in §9G. For LAPLACE, resting securely upon GAY-LUSSAC & WELTER's conclusion that $\gamma =$ const. for most gases, DELAROCHE & BÉRARD's result must have seemed to confirm all that he had guessed and calculated. As the range of variation of p and θ in all the experiments was small, he may have felt no need to ask what his formulae (3C.13) predicted for extreme conditions.

Be that as it may, LAPLACE's formulae stand, firmly drawn from his assumptions. The theorist cannot shut his eyes to what they imply for extreme conditions. The difficulty is not confined to LAPLACE's special choice (3C.12) of ψ; MEIKLE's formulae (3E.1), which we have shown to hold for any ψ if we choose to follow an adiabat, show that the same objectionable conclusions hold under all possibilities consistent with LAPLACE's theory. The simplest molecular pictures of a gas forbid the specific heats of a gas to vary greatly with temperature or density. It is not surprising that HERAPATH[11], who had proposed a kinetic theory, noticed this fact; his is the merit of being the first to criticize the fatal experiments of DELAROCHE & BÉRARD. That it should be he whom the tragicomic muse should choose as sole spokesman for the truth, and at that in a footnote to a translation, is an early example of her whimsy, for in his own day HERAPATH was dismissed as a crank. Anyway, he seems not to see where the basic fault of the theory lies.

Something is wrong. The LAPLACE-POISSON formula is not it. Indeed, to derive that we need use none of LAPLACE's apparatus or POISSON's unconvincing remarks. First, to obtain the basic differential equation

$$\frac{\dot{p}}{p} = \gamma \frac{\dot{\rho}}{\rho}, \qquad \text{(3C.10)}_r$$

we need only specialize (6) to an ideal gas; if we then assume γ to be constant

---

[11] P. 337 of his translation of the note of POISSON [1823, *2*], in the context of vapors:

> I cannot satisfy myself of the degree of confidence to be attached to the experiments of MM. Laroche and Berard. Calculations from the influence of currents of air do not impress me with the idea that such methods are susceptible of much accuracy. Besides, it certainly seems to be adverse to the theory of caloric itself, that so rarefied and expanded a body as vapour should have a less specific heat than its generating water; which is the case in the above philosopher's results.

on adiabatic processes[12], we may integrate (3C.10) just as POISSON did. That is all there is to it. *The LAPLACE-POISSON law of adiabatic change follows directly from the theory of calorimetry*[13], on the assumption that $\gamma = \text{const.}$ on each adiabat.

Therefore, either the sonorous motion is not adiabatic, or the Caloric Theory is wrong. Why did no-one in the early 1800s make this easy comparison of fact with theory? I think the reason lies in the dense tangle of LAPLACE's writing. Even today, with hindsight, it was no easy matter to penetrate the thicket and extract and interpret the equations given above in §3C, equations which make the comparison easy if not even obvious. POISSON's analysis, while not so mysterious, suggested a man's working backward to recover a result already recognized as correct.

In LAPLACE's *Œuvres* over 100 pages are filled by his publications on the theories of sound and heat yet do not reprint all of them. Everything positive in this work he could have developed in twenty simple equations, clearly explained and securely derived, along with four or five pages on the experimental data. Had he done so, his work might have set an example worthy for others to follow. Instead, he cast out his good ideas sporadically in the course of an orgy of expansions and substitutions and supplementary hypotheses and neglect of small terms—the sort of gyrations which Historians of Science and physicists often call "mathematics". It would be facile to apply to his work on sound his own estimate of NEWTON's: "His theory, although imperfect, is a monument of his genius."

LAPLACE is one of those mathematicians who won a great reputation in his own day and has held it ever since, safe within his forbidding eruption of formalism. The few who have had the courage and industry to follow through some of his teetering calculations have adopted a certain reserve toward him. As a physicist, he preferred contorted structural hypotheses; as a mathematician, he was unusually loose, even for his day; as a teacher, he wrote so as to dazzle rather than enlighten. The part of his work we have just analysed is typical of him.

LAPLACE's study of heat and sound falls within the Caloric Theory and does not attempt anything in the province of thermodynamics. Nevertheless, it frames the essential concept of an adiabatic process and provides a major relation which is obeyed in such processes. Also by its verbiage, mixture of scarcely compatible ideas, and preference for the complicated where the simple would have sufficed, it sets the tone of the ensuing tragicomedy.

---

[12] The LAPLACE-POISSON law does *not* hold unless γ is constant on each adiabat, but γ need not have the same value on all adiabats. *Cf.* the remarks following Corollary 4.3 in §4 of *Concepts and Logic*.

[13] This conclusion may have been clear to the pioneers of thermodynamics in the 1850s, but the earliest explicit statement and proof I have found are those of REECH [1868, §37].

4. Act I. Workless Dissipation: FOURIER

> Ma tanto più maligno e più silvestro
> si fa 'l terren con mal seme e non cólto,
> quant' elli ha più di buon vigor terrestro.
> DANTE, *Purgatorio* XXX, 118–120.

4A. FOURIER's Predecessor: BIOT

LAMBERT in his *Pyrometrie*[1] seems to have been the first man to attempt a precise treatment of the conduction of heat. He considered a long bar open to the air, resting upon thin wires, and with one end in a fire. "Thus the bar is heated at one end only. The heat penetrates by and by into the more distant parts but finally passes out through each part into the air. If the fire is maintained long enough and with equal heat, finally every part of the bar contains a certain degree of heat because it again and again receives just as much heat from the parts lying nearer to the fire as it communicates to the more distant ones and to the air." [As MACH[2] remarked, LAMBERT's analysis does not exhibit clarity corresponding to this description of the physical problem.] LAMBERT regards the "heat" y as a function of position x alone and writes down the expression for the subtangent $\mathcal{T}$ to the corresponding curve:

$$dy: y = dx: \mathcal{T} \ . \tag{4A.1}$$

Giving no reason at all, he assumes[3] that $\mathcal{T}$ is constant. Hence he concludes

---

[1] LAMBERT [1779, §§326–327]. Some isolated passages in the works of NEWTON, DANIEL BERNOULLI, and EULER may possibly refer to what we now regard as conduction of heat.

[2] E. MACH [1896, beginning of the *Historische Uebersicht der Lehre von der Wärmeleitung*].

[3] In §270 LAMBERT had treated "NEWTON's law of cooling" by radiation alone and had discussed the subtangent that appears there. Perhaps he expected his readers to apply the same sort of ideas to the more complicated case discussed here. It seems strange that he found for radiation alone an equation with one more term in it than the one he presents here for radiation and conduction combined. The matter is discussed by MACH [1896, *Historische Uebersicht der Lehre von der Wärmestrahlung*].

What NEWTON meant to say in connection with heat and temperature is not clear. Both his theory and his experiment have been discussed by J. A. RUFFNER, "Rethat the curve is "a logarithmic line". [Indeed,

$$y(x) = y(0)\, e^{-x/\mathscr{T}} \,, \tag{4A.2}$$

with the convention of sign nowadays normal.] LAMBERT analyses the experiments of NEWTON and AMONTONS, finds them not as discrepant as was then thought, and presents experimental results of his own. These comparisons show that LAMBERT interpreted the "heat" y as the temperature. From this time on the "logarithmic law" (2) was considered good. [No-one seems to have remarked that LAMBERT's differential equation had no solution for a bar of finite length, the ends of which were maintained at arbitrarily fixed temperatures.]

The next to take up the theory of conduction seems to have been BIOT[4], a generation later. In new experiments he verified LAMBERT's "logarithmic law". Thereupon he stated,

> It was not enough to conclude these results by experiment; it was necessary to find them by theory, for experiment alone shows only some isolated facts, while it is theory that makes us perceive the relations between them.
>
> For that, we must start from this law: when two bodies of different temperatures are put in contact, the quantity of heat that the hotter gives to the colder in a very short time, other things being equal, is proportional to the difference of their temperatures.... This law was assumed by Newton[5] in his essays on heat. Richman[n] confirmed it subsequently by his own experiments and those of Krafft, and afterward Count Rumford himself by new facts has added new weight to those authorized[6].

[LAMBERT had stated clearly that the bar would be subject to conduction in its interior and to radiation on its surface. Only the latter process may be governed by "NEWTON's law".] BIOT seems to think that "NEWTON's law" applies to conduction as well:

> To establish the calculation in accord with this law, we must bear in mind that each point of the bar receives some heat from that which

---

interpretation of the genesis of Newton's 'Law of Cooling'", *Archive for History of Exact Sciences* **2** (1962/6), 138–152 (1964). It seems to me that NEWTON's "degree of heat" was a temperature, not a flux of heat, and that NEWTON meant to state

$$\frac{d\theta}{dt} \propto (\theta - \theta_a) \,, \qquad \theta_a = \text{const.}$$

As we shall see below, this interpretation conforms with BIOT's and FOURIER's.

[4] BIOT [1804].

[5] *Cf.* Footnote 3, above.

[6] Both printed texts have "ces autorisés", perhaps a misprint for "ces autorités".

precedes it and communicates some heat to that which follows it. The difference is what remains in it on account of its distance from the furnace, and it loses a part of that to the air, be it by immediate contact with that fluid or be it by radiation.

Thus in the state of equilibrium, when the temperature of the bar is become steady, the increment of heat that each point of the bar receives in virtue of its position is equal to that which it loses through radiation, a loss which is proportional to its temperature.

And in a state of motion, when the temperature of the bar changes at each instant, the quantity of heat received by each point on account of its position, less the quantity it loses through radiation and contact with the air, equals the quantity by which its temperature increases in the same interval.

The first condition when reduced to calculation gives rise to a differential equation of second order between two variables: the increase of temperature of each point and its distance from the constant source of heat. This equation is linear, with constant coefficients, and it may be integrated by known methods.

The second condition, in which one more variable enters, namely the time, leads to a partial differential equation of second order. This equation, which gives the state of the bar at any instant, includes the preceding one implicitly.

BIOT considers only the case corresponding to LAMBERT's assumptions. "The differential equation related thereto contains in its integral two arbitrary constants multiplying two exponentials, and beyond that another constant but not arbitrary quantity which depends upon the ratio of the conductibility to the radiation."

[These words certainly seem to describe the partial differential equation

$$\rho C \frac{\partial \theta}{\partial t} = K \frac{\partial^2 \theta}{\partial x^2} - h\theta \qquad (4A.3)$$

and its steady case,

$$\frac{K}{h} \frac{d^2 \theta}{dx^2} = \theta \ . \qquad (4A.4)$$

C is a constant that will appear below in FOURIER's work, K is the thermal conductivity, and h is a coefficient of radiative transfer. The integral of (4) is

$$\theta(x) = A \exp\left(-\sqrt{\frac{h}{K}}x\right) + B \exp\left(\sqrt{\frac{h}{K}}x\right) \ ; \qquad (4A.5)$$

A and B are the two arbitrary constants BIOT mentions; for an infinitely long bar the temperature is bounded at ∞ if and only if $B = 0$], and the result confirms LAMBERT's "logarithmic law". [Moreover, the solution (5) can be adjusted to correspond with a bar of finite length, the ends of which are kept at given temperatures. BIOT seems to have solved the problem he set himself. Nevertheless, he writes out no equations. Did he obtain (3), (4), and (5)?]

In the course of a treatise published twelve years later BIOT prints the equations and tells us the story, partly in a footnote[7].

> [The three equations] were asserted and applied, I think, for the first time in a little memoir... which I read to the Institute in 1804 and which was printed in the Bibliothèque britannique [*i.e.*, the note we are presently discussing]. But not being satisfied then about the difficulty of analysis regarding homogeneity, I indicated the structure of the formulae without proof.

The "difficulty of analysis regarding homogeneity" BIOT explains as follows[8]:

> But when we come to form this equation, we find that the laws of homogeneity which govern differentials cannot be satisfied if we suppose that each material and infinitesimally small point of the bar receives heat only by contact with the point which precedes it and transmits heat only to the point which follows it. This difficulty can be set aside only by assuming, as Mr. Laplace did, that a particular point is influenced not only by those that touch it but also by those that are only a small distance away from it, ahead and behind. Then homogeneity is re-established, and all the rules of differential calculus are observed.

[What BIOT tells us here is that "NEWTON's law of cooling" refers to the surface of contact of two bodies, be they large or be they small. Somehow BIOT in 1804 had obtained the right partial differential equation and had found and interpreted its steady solutions, but *he was not able to derive it from "NEWTON's law of cooling"*.

[That is quite right. BIOT's equation is incompatible with "NEWTON's law" if that law is applied to interior parts of a homogeneous body. What LAPLACE disclosed to BIOT, if we may accept BIOT's statement, was *a new concept of heat transfer*: The conduction of heat arises not in response to *differences* of temperature at an actual dividing surface but to *gradients* of temperature within an undivided body. LAPLACE was thoroughly familiar with EULER's hydrodynamics, which represents the accelerating force in a fluid not as a pressure difference effected by a piston but as the result of a pressure gradient within an undivided, homogeneous mass of fluid. Both distinctions are just matters of simple physics. Nevertheless we shall encounter in later parts of this tragicomedy claims that NEWTON's law is the basis of the theory of heat conduction; such claims persist sometimes even today.

---

[7] BIOT [1816, Chapter VI, pp. 669–670].

[8] BIOT [1816, Chapter VI, pp. 667–668].

[How did BIOT arrive at the partial differential equation[9]? I have found in his works nothing else to suggest him capable of thought at this level in mathematical physics. For LAPLACE, whose bread and butter for much of his life's work was EULER's way of looking at physical problems, to correct LAMBERT's incomplete and unconvincing treatment would have afforded no great problem. Perhaps LAPLACE gave BIOT the equation and left him to sink or swim for a few years in trying to derive it. That would have been merely an instance of the way great mathematicians since the very beginnings of mathematical research have effortlessly maintained their superiority over ordinary mortals.]

BIOT's footnote continues:

> Later Mr. Fourier reproduced the partial differential equation in a large work which has received a prize of the Institut de France.

BIOT does not tell us the date at which LAPLACE disclosed to him the nature of heat conduction. [If it was in 1804, then BIOT and LAPLACE largely anticipated FOURIER in the physical aspects of his theory. GRATTAN-GUINNESS[10] states that BIOT sent his paper to FOURIER in 1804 and that the earliest surviving fragments of FOURIER's work on heat conduction date from 1805. There and in the draught of 1807 FOURIER's basic partial differential equation simply replaces $\partial^2/\partial x^2$ in BIOT's (3) by $\partial^2/\partial x^2 + \partial^2/\partial y^2 + \partial^2/\partial z^2$ and leaves intact the term representing radiation through the surface of the bar. *Cf.* the footnotes to §§4B and 4E, below. The blunder, which FOURIER corrected in 1808 and which does not appear in any abstract or text published during his lifetime, certainly suggests that FOURIER had seen and appropriated BIOT's equation without fully understanding the physical concepts on which it was based. The gigantic figure of LAPLACE stands in the background, too. Nevertheless, neither LAPLACE nor BIOT did anything further regarding the conduction of heat; while their voices offstage will be heard, neither will again tread the boards of this tragicomedy.]

4B. FOURIER's Program

Thermodynamics is the science of the power of heat to do work and of the dissipation of that power. The second aspect, although in a case so special as

---

[9] The matter is discussed by I. GRATTAN-GUINNESS in collaboration with J. R. RAVETZ, pp. 83–85 of *Joseph Fourier 1768–1830*, Cambridge and London, MIT Press, 1972. Apparently determined to defend FOURIER's originality at all costs, GRATTAN-GUINNESS explains why BIOT could not have gotten the right equations: because his "philosophical views on physics" forbade him to take a "genuinely 'continuous' view of heat diffusion", *etc. etc.*

[10] P. 84 of the work cited in the preceding footnote.

to be degenerate, was the first to be treated successfully. FOURIER's *Analytical Theory of Heat*[1] was published in 1822; its contents derive largely from previous years[2], many of the results having been announced[3] in print as early as 1808.

---

[1] FOURIER [1822]. Two difficulties, at the very least, attend the reader of this work. If he thinks to consult the original text when he opens Volume 1 of FOURIER's *Œuvres*, published in 1888, he will be grievously deceived. DARBOUX, the editor, after the standard words of mellifluous eulogy we expect in the published judgments of French savants upon each other, goes on to remark that FOURIER's book

> contains, we must admit, many careless slips, errors in calculation and detail.... Guided by the advice of our eminent publisher, M. Gauthier-Villars, we have worked hard to wipe out the inaccuracies in the printed text. We have repeated the calculations, corrected with greatest care the incorrect references, the errors in notation, and the misprints....

The preface to the translation into English by A. FREEMAN, first published in 1878, assures us that "the translator has followed faithfully the French original", but here too, the reader must be on his guard against silent improvements apparently intended to whiten the monument left by the popular hero of nineteenth-century utilitarian science.

All quotations in the text are in my translation from the first edition. The sign § denotes "Article".

[2] FOURIER's memoir of 1807, along with the surviving fragments of a draught written in 1805, are included in the volume by I. GRATTAN-GUINNESS which we have cited in Footnote 9 to §4A.

Some persons expert in rational thermodynamics have expressed astonishment that so simple a theory as FOURIER's was not proposed until a hundred years after the theory of elastica and fifty years after hydrodynamics, both of which are subtler in nature, in concept, and in product, especially since, as FOURIER himself makes plain, EULER's hydrodynamics was his model in establishing his basic equations. The reason seems to be that the conduction of heat was not recognized until it had been distinguished from radiation of heat in the experimental researches of INGEN-HOUSZ, LESLIE [?], and RUMFORD, toward the end of the eighteenth century. *Cf.* pp. 200–201 of E. HOPPE, *Geschichte der Physik*, Braunschweig, Vieweg, 1926. The first quantitative experiments, those of DESPRETZ, did not appear until after FOURIER's treatise; indeed, it seems that before FOURIER's ideas became known, experimentists did not have a clear basis for interpreting and reporting their results. *Cf.* the theory and experiments of LAMBERT described above in §4A.

The work of LESLIE is particularly confusing. The mischievous muse of thermodynamics made him inweave his simple statements about heat in a horrid mess of difficult, irrelevant, and unexplained calculations. His and other early theories of heat make much of entities as imperceptible as voids and vortices or, for that matter, angels. They belong not to physics but to what would now be regarded as speculative philosophy.

LAMBERT, LAPLACE, and FOURIER were the first men to publish even any rudimentary mathematical treatment; hence they were the first to offer the world anything about calorimetry or the conduction of heat sufficiently specific to be refutable by experiment.

[3] P[OISSON] [1808]. This abstract made the partial differential equation (4E.1) widely known.

Both in the Discours Préliminaire and in §429 FOURIER describes his manuscript deposited with the Academy in 1807 and his revised manuscript of 1811, to which a prize was awarded. The latter manuscript he caused to be published by the Paris

FOURIER prefixes to his book a long, [eloquent, boastful, and discursive] preliminary discourse.

> Primordial causes are totally unknown to us, but they are subject to simple and constant laws which may be discovered by observation, and the study of which is the object of natural philosopy.
>
> Heat penetrates, like gravity, all the substances of the universe; its rays occupy all the parts of space. The object of our work is to lay bare the mathematical laws which this element follows. This theory will form henceforth one of the most important branches of general physics.

Regarding heat, there are "three fundamental observations. Indeed, different bodies do not possess to the same degree the faculty of *containing* heat, of *receiving* it or of *transmitting* it across their surfaces, and of *conducting* it in the interior of their masses. These are the three specific qualities our theory clearly distinguishes, and which it teaches how to measure."

Pages of physics and rhetoric regarding the importance of certain special applications follow.

> The principles of this theory are deduced, like those of rational mechanics, from a very small number of primordial facts, the cause of which the geometers do not consider at all, but which they assume as resulting from common observations, confirmed by all the experiments.
>
> The differential equations of the propagation of heat express the most general conditions and reduce the physical questions to problems of pure analysis, which is justly the aim of theory. They are no less rigorously proved than are the general equations of equilibrium and

---

Academy many years later, in 1824, after he had become its secretary. The former was published in 1972 (*cf.* Footnote 2). Examination of the text does not confirm FOURIER's statement that therein "the general equation of the motion of heat in the interior of solids of any dimension and at the surface of the body" had been "rigorously derived": The differential equation at which FOURIER arrives in the manuscript of 1807 contradicts (4E.1) in general (*cf.* our discussion below in Footnote 5 to §4E), and the only boundary condition FOURIER considers there is that of prescribed temperature. P[OISSON] [1808] generously passes over in silence FOURIER's central, damning error, for in his summary published in March 1808 he reports only the correct general equation (4E.1) and gives the reader every reason to believe that that equation stood in the manuscript read before the Academy on 21 December, 1807. In the interim FOURIER must have seen and by some communication with the Academy corrected the potentially disastrous error. Both (4E.1) and the boundary condition (4E.2) appear in FOURIER's manuscript of 1811. Thus FOURIER obtained his basic theory in the form in which it has been preserved to this day some time between 21 December, 1807, and 28 September, 1811, these being the dates on which the memoirs of 1807 and 1811 reached the Academy.

motion. It is so as to render this comparison more perceptible that we have always preferred demonstrations analogous to those of the theorems which provide the foundation of statics and dynamics.

FOURIER goes on to emphasize the importance of *solving* the general equations in special cases corresponding to "all the given conditions". For this, "a special analysis" is needed, and indeed most of his treatise concerns aspects and details of this special analysis. FOURIER vaunts his originality here and the value of his results also for "the solution, long desired" of certain problems of "general analysis and dynamics". He includes a paean for the virtues of what has since come to be called applied mathematics, devoid of "vague questions and calculations to no effect"; he claims that his solutions, marred by "nothing vague or indeterminate", lead "all the way to numerical applications, a condition necessary in every research, and without which only some useless transformations emerge...."

FOURIER complains of the long delay in publishing his work and calls attention to various notes and abstracts which establish his priority in aspects of the theory.

According to FOURIER, "The new theories expounded in our work are joined forever to the mathematical sciences and like them rest upon invariable foundations; they will retain all the elements they have today, and they will continually gain in extent." He predicts that great terrestrial and cosmic discoveries will be made on the basis of them; for example, the constant temperature of interplanetary space will be determined. "The theory itself will direct all these measurements and will delimit their precision. From now on any considerable progress it can make must be founded upon these experiments, for mathematical analysis can deduce from general and simple phenomena the expression of the laws of nature, but the particular application of these laws to very compound effects requires a long sequence of exact observations."

So concludes FOURIER's preface. However, in the very first paragraph of the text of the book FOURIER repeats his boast that he has reduced "all physical researches on the propagation of heat to questions of integral calculus, the ingredients of which are given by experiment." [These claims are not so contradictory as their wording might make them seem. FOURIER believes that his basic principles are not only final and exact but complete; in order to apply them to cases, we must specify appropriate initial conditions and boundary conditions as well as the conductivities and heat capacities of bodies, and it is for these, not for the general laws, that we must turn to experiment so as to set up the particular problem of integral calculus we are to solve.

[After this verbiage, we expect a long and slow-moving book, and such it is. Most of it concerns analysis. Only the parts that bear upon thermodynamics call for our attention in the present essay.

[FOURIER's mathematical terminology is notoriously vague, his mathematical analysis notoriously loose, but in this essay I shall silently clear and straighten his presentation where it is easy to do so without violence to the physical principles toward which he seems to grope[4].]

4C. FOURIER's Premisses Regarding Specific Heat and Temperature

At §22 FOURIER begins to present his physical concepts and assumptions:

> About the nature of heat, nothing more than uncertain hypotheses could be formed, but knowledge of the mathematical laws to which its

---

[4] This being an essay on the history of physical theory, there is no point in making much of weak points of mathematical analysis except in cases when they led the physics astray. In the parts of FOURIER's work that concern us, the needed sanitary operations are more or less obvious to anyone competent in vector algebra and analysis, and often purely verbal or notational changes do the job. No reader should infer therefrom that FOURIER's own level of mathematical reasoning met the standards of the two centuries that preceded his. Two estimates, both by distinguished Frenchmen, should suffice:

1. In awarding FOURIER the prize in 1811 the judges, among whom were LAGRANGE, LAPLACE, and LEGENDRE, observed "that the way the author arrives at his equations is not exempt from difficulties, and his analysis still leaves something to be desired, be it in generality, be it even in rigor." (It is amusing to note that not one of the three famous judges was a stickler for tight proofs: Each of them had at least once been reproached in print by some great contemporary mathematician for failing to maintain the rigor even then considered sufficient.)

2. N. BOURBAKI writes in the chapter on integration in his *Eléments d'histoire des mathématiques*, Paris, Hermann, 1960, "... the proofs of Fourier are altogether devoid of rigor, and their range of validity is not clear...." (While a mathematician today who examines proofs by HUYGENS, NEWTON, LEIBNIZ, the BERNOULLIS, and EULER may find some gaps and errors as well as frequent imprecision, no competent judge has ever suggested that they were "*altogether* devoid of rigor". Among mathematicians commonly reputed great, FOURIER sets a record.)

Only the first of these quoted criticisms and of it only the first part concerns our tragicomedy. We need to mention here, once and for all, two particular points of weakness and possible ambiguity.

1. "Molecules". Almost always "molécule" and "une portion infiniment petite" are interchangeable terms, meaning what would today be called an "element of volume"; such "molecules" have volume, shape, *etc.* (*e.g.*, in the essential arguments in §127 and §§151–154). These are the only "molecules" to which mathematical analysis is applied anywhere in the book, and this fact, along with FOURIER's descriptions in words, justifies my stating his assumptions, *e.g.* (4D.6), in terms of integrals over volumes. On the other hand, in some places FOURIER speaks of a "material point". In others, an infernal confusion of points and small elements rules unchallenged, for example in §59.

2. Differences of temperature. FOURIER, well aware that units may be selected arbitrarily, nearly always takes increments of temperature as being from 0° to 1° in intentionally arbitrary units. Sometimes he regards this difference as being just what it is; more often he treats it as an infinitesimal quantity. This ambiguity, although it may confuse the reader, does not mirror any defect inherent in the mathematical arguments, which are cleared (insofar as this one matter is concerned) by expression in general units, and so cleared they are presented in the text of this essay.

effects are subject is independent of any hypothesis: It requires only attentive examination of the principal facts that common observations have indicated, facts which have been confirmed by precise experiments....

The action of heat tends to dilatate all bodies,—solid, liquid, or aeriform; it is this property that renders its presence visible.

The first basic assumption is stated in §26:

To raise a mass...of a certain weight...from the temperature 0 to the temperature 1, it is necessary to add a new quantity of heat to that which was already contained in this mass. The number C, which denotes this quantity of heat added, is the specific capacity of heat[1]...; ...C has very different values for different substances.

That is, C is a material parameter, what today would be called a constitutive coefficient. Except for a single comment to the contrary in §433, FOURIER seems to regard C as constant for any one material.

Next (§27):

If a body of specified kind and specified weight...occupies the volume V when the temperature is 0, it will occupy a greater volume $V + \Delta$ when it takes on the temperature 1, that is, when the heat that it contains at the temperature 0 has been increased by a new amount C, equal to its specific capacity of heat. But if instead of adding this quantity we add zC ..., the new volume will be $V + \delta$ instead of $V + \Delta$. But experiments show that if $z = \frac{1}{2}$, the increase of volume δ is only $\frac{1}{2}$ the total increase Δ, and that in general the value of δ is $z\Delta$ when the quantity of heat added is zC.

Next (§28):

This ratio z of the two quantities of heat added, zC and C, which is also that of the two increases of volume δ and Δ, is what is called the *temperature*....

FOURIER remarks (§29) that the linear rule of expansion he has just stated

is exact only in the cases in which the bodies in question are subjected to temperatures far from those which determine their change of state.

---

[1] FOURIER seems not to distinguish specific heats for different circumstances. Here and in §27 he calls two different quantities "la capacité spécifique de chaleur". *Cf.* also the distinction between "specific" quantities per unit mass ("weight") and per unit volume made in §159.

It would be unjustified to apply these results to all liquids; in regard to water especially, the dilatations do not always follow the increments of heat.

In general, temperatures are numbers proportional to the quantities of heat added; in the cases we shall consider, these numbers are proportional also to the increments of volume.

At only two places in all the rest of the book does FOURIER refer again to the changes of volume resulting from flow of heat. In §53:

> Heat is the principle of all elasticity; it is its repulsive force that conserves the shapes of solid masses and the volumes of liquids. In solid substances the neighboring molecules would yield to their mutual attraction if its effect were not destroyed by the heat that keeps them apart.
>
> This elastic force is greater in proportion as the temperature is higher; it is for this reason that bodies expand or contract when their temperatures are raised or lowered.

In the fourth from last paragraph of §433, the very last section in the book: "the faculty of solids for expansion... is not at all the same at different temperatures", but "for the great natural phenomena...it is justified to regard the values of the coefficients as constant."

4D. Critique of FOURIER's Premisses

It is strange that FOURIER, with his vaunted knowledge of experiment, could have claimed that addition of heat to a body would never make it contract. Ice floats upon water, and the Accademia del Cimento had reported in 1670 that the density of water at atmospheric pressure reached a maximum at a temperature above the freezing point. In 1804, a few years before FOURIER's first published announcements of his work on the conduction of heat, HOPE had read a description of a beautiful and precise experiment, from which he concluded that the dividing temperature was about 4°C. HOPE's results were published in 1805, not only in Scotland and England but also in France[1]. During the next year RUMFORD[2] read to the Institut an account of his experiments to show that water at 41°F sinks in water at 32°F. This paper was

[1] For a brief account of HOPE's work and references to twelve publications on the subject, by various authors, before the appearance of FOURIER's treatise, see §VIIIC2 of Volume 2 of J. R. PARTINGTON's *An Advanced Treatise on Physical Chemistry*, London *etc.*, Longmans, 1951.

[2] See Chapter 5 of S. C. BROWN, *Benjamin Thompson—Count Rumford*, Oxford *etc.*, Pergamon Press, 1967.

published in five different journals. It was not only FOURIER who ignored[3] this work. He was joined by every early theorist in thermodynamics[4], for every one of them, either explicitly or tacitly, at one point or another used the constitutive inequality

$$\Lambda_V > 0 \ . \qquad (2C.5)_{1r}$$

KELVIN's observation in 1854 that sometimes $\Lambda_V < 0$ will be discussed below in §9F.

FOURIER's assertion in §26 amounts to

$$Q = CM\dot{\theta} \ ; \qquad (4D.1)$$

M is the mass of the body in question, and C, the "specific capacity of heat", is characteristic of the substance of which the body is composed. [Here the body is rigid.] In §27 FOURIER considers an expansible body and asserts that

$$Q = k\dot{V} \ . \qquad (4D.2)$$

FOURIER regards C and k as being constant, or nearly so, for a given body. [What we here denote by k is not what FOURIER in §27 denotes by C, which DARBOUX silently emends to C_0.]

If C and k are both constants, and if both (1) and (2) hold, then obviously

$$\theta = \frac{k}{MC}(V + \text{const.}) \ . \qquad (4D.3)$$

Using subscripts 1 and 2 to denote quantities associated with the first and second additions of heat as described in §27, we see that

$$\frac{\int_{t_0}^{t_2} Q dt}{\int_{t_0}^{t_1} Q dt} = \frac{V_2 - V_0}{V_1 - V_0} = \frac{\theta_2 - \theta_0}{\theta_1 - \theta_0} \ , \qquad (4D.4)$$

as FOURIER claims there.

Although FOURIER uses frequently several terms involving the words "specific" and "heat", he never describes the circumstances to which they refer. Neither does he use the term "latent heat" except in reference to fusion. Nevertheless his claims, though absurdly special and unjustified by the experimental facts available to him, are perfectly consonant with the Doctrine of Latent and Specific Heats:

$$Q = \Lambda_V(V, \theta)\dot{V} + \mathrm{K}_V(V, \theta)\dot{\theta} \ . \qquad (2C.4)_r$$

---

[3] We cannot justly regard FOURIER's vague statement about water, quoted in the preceding section, as acknowledgment of anything but failure of linear response. LAVOISIER & LAPLACE refer to the isobaric maximum density of water at the very beginning of an experimental paper of 1781/2, reprinted in *Œuvres de Lavoisier* **2**, 739–764.

[4] In *Concepts and Logic* BHARATHA & I construct a thermodynamics along classical lines without assuming that ϖ be invertible for θ when V is held constant, but in order to do so we have to take far greater care in the mathematics than any classical author seems to have been able to afford.

Either Λ_V or K_V may be chosen at will, and then the other is determined by the relations

$$\Lambda_V = k\left(1 - \frac{K_V}{MC}\right) , \qquad K_V = MC\left(1 - \frac{\Lambda_V}{k}\right) , \tag{4D.5}$$

as the reader will easily verify.

FOURIER cannot have intended his slipshod remarks about temperature to serve as a definition or a searching inquiry into the concept. We may only conjecture that he thought any book on heat ought to explain temperature. In the mathematical theory which it is his book's purpose to present, FOURIER never makes use of these preliminaries.

Although FOURIER writes much about heat and its flow from place to place, he never tells what he means by it [5]. In accord with the opening sentence of his book, he has no need to do so. He is concerned only with its effects upon the temperature, not with its nature.

In effect FOURIER treats both temperature and heat as primitive concepts, in need of no definition. Except in his preliminary remarks about thermometers, he never makes use of any relation between volume, heat, and temperature; *a fortiori*, he never calculates any change of volume. He treats bodies as if they were rigid. His first operative axiom, which he never states openly, is that for a body occupying the region $\mathscr{V}$

$$Q = \int_{\mathscr{V}} \rho C \dot{\theta} dV . \tag{4D.6}$$

The field C is a function of θ, generally taken as constant. We may call it the *specific heat per unit mass* if we wish to.

Since θ, too, is a field, various sorts of time derivatives might appear where we have written $\dot{\theta}$. In the case of a rigid body, to which FOURIER devotes nearly all of his work, there is no possible ambiguity, and (6) must mean

$$Q = \int_{\mathscr{V}} \rho C \frac{\partial \theta}{\partial t} dV . \tag{4D.7}$$

Further axioms, of course, are necessary if a theory of conduction is to be constructed.

It is difficult not to conclude that FOURIER adopted the Caloric Theory of

---

[5] In §33 FOURIER mentions "rays of heat", and he develops this idea in §§40–47 in his analogy between heat and light; in §52 he says heat is not at all like a fluid; in §54 he asserts that the equilibrium of molecular attraction and heat (repulsion) "is stable; that is, when it is disturbed by any accidental cause, it re-establishes itself." All this is merely physics in the eighteenth-century sense of the word: speculation about the nature of things, with no quantitative interrelation of phenomena.

heat in the sense we have explained in §3C; on the other hand, FOURIER himself never says so, and many of his results and applications do not require it. Nevertheless, in §4H we shall see that his ideas are compatible with general notions if and only if the Caloric Theory is adopted.

I have not studied the memoirs of POISSON on conduction of heat. In his textbook[6] he is neater and perhaps somewhat clearer in regard to fundamentals than was FOURIER. He clearly adopts the Caloric Theory of heat, which he models in terms of a concept of "molecular radiation". Then old, he takes no note of the works of CARNOT and CLAPEYRON, which had been published in 1824 and 1834, respectively.

4E. FOURIER's Concept of the Flux of Heat, and his General Differential Equation and Boundary Condition

In §128 (and again in §142) FOURIER finally states his general partial differential equation for the *temperature field*[1]:

$$\rho C \frac{\partial \theta}{\partial t} = K \Delta \theta \ ; \qquad \text{(4E.1)}$$

the positive coefficient K, which (§161) bears the physical dimensions of heat per unit time, length, and temperature, is the *specific conductivity*[2] of the body. In §146–147 he at last obtains his general condition to be satisfied at a boundary having outer unit normal $\mathbf{n}$:

$$K\mathbf{n} \cdot \operatorname{grad} \theta + h(\theta - \theta_a) = 0 \ , \qquad \text{(4E.2)}$$

in which θ_a denotes the temperature of a steady stream of circumambient air, while h is the *superficial conductivity*[3] of the body in this specified environment.

---

[6] POISSON [1835, Chapters I–V].

[1] Here θ stands for a function of $\mathbf{x}$ and t, while earlier it stood for a function of t alone or an independent variable.

[2] §68: "the constant flux of heat that in a unit of time ... traverses a unit of surface in the ... solid, if it were formed of a given substance." §69: "la conducibilité spécifique", ..., "la conducibilité ou conductibilité dans les différentes substances" represents "the quantity of heat ... that flows in one minute across a surface one meter square ..." in certain specified, homogeneous conditions corresponding to a prescribed difference of temperature. §73: "la conducibilité spécifique intérieure". §430: "cette propriété que les physiciens ont appelée conductibilité ou conducibilité". §433: "mesure de la perméabilité". BIOT's term (§4A) was "la conductibilité".

From its introduction until §146 FOURIER denotes this coefficient by K; from §146 onward it is frequently misprinted as k, which FREEMAN once and DARBOUX always silently restored to K.

[3] §§30, 36: the quantity of heat that flows out through unit area of the surface in unit time, the difference $\theta - \theta_a$ being supposed unity. §60: "la mesure de la conducibilité extérieure". §146: "la conductibilité relative à l'air atmosphérique". §161: "la conducibilité de la surface". §433: "qui mesure la pénétrabilité de la surface".

[By the time the reader has progressed this far, he may find it difficult to remember where, in the endless preceding parade of special cases, experiments real or fancied, and boasts of vast importance for his work through which FOURIER creeps toward these few specific statements, are to be found the physical principles he claims to have established and upon which he bases these general equations.] At the very end of his book (§429) he adds some remarks about the nature and origin of his ideas. [Justly] he regards himself as the first to introduce the essential concept of *flux of heat*. "This notion of flux is fundamental. Insofar as a person has not acquired it, he cannot form an exact idea of the phenomenon and of the equation that expresses it." FOURIER never denotes the flux of heat by a symbol. [If we represent it by $q(\mathbf{n})$, we may say that the fundamental relation he uses again and again though never fully states is

$$q(\mathbf{n}) = \mathbf{q}\cdot\mathbf{n}\ , \qquad \mathbf{q} = -K \operatorname{grad} \theta\ , \tag{4E.3}$$

at points interior to conducting bodies. Here $\mathbf{n}$ is the outer unit normal to a closed surface. If $q(\mathbf{n}) > 0$ at a point on the surface, heat is flowing outward there. The surface is a boundary, either an actual boundary where some given body is in contact with its surroundings, or the conceived boundary of some part inside such a body and hence adjacent only to other parts of that same body.] FOURIER does not write $(3)_2$ as such, but he does convince himself that the values $q(\mathbf{n})$ for three orthogonal vectors $\mathbf{n}$ form what we should now call the rectangular Cartesian components of a vector field (§§149–150).

According to FOURIER (§58, also §429, 1°), experiment shows us that "all other things being equal, the quantity of heat received by one molecule from another [infinitesimally near one] is proportional to the difference of the temperatures of these two molecules...." [By "molecules" FOURIER means here elements of volume, as he usually does.] He goes on to make it clear that the "quantity" is in fact a time rate (§59): If m and n are "two equal molecules" an "extremely small distance p apart", then "the quantity of heat m receives from n" in the "infinitesimal duration dt of the instant" will be given by $(\theta' - \theta)\phi(p)dt$, in which "$\phi(p)$ is a certain function of the distance p. a function which in solid and liquid bodies becomes 0 when p has a sensible magnitude. This function is the same for all points of some one given substance; it varies with the nature of the substance." [Just as FOURIER's preceding statement of this, his *fundamental principle*, is misleading in that it fails to mention $\phi(p)$, so also is his final statement of it in §429, 1°.

[From this statement FOURIER comes nearer to the general law (3)] by appeal to special cases (§§65–72, 81–97), [and the reader must trudge through pages and pages of remarks about linear functions, dignified into] theorems on steady fields of temperature. In §§65–68 FOURIER convinces himself that for a steady distribution of temperature in a solid confined between planes a distance L apart and having the temperatures θ_1 and θ_2, the flux of heat across the planes is the constant $K(\theta_1 - \theta_2)/L$ [in conformity with (3)]. By the time he reaches §96 he can affirm that "one of the principal elements of the theory of heat" remains to be revealed: "to define and measure exactly the quantity of heat that flows at each point of a solid mass across a plane of given direction." In §127 he states (3), partly in words, for the faces of a cube. Again he resorts to special cases (§§132–138). He claims to prove as a theorem (§§140–141) that if $q(\mathbf{n})$ is the outward flux of heat across an interior surface whose outer unit normal is $\mathbf{n}$, then

$$q(\mathbf{n}) = -K\frac{\partial\theta}{\partial n} . \tag{4E.4}$$

By reduction to co-ordinates FOURIER then (§149) arrives at $(3)_1$. [We know that (4) is equivalent to $(3)_2$, but I find no such statement in FOURIER's book.]

The same sort of progress leads FOURIER to conclude (§§30–32, 60, 432) that on the bounding surface $\mathscr{S}$ of a body exposed to a steady current of air which is maintained at the temperature θ_a, the flux of heat $q_{\mathscr{S}}$ from the body is given by

$$q_{\mathscr{S}} = h(\theta - \theta_a) , \tag{4E.5}$$

provided $\theta - \theta_a$ be sufficiently small. It is intended to express the losses effected "either by radiation or by contact" (§433, also §36). The former can be made much smaller by polishing the surface; a piece of metal cools much faster if its surface is covered by a black coating (§32). Thus the coefficient depends upon "the various states of the surface" as well as the substance making up the body (§32); it depends also on the speed of the current of air (§30).

FOURIER regards (4) as his own discovery, without any precursor. As for (5), he states (§429, 3°), "Newton was the first to consider the law of cooling of bodies in air. That which he assumed for the case when the air is swept away with constant velocity squares the better with experiment, the less is the difference of temperatures; it would hold exactly if this difference were infinitely small." The extent of FOURIER's indebtedness to NEWTON, like his indebtedness to BIOT, will remain a matter of doubt, especially since NEWTON's statements are obscure and BIOT's final work is not precisely dated. *Cf.* §4A, above. FOURIER was thrifty in acknowledging the work of his predecessors[4].

[4] In §429, 3°:

> Amontons made a remarkable experiment on the establishment of heat in a prism, one end of which is subject to a specified temperature. Lambert ... was the first to give the logarithmic law of decrease of the temperatures in this prism. Messrs. Biot and de Rumford have confirmed this law by experiments.

The "logarithmic law" is (4A.2).

In a sloppy summary of FOURIER's manuscript of 1807 P[OISSON] [1808] stated that "the known principle of Newton" implied the "logarithmic law"; that the latter had been verified in a direct experiment by BIOT [1804] (*cf.* §4A, above); and that that experiment "can thus serve as demonstration of this principle, the only one that Mr. Fourier borrows from physics, and on which he rests all his analysis." Whether to prove

In some parts of his work FOURIER sees the essential difference between (4) and (5). The conductivities h and K have different physical dimensions (§161). FOURIER considers (4), which refers to infinitesimal differences of temperature, to be an exact and general law of nature. He is hesitant to recommend (5) except for small differences of temperature; he regards it as "observed" rather than "exactly known". In "several important cases" it is "replaced by a given condition which expresses the state of the surface, be it constant or variable or periodic" (§432, also §§31, 60). However, in the mathematical work FOURIER does not consider any such case. Elsewhere (§429, 2°, also §154) he seems to regard (5), too, as a natural law he himself has had the genius to formulate:

> It was not deduced from particular cases, as has been groundlessly supposed, and it could not have been; the proposition it expresses is not of the kind to be discovered by induction; it cannot be known for some bodies yet remain unknown for others; it is necessary for all, in order that *the state of the surface not undergo an infinite change in a determined time.*

[We may see[5] that (4) and (5) are special cases of the fundamental principle

---

so much by analysis after having "borrowed" so little from physics is a virtue, or whether it is a failing to let so much physics go unrepresented by the analysis, must remain a secret locked in POISSON's tomb. BIOT had let his readers think he had derived from "NEWTON's law of cooling" whatever it was he did derive, so POISSON's remarks may be meant to imply only that FOURIER's principles are at bottom those already sketched by BIOT. As we have seen above in §4A, it was only in 1816 that BIOT printed the lesson he had learned from LAPLACE some time before: "NEWTON's law of cooling" does *not* apply to the conduction of heat.

Be all that as it may, the widespread misconception that FOURIER's law of heat conduction expresses the same idea as "NEWTON's law of cooling" may derive from POISSON's summary.

[5] As the above quotations show, FOURIER was neither clear nor consistent in regard to his fundamental principle. In his first surviving draught, written about 1805 but not published until 1972 (*cf.* Footnote 2 to §4B), FOURIER obtained instead of (1) the equation

$$\rho C \frac{\partial \theta}{\partial t} = K \Delta \theta - h\theta \ .$$

He seems to have been misled by the accepted "logarithmic law" (4A.2) and its generalization (4A.5), which BIOT had obtained in 1804. As BIOT's paper clearly states, in a bar exposed to the air the radiation from the sides contributes to the balance of heat *at each point* if the bar is idealized as a line. Effects represented by boundary conditions in a three-dimensional theory must be absorbed into the conditions that hold at interior points of a corresponding thin body. Classic examples occur in the works of the BERNOULLIS and EULER: In the hydraulics of flow in tubes there is a term representing change of cross-section, in the theory of the elastica the effects of tension in the fibres of the cross-section are represented by a couple at each point, *etc.* FOURIER's blunder suggests that in the physics of heat he was more indebted to BIOT than he cared to acknowledge.

By 1808 FOURIER had seen and corrected his mistake.

FOURIER laid down in §59: $\phi(p) = 0$ when p has "a sensible magnitude", while for infinitely small p

$$\phi(p) = \begin{cases} -\dfrac{K}{p} & \text{for solids and liquids ,} \\ h & \text{for air .} \end{cases} \tag{4E.6}$$

The temperature of the air is supposed to be uniform in the interior of the current but to jump discontinuously on the boundary to the value θ of the solid or liquid body with which the air is in contact. This interpretation is consistent with the then accepted belief that air did not conduct heat.]

The fundamental principle to which FOURIER appeals to derive his fundamental equations (1) and (2) expresses the conservation of heat. He does not state this principle clearly, but he uses it again and again. It yields at once the general boundary condition (2), provided (5) be admitted. As FOURIER writes (§148), "the quantity of heat which tends to leave in virtue of the action of the molecules is always equivalent to that which the body must leak into the medium". In other words, the quantity $q(\mathbf{n})$ as calculated from (4), interpreted as a limit taken from inside the body, must equal $q_{\mathscr{S}}$ as given by (5). FOURIER applies this reasoning in special cases (§§115, 129) and finally in general (§§146–148). The same principle yields the partial differential equation (1) at interior points. FOURIER writes (§150), "if in the interior of the solid an element of any shape is conceived, the quantities of heat that penetrate this polyhedron through its different faces compensate each other reciprocally. More exactly, the sum of the terms of first order that enter into the expression of these quantities of heat received by the molecule is zero, so the heat that accumulates there in fact and causes its temperature to vary can be expressed only by terms infinitely smaller than those of first order." It is "the heat that accumulates there" that has already been determined in one way by FOURIER's basic assumption

$$Q = \int_{\mathscr{V}} \rho C \frac{\partial \theta}{\partial t} \, dV \; . \tag{4D.7$_r$}$$

To obtain another expression for it, FOURIER calculates the net flow of heat into a molecule. He does so thrice, once in §127 in the context of a cubical body, once in §142 for a body of any form, and again in §§151–154; the molecule is an infinitesimal cube or truncated prism with rectangular base, and the proofs are the same in concept[6], the first two being almost identical,

[6] The third proof is presented in the context of a boundary point. As DARBOUX remarks in a footnote, it does not generally apply at such a point, since FOURIER assumes that the point in question lies upon one of the faces of a convex polyhedron interior to the conducting body, which is always true of interior points but not generally of boundary points. DARBOUX's explanation of how to proceed at a boundary point rests on properties of the gradient and is therefore restricted to use of FOURIER's constitutive relation (3). In failing to see the difference between a constitutive relation and a generic principle, DARBOUX follows the tradition of nineteenth-century thermodynamics.

word for word. The result FOURIER obtains is the differential equivalent of

$$Q = \int_{\mathscr{S}} \operatorname{div}(K \operatorname{grad} \theta) dV \ . \tag{4E.7}$$

Comparison with (4D.7) yields

$$\rho C \frac{\partial \theta}{\partial t} = \operatorname{div}(K \operatorname{grad} \theta) \ , \tag{4E.8}$$

which reduces to (1) when K is constant[7].

FOURIER's final formulation of his theory provides three *material coefficients* to represent the variety of materials, bounding surfaces, and environments: ρC, K, and h. They are (§433) "variable magnitudes which depend upon the temperature or the state of the bodies. However, in applications to the natural questions of greatest interest to us, these coefficients may be given values which are sensibly constant." The specific conductivity K probably varies considerably with temperature; experiments indicate it to be "more variable" than ρC. The difficulties attendant upon the superficial conductivity h have been mentioned above.

Most of the rest of FOURIER's book concerns in effect questions of pure analysis which do not lie within the scope of our drama. An exception is furnished by §§160–161, which discuss physical dimensions and may well provide the earliest explicit reference to a dimension independent of mass, length, and time.

> It must now be remarked that each undetermined magnitude or constant has a *dimension* proper to itself and that the terms of any one equation could not be compared if they did not have the same *dimensional exponent*. We have introduced this consideration in the theory of heat so as to render our definitions firmer and to serve as a check upon the calculation. It derives from primordial notions about the quantities. For this reason, in geometry and mechanics, it is equivalent to the fundamental lemmas that the Greeks have left us without demonstration....
>
> If its own *exponent of dimension* is attributed to each quantity, [every] equation will be homogeneous, because each term will have the same total exponent.

Numbers representing areas and volumes have the exponents of length 2 and 3, respectively. Angles, sines, logarithms, and other abstract numbers do not change with choice of the unit of length and hence ought to have the

[7] In the body of the book FOURIER gives only (1) and special cases of it, but in §429, 2°, he states (8) in words.

dimension 0. FOURIER then provides a dimensional matrix, probably the earliest example:

	Length	Duration	Temperature
x	1	0	0
t	0	1	0
θ	0	0	1
K (specific conductivity)	-1	-1	-1
h (superficial conductivity)	-2	-1	-1
ρC (heat capacity per unit volume)	-3	0	-1

[FOURIER leaves the units of heat unmentioned, because they cancel out of all his general equations.] He remarks (§§69, 159) that the dimensions of mass, while they are used to express ρ and C, cancel out in the product ρC, which is all they contribute to the theory of heat, and from time to time he writes c for ρC.

4F. Critique of FOURIER's Concepts and Methods

Although FOURIER perceives that the fluxes of heat in three orthogonal directions form components of a vector field, his proof of this fact (§§149–150) is circular. It rests upon use of the constitutive assertion

$$q(\mathbf{n}) = -K\frac{\partial\theta}{\partial n} \qquad (4E.4)_r$$

and thus amounts to no more than verification of the then already known fact that the gradient of a scalar field is a vector field[1]. The physical concepts are thereby obscured. Logically, it is necessary first to show that $\mathbf{q}$ is a vector before one can justly set it equal to K grad θ. FOURIER did not see any such need.

FOURIER's basic but never clearly stated assumption about the flux of heat amounts to the following: *If $\mathscr{V}$ is a part of the body, on its bounding surface $\partial\mathscr{V}$ there is a scalar field $q(\mathbf{x}, \mathbf{n}, t)$, $\mathbf{n}$ being the outer unit normal to $\partial\mathscr{V}$, such that*

$$Q = -\int_{\partial\mathscr{V}} q(\mathbf{n})dA \ . \qquad (4F.1)$$

$\int \ldots dA$ denotes integration with respect to surface area, and the arguments

[1] FOURIER's mathematics is clumsy as well as loose even in this regard. DARBOUX remarks in a footnote to the derivation of (4E.4), "In deducing the consequences of this rule, Fourier could have simplified the exposition and avoided some uncertainties which we shall note below."

x and t are understood unwritten. The scalar field $q(\mathbf{n})$ upon $\partial\mathscr{V}$ is the *flux of heat* through $\partial\mathscr{V}$. Comparison with FOURIER's basic assumption

$$Q = \int_{\mathscr{V}} \rho C \frac{\partial \theta}{\partial t} \, dV \qquad (4D.7)_r$$

or with the more general idea that Q vanishes along with the volume of $\mathscr{V}$, makes it possible to prove that there is a vector field $\mathbf{q}(\mathbf{x}, t)$ over $\mathscr{V}$ such that

$$q(\mathbf{n}) = \mathbf{q}\cdot\mathbf{n} \ . \qquad (4E.3)_{1r}$$

The field $\mathbf{q}$ over $\mathscr{V}$ is the *heat-flux vector field*. This *fundamental theorem on the flux of heat*, which is in no way contingent upon FOURIER's assumption

$$\mathbf{q} = -K \operatorname{grad} \theta \ , \qquad (4E.3)_{2r}$$

was first stated and proved by STOKES[2] in 1851. STOKES merely adapted to the case of a scalar field defined on $\partial\mathscr{V}$ the argument invented by CAUCHY in 1823 so as to prove his great theorem of the existence of the stress tensor. In FOURIER's book there is not even a hint toward proof of $(4E.3)_1$, which is simpler than CAUCHY's theorem because it relates a scalar to a vector rather than a vector to a tensor. We cannot attribute to FOURIER the fundamental theorem, which is a cornerstone of rational thermodynamics today. We can only say that he introduced and developed two important special instances of the flux of heat.

The fundamental theorem $(4E.3)_1$ follows from the *generic principles*, common to all kinds of conducting bodies. In contrast, *Fourier's law of heat conduction* $(4E.3)_2$, which defines a particular kind of conductor, is a *constitutive relation*. Constitutive relations, which define particular materials within a general theory, serve to model the diversity of natural bodies[3]. FOURIER shows no sign of seeing this distinction.

---

[2] STOKES [1851]. The theorem has two parts: (1) the scalar function $q(\mathbf{x}, \mathbf{n}, t)$ is a linear function of $\mathbf{n}$, and (2) every linear scalar function of vectors has a unique representation as an inner product. While the second is now a standard result in linear algebra, it was not known in the days of CAUCHY and STOKES, so their arguments had to include a proof of it as well, in the appropriate special cases.

[3] In his note on p. 120 of Volume 1 of the *Œuvres de Fourier* DARBOUX writes "Fourier ... presumes tacitly that the solid body enjoys the properties we express today by saying that the body is isotropic." This remark is true but fails to reveal how enormous is the shortcoming. As we know from MAXWELL's second kinetic theory of gases, even in the case of an isotropic fluid the flux of heat need not be determined by grad θ or even by the temperature field alone.

When the linear theory is not restricted to isotropic materials, the constitutive relation is

$$\mathbf{q} = -\mathbf{K} \operatorname{grad} \theta \ ,$$

K being a tensor called the *thermal conductivity*. This more general relation was first obtained by DUHAMEL [1832], who employed a molecular model which STOKES [1851]

Once we have $(4E.3)_1$ and (1), by use of (4D.7) we obtain

$$\int_{\mathscr{V}} \rho C \frac{\partial \theta}{\partial t} dV = -\int_{\partial \mathscr{V}} \mathbf{q} \cdot \mathbf{n} dA \ . \tag{4F.2}$$

We recognize this statement as providing the formal basis upon which FOURIER's theory rests. It expresses two ideas: Heat is neither created nor destroyed, and the density of heat in a body is proportional to its temperature. The former reflects one aspect of the Caloric Theory of heat, another aspect of which we have specified and developed in §3C in connection with the work of LAPLACE.

As we know today, from (2) we may easily prove that at a given element of surface, the flux of heat into one side must equal that out of the other. If we denote limits from the two sides of a surface by + and −, then

$$q^+(\mathbf{n}) = q^-(\mathbf{n}) \ . \tag{4F.3}$$

That is, the component of $\mathbf{q}$ normal to $\partial \mathscr{V}$ is continuous at the surface, while the component tangential to $\partial \mathscr{V}$ need not be. Because FOURIER asserts and uses (3) in §148 in connection with his boundary condition (4E.2), we might easily be misled[4] to attribute proof of it to him, but proof he never attempted. He seems to have regarded the statement as obvious, denied by no-one. Such it is if heat is neither created nor destroyed, but not more generally. The ritualistic prestidigitation in which FOURIER habitually took refuge was ill suited to a conceptual problem such as the proof of (3) as a theorem. It

---

was to call "the hypothesis of molecular radiation" and to reject emphatically. This hypothesis led DUHAMEL to conclude that $\mathbf{K}$ was symmetric: $\mathbf{K}^T = \mathbf{K}$. STOKES gave a phenomenological reason in favor of this conclusion; controversy regarding it continues to the present day; the position has not changed since the summary of it by TRUESDELL [1969, Chapter 7].

[4] TRUESDELL & TOUPIN [1960, Footnote 3 on p. 610] succumbed to the temptation. Retrospective generosity is a sin easy for the creating scientist, if he is honest, to fall into. Certainly he shrinks from letting his readers infer that he regards as his own some idea that may have been familiar to IMHOTEP or Labor-loving JOHN. As Historians of Science run no such risk (the subsets of the null set being the null set), for them the Devil has other shifts.

Another instance which may seem one of retrospective over-generosity but is not is the modern name "Fourier inequality" for the condition

$$\mathbf{q} \cdot \operatorname{grad} \theta \leqq 0 \ ,$$

namely, the flow of heat is always from a hotter to a colder place, never from a colder to a hotter. Because FOURIER never treated $\mathbf{q}$ except when it was subject to the constitutive relation (4E.4), he could not have stated this inequality except in the form

$$K \geqq 0 \ .$$

Although I cannot find any place where he does affirm this constitutive inequality outright, certainly throughout his work he takes K as positive and never entertains any other possibility. In this sense we may justly associate the general inequality $\mathbf{q} \cdot \operatorname{grad} \theta \leqq 0$ with FOURIER's name, though we should stop short of attributing it to him.

was left to CAUCHY to face the concepts as such, and in the more difficult context of tensor fields. I refer to the condition asserted by POISSON in 1829 and proved by CAUCHY in 1841: At a weak singular surface the balance of linear momentum requires that the traction vector be continuous.

If $q(\mathbf{n})$ is a continuous function of $\mathbf{x}$—something that at a point on the boundary of a conducting solid and the circumambient air it certainly is not, at least in FOURIER's view—we may prove from (2) that

$$q(-\mathbf{n}) = -q(\mathbf{n}) \ . \qquad (4F.4)$$

The modern student regards this condition, a counterpart of CAUCHY's fundamental lemma in continuum mechanics, as an overriding requirement, which every constitutive relation must satisfy. Since FOURIER's constitutive equation (4E.4) does satisfy it, we may presume that he saw no need to isolate it from his several specializing assumptions. The one critical point where we might expect him to have used (4) is in §§146–148, but it is not needed there[5]: FOURIER assumes that the flux of heat *out* of the boundary *into* the air is given by

$$q_{\mathscr{S}} = h(\theta - \theta_a) \ , \qquad (4E.5)_r$$

while the flux into that boundary is $q(-\mathbf{n})$ on the understanding that $q(\mathbf{n})$ is given by FOURIER's usual formula

$$q(\mathbf{n}) = -K\frac{\partial\theta}{\partial n} \ . \qquad (4E.4)_r$$

FOURIER uses his constitutive relation (4E.4) in every instance where q appears. In not all, however, does it play an essential part. In his three derivations (§§127–128, 142, 151–154) of his partial differential equation

$$\rho C\frac{\partial\theta}{\partial t} = K\Delta\theta \qquad (4E.1)_r$$

it serves only to lengthen the formulae. What he actually proves, crudely and in terms of infinitesimal elements, is the result we should now write in the form

$$\int_{\partial\mathscr{V}} \mathbf{q}\cdot\mathbf{n}dA = \int_{\mathscr{V}} (\operatorname{div}\mathbf{q})dV \ , \qquad (4F.5)$$

that is to say, the divergence theorem. The argument is step by step parallel to the one given in the 1750s by EULER to calculate the resultant force of a field of pressures:

$$\int_{\partial\mathscr{V}} p\mathbf{n}dA = \int_{\mathscr{V}} (\operatorname{grad} p)dV \ , \qquad (4F.6)$$

[5] So far as I can see, the only way to obviate DARBOUX's objection to FOURIER's treatment of boundary points (Footnote 6 of §4E) is to begin from a proof of (4) as a consequence of generic principles, independent of constitutive relations.

of course stated by EULER likewise in terms of differential elements; to derive (5) when (6) is known is no more than a student's exercise. Elsewhere[6] FOURIER refers to EULER's hydrodynamical equations: "It would be useless to recall here the well known demonstrations of these equations. We presume the reader bears in mind the elements of this question as they are presented in the works of EULER (*Berlin Mémoires* for 1755)." Possibly FOURIER, always pitiably eager to claim something new even when there is naught, refers obliquely to this debt when he writes (§429, 1°) that his proofs are "no less exact than those of the elementary propositions of mechanics", though he does not specify which propositions he has in mind.

The foregoing remarks suggest that while FOURIER could handle well the rather simple special cases he introduced, he fell short of the foresight, the depth of the greatest mathematicians, for whom a key special case solved served to set a new problem. Unlike the giants of mathematical science from HUYGENS to MAXWELL, FOURIER is not indefatigable in the search for logical threads and essential hypotheses. In general, he silently adapts to his ends the mathematics created by EULER and his predecessors for continuum mechanics. Not only does FOURIER fail to create any new arguments or concepts, but also he is vague and sometimes awkward in his use of known methods.

To complete this critique, we must return to FOURIER's physical principles, already discussed in part in §§4C–4E. FOURIER's basic concept is the *flux of heat*. No absolute amount of heat appears anywhere in his book. Thus, it seems, his theory implies not only no limit to the amount of heat a body may receive, but also none to the amount that may be extracted from a body.

As we saw in §§4B and 4C, FOURIER regarded an increase of volume as the inevitable companion of an increase of temperature. Nevertheless the theory that he constructs allows the temperature field to be calculated without taking account of the expansion or deformation to which heating gives rise. If the temperature changes in the course of time, as it generally will according to FOURIER's theory, the body's size and perhaps also shape will change. The boundary of the body, therefore, will move. It cannot then be correct to apply the boundary condition at the points where the boundaries were initially, for generally those points now lie outside of the body or in its interior, no longer on its present boundary. Because we do not know where the present boundary is, we do not know where to apply FOURIER's boundary condition. Thus, except when the body is rigid, FOURIER's theory provides only an approximation of unspecified and inassessable nature, or a plain hoax.

Though FOURIER nowhere says a word about the matter, in using his theory we are pretty well forced to forget all his preliminary statements about heat and temperature and to assume that the bodies he deals with do *not* expand when heated. We may charitably presume that at the beginning of his book he was describing thermometers, and that it is these thermometers

---

[6] FOURIER [1833] (posthumously published work of 1820).

he expects us to use when confirming his predictions about the differences of temperature in a rigid body.

If we agree to regard FOURIER's theory as appropriate to rigid bodies only, we can easily find its place on the framework of later concepts of heat. We write the axiom expressing balance of energy for a particular body in the form

$$\dot{\mathrm{E}} + \dot{K} = P + JQ \ . \tag{4F.7}$$

The four quantities Q, P, K, and E are functions of time alone, and J is the mechanical equivalent of a unit of heat (see §8A, below). Q we have encountered before; in a major special case it is given by (1). P is the *power* of the forces applied to the body, K is the *kinetic energy* of the body, and E, the *internal energy* of the body, is the value of an absolutely continuous function of mass:

$$\mathrm{E} = \int_{\mathscr{V}} \rho \epsilon dV \ , \tag{4F.8}$$

ϵ being the field of specific internal energy at points within the body. Then for a stationary body

$$\dot{\mathrm{E}} = \int_{\mathscr{V}} \rho \frac{\partial \epsilon}{\partial t} dV \ . \tag{4F.9}$$

If ϵ is the value of a function of θ and possibly also of variables which do not change when the body is at rest, then

$$\frac{\partial \epsilon}{\partial t} = C \frac{\partial \theta}{\partial t} \ , \qquad C \equiv \frac{\partial \epsilon}{\partial \theta} \ , \tag{4F.10}$$

and (7) reduces to FOURIER's effective starting point

$$Q = \int_{\mathscr{V}} \rho C \frac{\partial \theta}{\partial t} dV \ . \tag{4D.7)$_r$}$$

Of course FOURIER's second effective basic principle (1) remains valid today for the special case in which all transfer of heat is by conduction. If, as we may do, we adopt FOURIER's constitutive relation

$$\mathbf{q} = -K \operatorname{grad} \theta \qquad \text{(4.E3)}_{2r}$$

at points inside an isotropic linear conductor, we recover FOURIER's theory in full. Boundary conditions, of course, are adjoined, and FOURIER's (4E.2) remains acceptable, in fact a major example. That is, for us FOURIER's theory furnishes a description of *workless dissipation*, and all his solutions in special cases illustrate this fact.

As we shall see in the following section, FOURIER did not have anything specific to offer in regard to deformable bodies, not only in general but even as described by the already well known Doctrine of Latent and Specific Heats.

His theory simply sets aside the phenomenon by which, certainly, we are able to measure temperatures.

For a century before FOURIER's time it had been customary in one theory after another to calculate the speed of propagation of small disturbances and to attempt to compare the calculated values with experimental data. According to FOURIER's theory, differences of temperature are diffused instantly through infinite distances. Although this fact may be illustrated by several of FOURIER's solutions for special problems, he does not remark upon it[7]. *A fortiori*, he does not attempt to explain or justify this physically implausible consequence of his theory. In later times it was to be seen as a major flaw, even a "paradox" repugnant to physics.

4G. FOURIER's Theory of the Conduction of Heat in Fluids

At the end of his book (§429, 5°) FOURIER refers to the "various and difficult questions" concerned with the effects of changes of shape. He states that in 1820 he found the differential equations "which express the distribution of heat in liquids in motion..., combined with changes of temperature." The memoir concerned incompressible fluids. An extract from it, followed by a draught, was published thirteen years later[1]. For FOURIER an "incompressible" fluid is one whose density ρ is determined by the temperature alone and is unaffected by changes of pressure:

$$\rho = \rho_0[1 + h(\theta - b)] \; ; \tag{4G.1}$$

ρ_0 is the density when $\theta = b$, and h is a constant which "expresses...the dilatability of the fluid mass. It is regarded as known from experiments." In his calculations and in most of his statements FOURIER chooses a scale of temperature such that $b = 0$. FOURIER regards (1) as being approximate for small changes of temperature. [Indeed, it is a linear approximation to the relation $\rho = f(\theta)$, which EULER in his paper of 1764 on the convection of heat in fluids had proposed and used as being appropriate to water but not to air]. Since generally ρ decreases when θ increases, we expect that $h < 0$ for most choices of b.

---

[7] Reading and rereading the mysterious §278, I have tried to find in it something about the matter, but I conclude that FOURIER's refusal there to let the diffusion of heat be infinitely slow is no more than an excuse for passing from the discrete ring to a continuous one in LAGRANGE's formal way. For me this kind of argument, while possibly suggestive toward conjecture, is neither physics nor mathematics. The earliest statement I can find in print that FOURIER's theory gives disturbances of temperature infinite velocity of propagation is by J. STEFAN [1863].

[1] FOURIER [1833].

FOURIER adopts without change the differential equations for the density and the pressure that EULER had published in 1757:

$$\frac{\partial \rho}{\partial t} + \operatorname{div}(\rho \mathbf{v}) = 0 \ , \tag{4G.2}$$

$$\rho \left[\frac{\partial \mathbf{v}}{\partial t} + (\operatorname{grad} \mathbf{v})\mathbf{v} \right] = -\operatorname{grad} p + \rho \mathbf{b} \ ; \tag{4G.3}$$

$\mathbf{v}$ is the velocity field, p is the pressure field, and $\mathbf{b}$ is the field of body force per unit mass. Although FOURIER refers his reader to EULER's memoir of 1755, in which "he gives these equations in a simple, clear form which includes every possible case, and he demonstrates them with that admirable clarity which is the principal trait of all his writings," FOURIER [manages to confuse matters] by using the word "molecule" in two senses[2] [which contradict one another]. FOURIER's final field equation is

$$\frac{\partial \theta}{\partial t} + \operatorname{div}(\theta \mathbf{v}) = \frac{K}{c} \Delta \theta \ , \tag{4G.4}$$

[This differential equation reduces to (4E.1) when $\mathbf{v} = \mathbf{0}$; however, it does not do so when $\mathbf{v} = \text{const.}$]

To derive (4), FOURIER[3] gives an argument [which is copied step by step from one of EULER's derivations of (3)]. What the argument shows is that for a material region

$$\frac{d}{dt}\left(\int_{\mathscr{V}} f dV\right) = \int_{\mathscr{V}} (\dot{f} + f \operatorname{div} \mathbf{v}) dV \ . \tag{4G.5}$$

FOURIER assumes that

$$Q = \frac{d}{dt}\left(\int_{\mathscr{V}} c\theta dV\right) \ ; \tag{4G.6}$$

apparently he regards this statement as an obvious extension of (4D.6) to a deformable body. It expresses the idea that the quantity of heat associated with an element of volume dV is $c\theta dV$. FOURIER writes that he "regards as constant the quantity of heat that the mass contains when it is at the temperature zero of melting ice..." and that his theory merely balances the differences from "this common constant". [Thus he has the Caloric Theory of heat in mind.] His calculation shows that he regards c as constant. Therefore, application of (5) to (6) yields

$$Q = \int_{\mathscr{V}} c(\dot{\theta} + \theta \operatorname{div} \mathbf{v}) dV \ . \tag{4G.7}$$

---

[2] The components of $\mathbf{v}(\mathbf{x}, t)$ are the "partial velocities of the molecule", so the molecule is a material *point*, yet p is "the pressure which is exerted against the molecule", so the molecule occupies a small *region*. In EULER's work on fluids there is no such confusion; EULER refers consistently to "un élément".

[3] FOURIER [1833, p. 603 of the *Œuvres*]. In this paper FOURIER uses the letter C for what he had denoted by ρC or c in his book. For clarity we replace it by c.

With (7) in hand, FOURIER can copy his old derivation of (4E.1), and the result is (4).

FOURIER suggests that usually $h(\theta - b)$, which of course is a dimensionless increment of temperature from the constant value b, will be negligibly small, so the density will be uniform. In this case, which corresponds to what in hydrodynamics has always been called an incompressible fluid, FOURIER's [somewhat uncertain] ideas about the relations between temperature and mechanical phenomena drop out of the picture. Since $\operatorname{div} \mathbf{v} = 0$, FOURIER's field equation (4) reduces to

$$c\dot{\theta} = K\Delta\theta \ , \tag{4G.8}$$

[an equation nowadays still used to describe the conduction of heat in fluids of uniform specific heat and thermal conductivity, but of course *not* subject to FOURIER's assumption that $\rho = f(\theta)$].

At the end of the note FOURIER writes a few words about gases. Although the relation between p, ρ, and θ is "known exactly" from experiment, experimental physics is not yet perfected sufficiently to determine "the relations between the densities and specific capacities of aeriform substances, and the property of receiving radiant heat." On the basis of remarks about the penetrability of gases by rays of heat, FOURIER claims that for aeriform bodies the governing equation "has a form very different from that we have found for solid substances. It is of an indefinite order, or, rather, it relates to that class of equations which include both finite differences and differentials."

4H. Critique of FOURIER's Theory of the Conduction of Heat in Fluids

FOURIER's theory rests on the assumption that $\rho = f(\theta)$, f being a constitutive function; while FOURIER adopts the special form (4G.1) for this function, we do not need to use it here. FOURIER's assumption allows us to reduce (4G.2) to the statement

$$\frac{f'(\theta)}{f(\theta)}\,\dot{\theta} = -\operatorname{div} \mathbf{v} \ . \tag{4H.1}$$

Therefore we may write (4G.4) in the form

$$\dot{\theta} = \frac{K}{c^*}\,\Delta\theta \ , \qquad c^* = c\left(1 - \frac{\theta f'(\theta)}{f(\theta)}\right) \ . \tag{4H.2}$$

If the velocity field is known, we thus obtain two scalar equations to determine θ. We may suspect that two equations for determining one scalar field are one too many. Certainly some strange conclusions result. One is that if $\mathbf{v}$ vanishes, θ must be a steady harmonic field: $\partial\theta/\partial t = 0$, $\Delta\theta = 0$. The sort of decay of temperature through conduction that is the dominant feature of FOURIER's theory for solids is thus excluded. The temperature of FOURIER's fluid cannot change at a given point unless the fluid is in motion. If the flow is isochoric, div $\mathbf{v} = 0$, so the temperature is both convected: $\dot{\theta} = 0$, and harmonic: $\Delta\theta = 0$.

FOURIER's concept of heating as expressed by (4G.6) is strange. Perhaps it is motivated in byways of the Caloric Theory which I have not explored. Certainly it makes no use of the conceptual apparatus provided by the then widely accepted Doctrine of Latent and Specific Heats. Of course we do not expect anyone in FOURIER's day to have used the structure based on the balance of energy, which we have sketched at the end of §4F; neither do we expect that the results for compressible substances obtained by anyone at that time will turn out to be compatible with that theory. On the other hand, the ideas represented by the Doctrine of Latent and Specific Heats suggest at once the proper generalization of (4D.6) to compressible fluids:

$$Q = \int_{\mathscr{V}} \rho(\lambda_v \dot{v} + \kappa_v \dot{\theta}) dV \ , \tag{4H.3}$$

in which $v \equiv 1/\rho$ and the coefficients κ_v and λ_v are given functions of v and θ. In the Caloric Theory, to which FOURIER seems to adhere, there would be a *heat density* $k(v, \theta)$ such that (*cf.* LAPLACE's formula (3C.7))

$$\begin{aligned} Q &= \frac{d}{dt}\int_{\mathscr{V}} \rho k dV = \int_{\mathscr{V}} \rho \dot{k} dV \ , \\ \lambda_v &= \frac{\partial k}{\partial v}, \qquad \kappa_v = \frac{\partial k}{\partial \theta} \ . \end{aligned} \tag{4H.4}$$

Be that as it may, directly from (3) and (4G.2) we conclude that

$$Q = \int_{\mathscr{V}} (\rho\kappa_v \dot{\theta} + \lambda_v \operatorname{div} \mathbf{v}) dV \ , \tag{4H.5}$$

which is similar to FOURIER's formula (4G.7). We cannot compare the two expressions term by term, for FOURIER always assumes that $\rho = f(\theta)$. Putting (1) into (5) and (4G.7), we see that to make the two agree for all $\mathscr{V}$ it is necessary and sufficient that

$$c^* = \rho\kappa_v - \frac{f'}{f}\lambda_v \ , \tag{4H.6}$$

c^* being defined by $(2)_2$. Thus FOURIER's proposal is not in contradiction with the then widely received ideas about heat, although it is far from being a consequence of them.

If we return to (3) and cast aside FOURIER's assumption that $\rho = f(\theta)$, we may otherwise follow his reasoning and obtain instead of his (4G.4) the field equation

$$\rho(\kappa_v \dot{\theta} + \lambda_v \dot{v}) = \operatorname{div}(K \operatorname{grad} \theta) \ . \tag{4H.7}$$

This is the differential equation for conduction of heat in fluids to which the ideas current in FOURIER's day would have led. With κ_v and λ_v suitably obtained from the specific entropy function, it remains today correct for inviscid fluids. We cannot fairly attribute it, even in a major special case, to FOURIER.

4I. FOURIER's Bequest

Not only what FOURIER did to promote the theory of heat must be noted here, but also what he did that was irrelevant to it, and what he failed to do.

FOURIER's "definition" of temperature rests on a supposed fact of experiment which he himself admits to be inexact; what it asserts is not only "inexact" but sometimes diametrically opposed to the results of experiment. Indeed, in another passage he recognizes that units of temperature are independent of those of mechanical quantities (and hence, obviously, temperature cannot be "defined" mechanically in a fashion independent of the properties of some particular body such as an ideal gas). To his own work this "definition" does no harm, for he never uses it, and it seems to serve only so as to give an impression of contact with what might be called real physics. In confusion of this kind FOURIER was to be followed by the whole tribe of thermodynamicists, but not until after the end of the period that this history describes.

As, again and again, FOURIER lauds physics, experiments, and "useful" mathematics in preference to abstraction, it is become a tradition to regard him as having been a great physicist. Certainly he avows the right party line, but I doubt if any critical reader today would find in his book much evidence of insight into nature beyond that common in works of his predecessors and contemporaries.

FOURIER did not face the difficulties afforded by conduction of heat in a deforming body of gas, in which the pressure, and hence the work done, depends upon both the temperature and the density. While neglecting the very parts of the theory of heat that others were soon to cultivate, he boasted[1] that he had "demonstrated all the principles of the theory and solved all the fundamental questions." FOURIER began the tradition of the theory of heat, a tradition we recognize as being still alive today in the claims of those many physicists who regard thermodynamics as a subject inherently limited yet arrogate to it a vast generality, in fact illusory.

FOURIER's great contribution is the concept of flux of heat. It has come into its own in researches of the last quarter century, but before then it had no effect upon the development of thermodynamics. One reason for neglect of it may be the confusing, special way FOURIER introduced it. It never appears anywhere in his book except in indissoluble union with the constitutive equation $(4E.3)_2$ of an isotropic linear conductor. Throughout FOURIER's work the general ideas governing the transfer of heat are jumbled with that particular constitutive relation. Failure to separate the generic principles of the theory of heat from constitutive relations of particularly simple bodies, which can serve as key examples but no more, was to become common in

[1] FOURIER [1822, p. xxiv in the *Œuvres*].

thermodynamics as in no other part of physics, and indeed characteristic of it until recently.

The distinction was not new: It had been made and used again and again in the rational continuum mechanics of the preceding century. The tradition of rational mechanics still flourished, and especially in France. CAUCHY, NAVIER, and POISSON knew full well the difference between a constitutive relation and a generic principle. When, a decade later than FOURIER's works of discovery and in the year after FOURIER's book appeared, CAUCHY came upon the counterpart of the flux principle for mechanics, where it is more difficult to perceive and develop, he recognized it at once as the pillar that it was and is. He stated his fundamental theorem clearly and proudly; he gave a splendid proof of it, which has been reproduced in every book on continuum mechanics from that day to ours. Of creative, conceptual mathematics like this, there is no example in FOURIER's book, nor anywhere in thermodynamics.

Although from the beginnings of mathematical science mathematicians have weighed assumptions and have *discarded the unnecessary ones*, such being the very essence of logical thought, in thermodynamics the unnecessary assumptions have been treasured, repeated, and inflated to the point that they are come to conceal the whole conceptual structure of the science. This unhappy quality is the tough and tortuous thread of the plot of the tragicomedy. It began with LAPLACE and was spun out by FOURIER.

Most of FOURIER's book concerns solutions of his linear partial differential equation (4E.1). In this part, too, the tragicomic muse ruled the scene. The fundamental "theorem", that "every" function can be expanded in a trigonometric series, is untrue, and FOURIER "proved" it through a mass of divergent gobbledegook which every competent mathematician of his own day rejected [2]. Up to this point the analysis, both in formulation of a theory and in mathematical proof, does not meet the standards of the preceding century. Then the magic of linearity takes hold. FOURIER is not the first to hit upon a theory governed by a linear partial differential equation with constant co-

---

[2] The reactions of LAGRANGE, LAPLACE, and MONGE to FOURIER's work show that they recognized its promise and fertility but found nothing basically new ("revolutionary") in it and could not accept the alleged proofs. Perhaps still fresh in LAGRANGE's mind was his own youthful defeat in the attempt to pass from a discrete system to a corresponding continuous one, a defeat he never admitted but chose to ignore by blandly republishing in 1788, after D'ALEMBERT's death, his old argument, which D'ALEMBERT and others immediately upon its first publication in 1759 had shown to be fallacious. That work came about as close to "FOURIER's theorem" as did FOURIER's own; LAGRANGE's argument rested at bottom upon an "approximation" at ∞ which is not only unproved but also incorrect, but in contrast with FOURIER's mountain of divergent formalism it is simple. (*Cf.* §10 of H. BURKHARDT's "Entwicklungen nach oscillirenden Funktionen und Integration der Differentialgleichungen der mathematischen Physik", *Jahresbericht der Deutschen Mathematiker-Vereinigung* $\mathbf{10}_2$, 1908.) Certainly LAGRANGE could find nothing new in "FOURIER's theorem" except the sweeping generality of its statement and the preposterous legerdemain advanced as a proof.

efficients, but he is the first to see how easy such a theory can be. Using mainly facts already known but up to then scattered and insufficiently appreciated, he weaves them into an elegant general method that pours forth solutions of boundary-value and initial-value problems in abundance never before seen. At the time, it was algebraic neatness and generality that struck home, the Eulerian brilliance in manipulation, especially the almost effortless methods for converting a formal explicit representation of boundary data or initial data into a corresponding representation of the solution. Of deeper and more lasting importance is the basic concept of the problems a linear theory can solve, leading in later years, in the hands of mathematicians superior to FOURIER, to searching studies of existence, uniqueness, smoothness, stability, and numerical calculation.

But all this had no influence on thermodynamics. Until about 1965 there was no definite field theory of the interaction of heat and work in general. Except for the very special cases of linear thermo-elasticity and linearly viscous fluids, thermodynamics had no field equations and hence no problems of boundary values or initial values to solve. The influence of FOURIER, and it was vast, was upon mechanics, acoustics, electromagnetism, the theory of approximation, probability, and functions of a real or complex variable—in two words, upon the linear field theories and upon pure mathematics—but not upon thermodynamics. FOURIER's title is belied by the contents: A theory of temperature differences he did give, but not a theory of heat.

Alas, FOURIER's brilliance in his own bailiwick had, indirectly, a negative effect upon thermodynamics. Glorying in his strictly linear theory, he taught a century of physicists and ancillary mathematicians that to ascend Parnassus they needed only by use of no more than sines, cosines, and a few other special functions, linked by the symbols $+ - \times \div \sum \partial \int$, spew out explicit and detailed solutions for special cases. From FOURIER's work grew that meaningless and popular Germanic shibboleth, "solution in closed form", which in the relatively blessed period before teams of apes twiddling costly machines reduced mathematics to the Cinderella of the sciences was sometimes used to depict the social difference between "applied" and "pure" mathematics. Indeed, the simpler the constitutive relation, the vaster the class of easy problems amenable to essentially routine mathematics! By paring the physical model to the bone, the theorist may extract from it incredibly precise predictions about the most intimate detail in bodies of the most complicated shapes. In FOURIER's theory the student must pay as the price for such detail and such precision a willingness to admit that differences of temperature propagate at infinite speed through material bodies and that the temperature within a body can be determined without taking account of the changes which the flow of heat into it may effect upon its size and shape.

Fortunately or unfortunately, according to taste, the constitutive relations natural and useful in thermodynamics are not linear. The kind of problem FOURIER deftly codified does not exist the moment his theory is enlarged so as to take just account of work done.

5. Act II. Dissipationless Work: CARNOT

O voi, ch'avete li 'ntelletti sani,
mirate la dottrina che s'asconde
sotto 'l velame de li versi strani.
DANTE, *Inferno* IX, 61–63.

5A. The General Quality of CARNOT's Treatise

In 1824, two years after the long delayed printing of FOURIER's theory of heat in its final form, appeared CARNOT's booklet called *Reflections on the Motive Power of Fire, and on Machines Fitted to Develop that Power*[1]. [Little of any consequence regarding this subject was then known. Anyone skeptical here need not resort to the writings of engineers, inventors, and constructors. Just eight years before CARNOT's work was published, a leading physicist[2] of the day could give his readers in a whole chapter on steam engines no more than an illustrated description of the machines, embellished by a few scientific terms and some numerical data regarding them, followed by a sketch of their evolution during the preceding 111 years, and finish with a discussion of how much work a horse of mean strength can do in a day.

[CARNOT's approach is entirely new.] After some opening remarks on the usefulness of steam engines he states (pp. 7–9) that "the phenomenon of the production of motion by heat has not been studied from a sufficiently general point of view." It is necessary, he writes, to "establish arguments applicable not only to steam engines but also to every imaginable heat engine...." Purely mechanical machines, CARNOT recalls, can be "studied down to their smallest details" by means of "the mechanical theory."

> All cases are foreseen, all imaginable motions obey certain general principles, firmly established and applicable under all circumstances. Such is the character of a complete theory. A comparable theory plainly

[1] CARNOT [1824]. Citations of pages refer to the first edition, and the translations into English are mine.
[2] BIOT [1816, Chapter VI].

wants for heat engines. We shall have it only after the laws of physics are extended enough, generalized enough, to make known beforehand all the effects of heat acting in a determined manner on any body.

[Despite his schooling at the Ecole Polytechnique and despite his early training by his father, who was a thoughtful mathematician, CARNOT does not follow the tradition of eighteenth-century rational mechanics he has just praised for its generality and extent. Instead, the sardonic muse directs him to write in a medium that anybody can understand. An obvious necessary condition is that no mathematics be used in the main text. This condition did not turn out to be sufficient. Among all writers on natural philosophy only CARNOT equals the pre-Socratics in ability to provoke an infinite sequence of cyclic quandary, acute and painful ponderation, conjecture, gloss, controversy, and quandary again, which bears witness that the outcome is comprehensible by nobody. It is easy to say, and it has often been said, that CARNOT wrote for engineers, but I can find no evidence that any engineer ever read and applied his results, which remained altogether neglected for twelve years, until CLAPEYRON put them into a semi-mathematical and hence at least semi-concrete form (see below, §6A), whereupon they soon began to attract notice. CARNOT's physical principles and such logic as he chose to bring to bear upon them he blurred through a veil of popular science. He is reported to have insisted that his brother, untrained in the subject, read and criticize the work; according to the legend, his brother understood it perfectly. Later students, unable to seek help from that brother, have puzzled, are puzzling, and forever will puzzle over it. In CARNOT's treatise we encounter that fuzziness which was to become and remain a distinguishing feature of thermodynamics for bewildered outsiders. He rivals HERAKLEITOS the obscure.

[That does not mean that CARNOT wrote nonsense. Far from it. In the following pages I seek to interpret CARNOT's claims as reflecting some mathematical statement, capable of dissection and proof or disproof by mathematical reasoning. This may not be just to CARNOT. Certainly, as we shall see, he comes off very well by this test.

[CARNOT's preference for reasoning in words rather than by mathematics is not the only obstacle to our understanding what he wrote.] When he comes to evaluate the motive power of an engine, he uses only specific numbers, and these numbers rest in part on properties of coal and steam. His main conclusions of this kind are (pp. 82, 84, 114–115):

1. 1000 units of heat[3] passing from a body maintained at the temperature of 1° to another maintained at 0° would produce, in acting upon air, 1.395 units of motive power (*i.e.* work[4]).

---

[3] A "unit of heat" is the quantity of heat required to raise by 1°C the temperature of 1 kg of water (p. 81).

[4] A "unit of motive power" is the work required to carry one cubic meter of water upward one meter (footnote, pp. 6–7).

2. 1000 units of heat passing from a body maintained at the temperature 100° to another maintained at 99° would produce, in acting upon steam, 1.112 units of motive power.

3. On the basis of the second figure, the motive power of 1 kg of carbon is 3920 units.

To justify his first and second figures, he presents strings of additions and multiplications of particular numbers. He regards the third figure as precarious. To obtain it, he starts from the second figure, written as 1.12, states that "if the motive power were proportional to the fall of caloric,..., nothing would be easier than to estimate it for 1000° to 0°." It would then be 1120. "But as this law is only approximate and perhaps deviates very much from the truth at high degrees, we can do no more than make an entirely crude evaluation. We shall suppose the number 1120 reduced by a half, that is 560. Since a kilogram of carbon produces 7000 units of heat and since the number 560 is relative to 1000 units, we must multiply it by 7, which gives 7.560 = 3920. There is the motive power of a kilogram of coal."

CARNOT's thermodynamics is purely and frankly phenomenological. He states his position clearly (p.15, footnote):

> I regard it useless to explain here what is quantity of caloric or quantity of heat (for I employ these two expressions indifferently[5]), or to describe how to measure these quantities by the calorimeter. Neither will I explain what is meant by latent heat, degree of temperature, specific heat, *etc.* The reader should be grown familiar with these terms through study of the elementary treatises of physics or chemistry.

While CARNOT regularly speaks of "all bodies", he regards the volume V and the temperature θ and certain functions thereof as sufficient to describe the condition of such bodies. For example (p. 37, footnote), he speaks of the "state considered relatively to the density, the temperature, and the kind of aggregation [*i.e.*, phase]...." [The "thermodynamic state", that king cobra in the pit of thermodynamic vipers, here first with innocuous semblance darts his forked tongue. Of course the word "state" as a general, vague term of ordinary speech had existed for a long time. The passage from a paper by

[5] Some persons have claimed that CARNOT intended a difference between "calorique" and "chaleur", using the former to denote what came later to be called "entropy". For references see Footnote 1 to §27 in the corrected reprint of my *Mechanical Foundations of Elasticity and Fluid Dynamics* (1952/3), N.Y., Gordon and Breach, 1966. I regret to have to confess that I once shared this superficial opinion, which MENDOZA in his edition of CARNOT [1824] has demolished. Indeed, as we shall see in §10A, the heat functions of CARNOT and LAPLACE on the one hand, and CLAUSIUS' entropy on the other, are different special cases of a general function introduced by REECH, and although this fact accounts for their having certain formal properties in common, if one exists the other does not.

Some specific comments in this regard may be found below in Footnote 3 to §5S.

LAVOISIER & LAPLACE quoted above in Footnote 9 to §2C uses it in this popular, conversational sense. What is new here and is destined to be deadly later is the illusion that the "state" may be identified with a finite number of variables such as ρ, θ, and "the kind of aggregation", an illusion that was to keep the theory of heat and temperature mainly in the kindergarten until recently[6]. In my comments I will not use the term again, since it represents no rational concept, beyond, of course, the trivial remark that sometimes a small number of independent variables suffices for an adequate description of a deformable body.]

For all of his calculations, CARNOT presumes the ideal gas law[7] $pV = R\theta$.

[Like FOURIER, CARNOT cannot separate a general idea from its application to a special constitutive relation. He limits the working substance to that most special of materials, an ideal gas, in the most special of circumstances, uniform fields of temperature and density. While LAPLACE's picture of the action of heat had been dismayingly complex, so much so that for him p and ρ did not suffice to describe the condition of anything but a dilute gas, CARNOT goes to the opposite extreme of oversimplification. CARNOT considers the effects of change of volume in circumstances such as to render the field of temperature uniform. FOURIER effectively neglects changes of volume and considers only the effects of conduction and radiation of heat upon the temperature field of a rigid body. The constitutive relations selected by the two principal founders of thermomechanics have as their common domain only the trivial case in which nothing at all happens[8]. From the dates of the two founders'

---

[6] For a specimen of the infinite confusion to which this simple word may give rise when someone tries to make it a scientific term, I quote BRIDGMAN [1941, pp. 58–59 of the reprint of 1961]:

> What is to be understood by "state" of the body? There is danger that it may degenerate into a tautology, because we might say that by definition the body has returned to its initial state when the total work and heat of the sequence of processes is zero. There must be more physical content to the concept of state than this. Actually, as already suggested in the discussion of temperature, "state" has an independent significance, and connects with the "properties" of the body—a body has resumed its initial state when all its properties have resumed their initial values. The first law states that there is some function of the parameters that determine the state (or the properties) of the body such that the difference of the function for two different states is the sum of the net heat and the work entering the body during the change from one state to the other, no matter what the details by which heat and work are imparted to bring about the change of state.

[7] *Cf.* Footnote 1 to §5G, below.

[8] The processes considered by CARNOT are functions of time alone. According to FOURIER's equation (4E.1), if the temperature field is uniform in space, it is also constant in time. CARNOT's General Axiom (5I.1) refers only to processes in which there are differences in temperature. Thus it cannot contradict what we have just concluded from FOURIER's theory. If we interpret FOURIER's work strictly as applying only to bodies whose volumes do not change, it is trivially consistent with CARNOT's, for then V, θ, and p are all constant, no work is done, and no heat is absorbed or emitted.

books, 1822 and 1824, the two main phenomena associated with heat were divorced. Two separate plots, with two different casts, took their turns upon the stage thenceforth.]

5B. Standard Concepts and Assumptions Used by CARNOT

CARNOT makes it plain, though he does not explicitly say so, that he accepts the Doctrine of Latent and Specific Heats[1] [which in this history we have expressed as follows:

$$Q = \Lambda_V(V, \theta)\dot{V} + \mathrm{K}_V(V, \theta)\dot{\theta} \,, \qquad (2C.4)_r$$

$$\Lambda_V > 0 \,, \qquad \mathrm{K}_V > 0 \,. \qquad (2C.5)_r$$

CARNOT does not write out these relations.] He is perhaps the first to introduce explicitly the concepts of *process*[2], *heat absorbed*, and *heat emitted*[3]. [We have defined these formally above:

$$C^+ \equiv \tfrac{1}{2}\int_{t_1}^{t_2} (Q + |Q|)dt \geqq 0 \,,$$

$$C^- \equiv \tfrac{1}{2}\int_{t_1}^{t_2} (|Q| - Q)dt \geqq 0 \,. \qquad (2C.22)_r$$

Like LAPLACE, whom he cites in this regard, he conceives an adiabatic

---

[1] *Cf.* the passage on p. 15 which we have quoted in the preceding section. On pp. 31–32 he writes, "The caloric used [when] no change of temperature occurs we shall call 'caloric due to change of volume'." In our notation that is

$$\int_{V_1}^{V_2} \Lambda_V(x, \theta)dx$$

when $\theta =$ const. *Cf.* also Footnotes 6 and 12 to §2C, above.

[2] "Les opérations", p. 10 *et passim.*

Dissent might be justified here. CARNOT in some passages may mean by "une opération" that which happens not only to the body *but also to its surroundings.* Such passages are quoted and cited in §§5D–5J. My critique there reflects my own preference for explicit mathematical treatment upon a specific conceptual framework of ideas with specific, explicit interpretation. That critique and the later parts of this chapter show that *such an interpretation of CARNOT's ideas is tenable* in nearly all cases. I do not mean to imply that no other interpretation is tenable. Perhaps a muddy author deserves muddy critics. If so, CARNOT has received his just deserts already and in abundance.

[3] "Les quantités de chaleur absorbées ou dégagées", p. 37, footnote; p. 52; p. 55. On p. 76, footnote, CARNOT denotes by *e* the heat absorbed in a Carnot cycle.

process[4] as describing very rapid changes. He takes for granted the *reversibility*[5] of work done and heat added, [which we have expressed above by formal reversal theorems:

$$C(-\mathscr{P}) = -C(\mathscr{P}) , \qquad (2C.7)_r$$

$$L(-\mathscr{P}) = -L(\mathscr{P}) .] \qquad (2C.21)_r$$

Essential to most of CARNOT's arguments, [whether obviously so or through implications which to him lay hidden,] is the *Caloric Theory of heat* [which we have specified in §3C. Like LAPLACE before him,] CARNOT assumes that the quantity of heat in a body is determined, to within an additive constant, by the volume and temperature of that body. He says so again and again: "Thus the production of motive power is due...not to any real consumption of caloric, *but to its transport from a warm body to a cold body*..." (pp. 10–11). Also[6], in regard to the famous argument alleged to prove that all bodies

---

[4] On p. 29 CARNOT remarks that the effects of a rapid change of volume are different from those of a slow one:

> When a gaseous fluid is rapidly compressed, its temperature rises; on the contrary, [the temperature] falls when the fluid is rapidly expanded. There we have one of the facts best of all established by experiment. I shall take it for the basis of my proof.

A footnote (pp. 29–30) adduces four facts of this kind; one of these is

> The results of experiment on the speed of sound. Mr. de Laplace has shown that to subject these results exactly to the theory and the calculation, heating of the air by a sudden compression must be assumed.

[5] He asserts this reversibility only for Carnot cycles. P. 19 (repeated almost verbatim on p. 35): "The operations I have just described could have been effected in an inverse sense and inverse order." Also, p. 36:

> The result of the first operations was to produce a certain quantity of motive power and to transport some of the caloric of body A to body B; the result of the inverse operations is to consume the motive power produced, and to return some caloric from body B to body A, in such a way that the two sequences of operations annul each other, in some way neutralize each other.

[6] P. 37, footnote. The quotation continues, "On the other hand, just to mention the matter in passing, the main foundations upon which the theory of heat rests ought to be subjected to the most careful examination. Several facts of experiment seem almost inexplicable in the present state of that theory."

From notes published after CARNOT's death we know that he came to reject this basis of "the whole theory of heat". Indeed, MENDOZA in his edition of CARNOT [1824] calls attention to the differences between the manuscript and the text as printed. For example, on p. 89 of the book (p. 46 of MENDOZA's edition), we read "Nevertheless, the fundamental law I desired to confirm seems to me to require new tests if it is to be set beyond doubt; it rests upon the theory of heat as that theory is conceived today; and I must admit that to me this foundation does not seem to be unshakeable. Only new experiments could decide the question ...." At this point in his manuscript CARNOT had expressed the very opposite opinion: "The fundamental law that we proposed to

have the same motive power when they undergo Carnot cycles corresponding to the same temperatures:

> I suppose implicitly in my proof that if a body has suffered any changes whatever and that if after a certain number of transformations it is brought back to its original state,... [it will be] found to contain the same quantity of heat as it contained at first, or, in other words, that the quantities of heat absorbed or emitted in its various transformations are exactly compensated. This fact has never been subjected to doubt; it has been assumed straight off without a thought and has been verified thereafter in many cases by experiments with the calorimeter. To deny it would be to overturn the whole theory of heat, which is based upon it.

[Not only that, we shall see in §5J that his calculations which refer to finite differences of temperature, while some of them seem to be in part independent of that assumption, are not so, but in fact require it.] Formally, CARNOT introduces a *heat function* $\mathrm{H_C}(V, \theta)$ (p. 63 and p. 77, footnote, denoted by s), the value of which is the quantity of heat in a fluid body, to within an additive constant. Thus in a process along any path $\mathscr{P}$ leading from (V_1, θ_1) to (V_2, θ_2) the heat added is given by[7]

$$C(\mathscr{P}) = \mathrm{H_C}(V_2, \theta_2) - \mathrm{H_C}(V_1, \theta_1). \tag{5B.1}$$

Henceforth in this tragicomedy the term *Caloric Theory* will refer to the existence of a heat function, not to the broader idea stated in §3C and illustrated in §4F.

CARNOT is thoroughly familiar with the concept of *work*, positive or negative or null, done by a body of gas in a process $V(t)$, $\theta(t)$ whose path is $\mathscr{P}$:

$$\begin{aligned} L(\mathscr{P}) &= \int_{t_1}^{t_2} \varpi(V(t), \theta(t))\dot{V}(t)dt, \\ &= \int_{\mathscr{P}} \varpi(V, \theta)dV. \end{aligned} \tag{2C.20$_{2,3r}$}$$

---

confirm seems to us to have been placed beyond doubt, both by the reasoning which served to establish it, and by the calculations which have just been made."

I relegate this matter to a footnote because I write here, not the history of ideas about heat, but the history of thermodynamics, which is a mathematical theory, and all of CARNOT's reasoning about finite differences of temperature, whether in his notes written after the *Réflexions* or in the book as published, is inextricably entwined with the Caloric Theory of heat, as my text below makes clear. For infinitesimal differences of temperature the whole matter is blurred, as we see below in §5J.

[7] CARNOT's heat function is related as follows to LAPLACE's $\mathrm{H_L}$, which we have discussed in §3C:

$$M\mathrm{H_L}(\varpi(V, \theta), M/V) = \mathrm{H_C}(V, \theta) + \text{const.}$$

5C. The Carnot Cycle

To represent the action of a heat engine CARNOT introduces a cycle[1], in which a gaseous body is brought back at the time t_2 to the volume and temperature it had at the earlier time t_1.

[We may picture a cycle as an oriented closed curve $\mathscr{C}$ in the V–θ quadrant (Figure 2, *cf.* Figure 1); no such diagram was used by CARNOT. A cycle des-

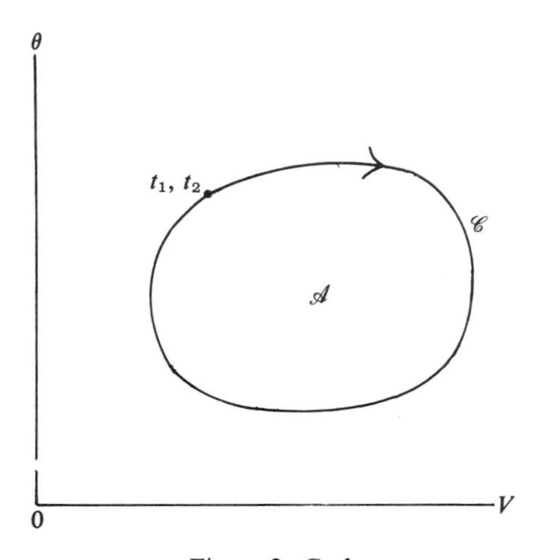

Figure 2. Cycle

cribed several times is again a cycle. To avoid confusion, we sometimes call *simple* a cycle described exactly once.

[The existence of a heat function H_{C} such as to satisfy (5B.1) is equivalent to the following statement: In every cycle $\mathscr{C}$,

$$C^+(\mathscr{C}) = C^-(\mathscr{C}). \tag{5C.1}$$

This relation, which CARNOT uses frequently, characterizes the Caloric Theory of heat as we shall use that term henceforth.]

The only cycle CARNOT considers explicitly is the one since named after him[2]. A body capable of expansion and contraction, and hence capable of

---

[1] P. 36, "cercle d'opérations"; p. 56, "un cercle complet d'opérations".

[2] Pp. 32–34, after an incomplete description on pp. 17–18. CARNOT's terms are "foyer" and "réfrigérant". His specification does not presume that heat be indestructible. In the reformulation by THOMSON [1849] one of the steps was stated in terms of a heat function, giving rise to some misunderstanding of what CARNOT himself had done. The note by KELVIN's brother J. THOMSON [1849] in reality corrects KELVIN, not CARNOT.

We have seen above that CARNOT thinks it is the rapidity of some natural processes that makes them virtually adiabatic. He does not mention that idea when he describes his cycle. At one point (p. 33), and at that point only, he states that the isothermal parts of his cycle should be "gradual", presumably to avoid "useless re-establishment of equilibrium in the caloric" (p. 35).

doing work, is caused to expand by absorbing heat from a *furnace* at the temperature θ^+; then insulated, so that it continues to expand while its temperature falls adiabatically to the temperature θ^- of a *refrigerator*; then put in contact with the refrigerator so as to give up heat and thereby contract; finally insulated again and allowed to contract to its initial volume while its temperature rises adiabatically to the temperature θ^+. The contact between the working body and the furnace (p. 33) is through a "wall which we shall suppose transmits caloric easily."

[It seems to me that CARNOT always envisions an ideal body which he conceives as having at any one time the same temperature at all of its points, no matter how it be heated or cooled. When two such bodies are put in contact, CARNOT expects that heat will flow from the one to the other through their common boundary, which "transmits caloric easily", in such a way as to make the cooler body become uniformly hotter all at once, and the hotter body likewise uniformly cooler. Except, possibly, for a vague remark here and there, CARNOT seems to suppose his ideal bodies incapable of conducting heat in the sense that the theories of LAMBERT, BIOT, and FOURIER represent.]

Formally, a Carnot cycle[3] consists in two distinct isotherms alternating with two distinct adiabats, [as represented in Figure 3, in which the form of the

---

[3] The historical literature has concerned itself with such things as whether CARNOT really described the whole cycle, and where he chose to begin it.

Do Carnot cycles exist? Surely the answer is yes if through every point of the V–θ quadrant passes one and only one adiabat, which is nowhere tangent to an isotherm, and if every pair of isotherms is connected by some adiabat. A modern textbook bludgeons the student into acquiescence by exhibiting the adiabats $\theta V^{\gamma-1} = \text{const.}$ of an ideal gas (*cf.* (3D.5)), but invariably the paedagogic author forgets to warn the innocent reader that *unless* $\gamma = \text{const.}$ *upon these curves, they are not adiabats*, for otherwise they do not satisfy (3C.10). Since CARNOT is to reject the possibility that $\gamma = \text{const.}$ (§§5S–5T, below), the issue is not one of "pure mathematics".

A differential equation for determining the adiabats follows from (2C.4) and (2C.5):

$$\frac{d\theta}{dV} = -\frac{\Lambda_V}{K_V} .$$

The nature of the adiabats is thus a constitutive property of the gaseous body. If Λ_V/K_V is Lipschitz-continuous, one and only one adiabat passes through each point of every *sufficiently small region*, but we cannot generally conclude that adiabats exist in the large. For many purposes, sufficiently small Carnot cycles suffice.

By appeal to (2C.5) we see that when an adiabat exists, θ is a strictly decreasing function of V along it.

Like this last conclusion, Figure 3 fails to be correct if (2C.5) is violated. In a region where Λ_V may change sign (as in modern thermodynamics it certainly may and in some cases must), Carnot cycles looking entirely different from Figure 3 are possible. One such is sketched below in Footnote 4 to §5M. *Cf. Concepts and Logic*, §7.

If $\Lambda_V = 0$ in a region, the adiabats and isotherms coincide there, so no Carnot cycles exist; CARNOT's General Axiom is left impotent then for want of the objects to which it refers. CARNOT's own statements to the effect that $\Lambda_V > 0$ we have quoted in Footnote 6 to §2C. We have noted there that consequently $K_p > K_V$.

Finally, it is worth mentioning that if $K_p = K_V$, LAPLACE's celebrated explanation of the speed of sound (§3F) is voided.

adiabats and the sense of description reflect the standing constitutive inequalities

$$\Lambda_V > 0 , \qquad K_V > 0 . \tag{2C.5)$_r$}$$

[The temperatures θ^+ and θ^- are the *operating temperatures* of the cycle. Since the adiabats of a given body are known curves, obtained from constitu-

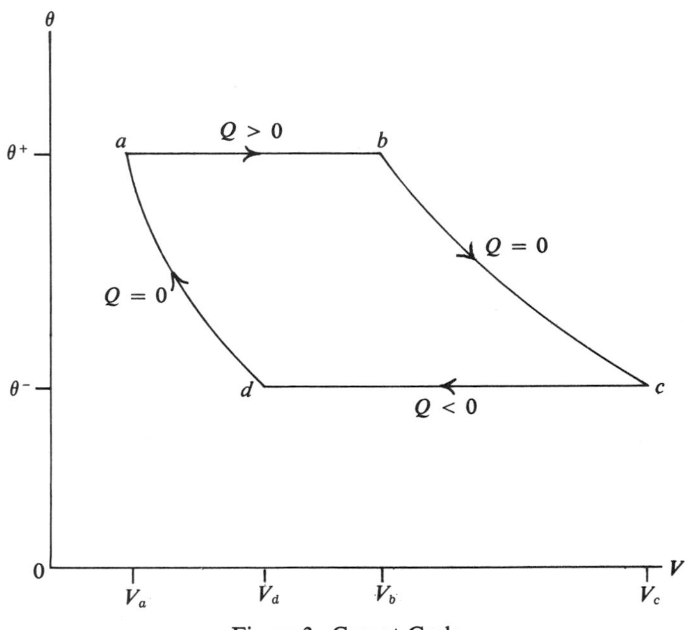

Figure 3. Carnot Cycle

tive properties of that body by integrating the differential equation given in Footnote 3, a Carnot cycle for that body is determined uniquely by the operating temperatures θ^+ and θ^- along with the volumes V_a and V_b which determine its upper isothermal segment. For definiteness in what follows we shall make it a *part of the definition* of a Carnot cycle that

$$\theta^+ > \theta^- , \qquad C^+(\mathscr{C}) > 0 , \tag{5C.2}$$

thus excluding trivial exceptions to general assertions.

[If a simple Carnot cycle is described repeatedly, the result is again a Carnot cycle, but the reverse of a Carnot cycle is not a Carnot cycle.

[The definition of a Carnot cycle uses only the concepts of the Doctrine of Latent and Specific Heats. It does not presume any relation between heat and work. In the context of the Caloric Theory of heat, which CARNOT invariably uses, we can calculate $C^+(\mathscr{C})$ and $C^-(\mathscr{C})$ as differences of values of the heat function $\mathrm{H_C}$:

$$\begin{aligned} C^+(\mathscr{C}) &= \mathrm{H_C}(V_b, \theta^+) - \mathrm{H_C}(V_a, \theta^+) \ , \\ C^-(\mathscr{C}) &= \mathrm{H_C}(V_c, \theta^-) - \mathrm{H_C}(V_d, \theta^-) \ , \end{aligned} \tag{5C.3}$$

and these two differences are equal.]

5D. CARNOT's Claim that Carnot Cycles Attain Maximum Efficiency

CARNOT writes (pp. 28–29)

> According to the ideas established so far we can with sufficient accuracy compare the motive power of heat to that of a fall of water: Both have a maximum that cannot be exceeded, whatever be...the machine used to receive the action of the water, and whatever be...the substance utilized to receive the action of the heat. The motive power of a fall of water depends upon its height and the quantity of liquid; the motive power of heat likewise depends upon the quantity of caloric employed and what we can and shall indeed call the height of its fall, that is to say, the difference of temperatures of the bodies between which is effected the exchange of caloric. In the fall of water the [maximum] motive power is rigorously proportional to the difference of the levels of the upper and lower reservoirs. In the fall of caloric the motive power no doubt increases with the difference of the temperatures of the hot body and the cold body, but we do not know whether it be proportional to this difference. We do not know, for example, if the fall of caloric from 100° to 50° produces more or less motive power than the fall of this same caloric from 50° to 0°. That is a question I plan to examine later on.

Cf. also p. 96. As KELVIN[1] was to put it, by "letting down" heat to a lower temperature, we can cause work to be done. For CARNOT the difference of temperatures is essential (p. 16, *cf.* also p. 12): "*Wherever a difference of temperature exists, there motive power can be produced.*" CARNOT accepts the Caloric Theory of heat, $C^+(\mathscr{C}) = C^-(\mathscr{C})$ for every cycle $\mathscr{C}$. This assumption is reflected on pp. 10–11:

> Thus the production of motive power is due...not to any real consumption of caloric, *but to its transport from a warm body to a cold body*....
>
> According to this principle, in order to produce motive power it is not enough to produce heat: Cold, too, must be procured; without it, heat would be useless.

[1] THOMSON [1848], W. THOMSON [1849, §§ 10,16,29].

[This statement has been misinterpreted[2]. It does not refer to C^+ or to C^-; it means only that a furnace alone will not suffice, we must provide also a refrigerator.

[Without first explaining what he means by "maximum",] CARNOT refers to "the maximum motive power of steam" and "the maximum motive power that can be effected by any means whatever" (pp. 21–22). He continues (pp. 22–23),

> It would be just to ask..., What is the sense of the word *maximum*?...
>
> Since any re-establishment of equilibrium in the caloric can be the cause of motive power, any re-establishment of equilibrium that occurs without producing this power should be considered a true loss. Now, very little reflection suffices to show that every change of temperature which is not due to a change of volume...can be nothing but a useless re-establishment of equilibrium in the caloric. Thus the necessary condition for the maximum is that *in the bodies employed to effect the motive power of heat there be no change of temperature not due to a change of volume*. Conversely, whenever this condition is fulfilled, the maximum will be attained.

---

[2] The misinterpretation: In any cyclic process that does positive work by absorbing heat, some heat must be emitted. These words, indeed, assert one of the many "Second Laws", but CARNOT had no need of that one, since the Caloric Theory makes it trivial. He refers here to another one, one which GIBBS [1873, *1*, p. 10 of the reprint in his *Collected Works*] expressed with his usual insight:

> ...heat received at one temperature is by no means the equivalent of the same amount of heat received at another temperature.... But no such distinction exists in regard to work. This is a result of the general law, that heat can only pass from a hotter to a colder body, while work can be transferred by mechanical means from one fluid to any other, whatever may be the pressure.

CARNOT had written something of this kind (pp. 34–35) in regard to his cycle:

> ...but we must notice that at equal volumes ... the temperature is higher during motions of dilatation than in motions of compression. During the former the elastic force of air is higher and consequently the quantity of motive power produced by the motions of dilatation is more considerable than that which is consumed to produce motions of compression. Thus an excess of motive power results....

CARNOT's statements presume that $\Lambda_V > 0$. If $\Lambda_V < 0$, the roles of compression and dilatation are exchanged.

MAXWELL [1871, p. 148] was to misinterpret CARNOT's "elastic force" as referring to "energy":

> Carnot, therefore, was wrong in supposing that the mechanical energy of a given quantity of heat is greater when it exists in a hot body than when it exists in a cold body. We now know that its mechanical energy is exactly the same in both cases, although when in the hot body it is more available for the purpose of driving an engine.

Thus, CARNOT concludes, the maximum work a given quantity of heat can do in a cycle is achieved when all that heat is absorbed at the higher temperature and then emitted at the lower temperature. This condition describes a Carnot cycle (§5C). Also later (p. 35), again in regard to a Carnot cycle,

> Thus the air has served as a heat engine; we have even employed it in the most advantageous way possible, for there has been no useless re-establishment of equilibrium in the caloric.

A "true loss" occurs (p. 24) "in the direct passage of the caloric from a more or less heated body to a colder body. This passage takes place mainly upon contact of bodies of different temperatures...." It is avoided in the Carnot cycle, because there only bodies at the same temperature come into contact with each other[3]. CARNOT explains (p. 25) how he conceives the transfer of heat in a Carnot cycle to occur:

> In truth, rigorously, things cannot happen as we have assumed. To cause caloric to pass from one body to another, the former must exceed in temperature, but by as little as we please. We may take the excess as null in theory without diminishing the exactness of the reasoning.

5E. Formal Statement and Critique of CARNOT's Claim of Maximum Efficiency

CARNOT himself does not make clear just what it is he claims his cycle renders a maximum. In the terms introduced by later physicists, two classes of competitors are implied:

1. Reversible cycles.
2. Both irreversible and reversible ones.

[3] CARNOT refers (p. 16) to percussion and friction as mechanical agents for effecting a change of temperature. He nowhere mentions the possibility that a body may change its shape without changing its volume, and he nowhere refers to a non-uniform field of temperature within the body. Two passages have been adduced by those who maintain the contrary, namely pp. 34–35 in regard to "useless re-establishment of equilibrium in the caloric" and pp. 89–90 in regard to motions of a metal bar heated at one end. I have studied these passages many times, and I cannot find in them anything more definite than the idea that when two bodies at different temperature come into contact, heat flows from the hotter to the colder and in so doing is partly wasted. The second passage with its repeated explicit reference to *contact* is the clearer. It is *we* who when we think of two such bodies in contact expect to find in each of them a field of temperature varying with position. We are not justified in attributing our ideas to CARNOT unless he himself has left at least some sign that he, too, employed them. In §5C I have explained how I think CARNOT conceived the flow of heat incident upon contact.

Only the former are compatible with the Doctrine of Latent and Specific Heats, and that axiom describes only such processes. The following claim may be stated upon the conceptual armature provided thereby. Not knowing what CARNOT really meant, I offer it as a possible interpretation, one that lends itself to precise statement and mathematical treatment.

Claim I (Maximum Efficiency for a Given Body). *Let a cycle $\mathscr{C}$ and a Carnot cycle $\mathscr{C}_{\mathrm{C}}$ have the same extreme temperatures and absorb the same amount of heat. If $\mathscr{C}$ is not a Carnot cycle, then*

$$L(\mathscr{C}) < L(\mathscr{C}_{\mathrm{C}}) \ . \tag{5E.1}$$

It is not claimed here that all Carnot cycles which absorb the same amount of heat and have the same operating temperatures do the same amount of work. Rather, each does more work than any corresponding cycle of any other kind.

The modern reader sees an important corollary of (1). Let $\mathscr{C}'_{\mathrm{C}}$ be any Carnot cycle that has the same operating temperatures as $\mathscr{C}_{\mathrm{C}}$ and absorbs the same amount of heat. Then we can construct a sequence of cycles $\mathscr{C}_{\mathrm{n}}$ that are not Carnot cycles but have the same extreme temperatures as $\mathscr{C}'_{\mathrm{C}}$, absorb the same amount of heat as it does, and as $n \to \infty$ become arbitrarily near to $\mathscr{C}'_{\mathrm{C}}$. For example, we may cut an arbitrarily small Carnot cycle out of the lower right-hand corner of $\mathscr{C}'_{\mathrm{C}}$. To each of these new cycles $\mathscr{C}_{\mathrm{n}}$ we apply (1):

$$L(\mathscr{C}_{\mathrm{n}}) < L(\mathscr{C}_{\mathrm{C}}) \ . \tag{5E.2}$$

Letting n tend to ∞, from the continuity of the integral defining L we conclude that

$$L(\mathscr{C}'_{\mathrm{C}}) \leqq L(\mathscr{C}_{\mathrm{C}}) \ . \tag{5E.3}$$

Interchanging the roles of $\mathscr{C}'_{\mathrm{C}}$ and $\mathscr{C}_{\mathrm{C}}$, we conclude that $L(\mathscr{C}_{\mathrm{C}}) \leqq L(\mathscr{C}'_{\mathrm{C}})$. Therefore

$$L(\mathscr{C}_{\mathrm{C}}) = L(\mathscr{C}'_{\mathrm{C}}) \ . \tag{5E.4}$$

We have established[1] the following ***corollary*** of Claim I:

For a given body, all Carnot cycles that have the same operating temperatures and absorb the same amount of heat have also the same motive power.

While CARNOT does not present this argument, it easily falls within the scope of his mathematical apparatus. Perhaps, since he stated the fact, he thought it so obvious a consequence of (1) as to need no proof. We may phrase the corollary as follows:

[1] This argument uses suggestions made by Dr. RICHARD JAMES.

For each body B *there is a function* G_B *such that for each Carnot cycle* $\mathscr{C}$ *that* B *may undergo*

$$L(\mathscr{C}) = G_B(\theta^+, \theta^-, C^+(\mathscr{C})) . \tag{5E.5}$$

The foregoing statement is not so astonishing as it might at first glance seem. As we have remarked in §5C, a Carnot cycle $\mathscr{C}$ is determined uniquely by its operating temperatures θ^+ and θ^- along with the extremities, say V_a and V_b, of its upper isotherm (Figure 3 in §5C). Anything determined by a CARNOT cycle for a given body B is therefore determined by θ^+, θ^-, V_a, and V_b. In particular, there is a function H_B such that

$$L(\mathscr{C}) = H_B(\theta^+, \theta^-, V_a, V_b) . \tag{5E.6}$$

The Doctrine of Latent and Specific Heats:

$$Q = \Lambda_V(V, \theta)\dot{V} + \mathrm{K}_V(V, \theta)\dot{\theta} , \tag{2C.4$_r$}$$

$$\Lambda_V > 0 , \qquad \mathrm{K}_V > 0 , \tag{2C.5$_r$}$$

shows us that

$$C^+(\mathscr{C}) = \int_{V_a}^{V_b} \Lambda_V(x, \theta^+)dx . \tag{5E.7}$$

Appealing to $(2C.5)_1$ again, we see that $C^+(\mathscr{C})$ is an increasing function of V_b when V_a is fixed. Thus (7) may be inverted to yield

$$V_b = g_B(V_a, \theta^+, C^+(\mathscr{C})) . \tag{5E.8}$$

Now we may eliminate V_b from (6) and so obtain

$$L(\mathscr{C}) = H_B(\theta^+, \theta^-, V_a, g_B(V_a, \theta^+, C^+(\mathscr{C}))) . \tag{5E.9}$$

This much follows from the Doctrine of Latent and Specific Heats alone. The corollary expressed by (4) asserts that the argument V_a drops out: The work done by a Carnot cycle absorbing a given amount of heat is independent of the position of the cycle on the two isotherms it employs.

CARNOT's argument to support Claim I is difficult to render precise. I refer to his remarks about a "true loss" and "useless re-establishment of equilibrium in the caloric", quoted in the preceding section. If $\mathscr{C}$ is described by the Doctrine of Latent and Specific Heats, all points of the body which undergoes it have at each time t a common temperature $\theta(t)$. If $\mathscr{C}$ is not a Carnot cycle, some heat is absorbed when $\theta(t) < \theta^+$, or some is emitted when $\theta(t) > \theta^-$. To avoid a difference of temperature between the working body and its surroundings (give them whatever names we desire), we must regulate those surroundings so that their temperature at each time is exactly $\theta(t)$. But then we are not using the same furnace and the same refrigerator as the Carnot cycle. There is now no basis for comparison, and the whole argument collapses.

Very well, then, let us stay with the given furnace and given refrigerator at constant temperatures. How do they affect a body that undergoes a cycle other than a Carnot cycle? Suppose that during part of the interval of time during which the working body is absorbing heat from the furnace its temperature is less than the furnace's: $\theta(t) < \theta^+$. If the interpretation of CARNOT's ideal bodies I have presented in §5C is just, heat from the furnace at the temperature θ^+ will flow instantly across the wall that "transmits caloric easily" (p. 33) and so will instantaneously become heat at the temperature $\theta(t)$. As CARNOT's many statements about "the fall of caloric" make plain, he well knows that *a given amount of heat is capable of doing more work at a higher temperature than at a lower temperature*. Thus when a body undergoes a cycle that is not a Carnot cycle, some of the heat it receives no longer has as much capacity to do work as it had while still in the furnace. "A true loss" results. The cycle cannot do as much work as a corresponding Carnot cycle. The same thing happens if the working body discharges to the refrigerator a part of its heat that is at a higher temperature than θ^-, for that heat has not done all the work it could have, had it remained in the working body until it had fallen to the temperature θ^-.

The Doctrine of Latent and Specific Heats makes *all processes reversible*. The cycle $\mathscr{C}$ we have just considered is one which does work. The reverse cycle $-\mathscr{C}$ consumes work. CARNOT knows full well that heat flows spontaneously only from the hotter body to the colder. That leaves us two ways to try to interpret what CARNOT means by his claim, which here we provisionally accept, that Carnot cycles are the most efficient for given extremes of temperature.

First, holding to the qualification *spontaneously*, we can conclude that the furnace and the refrigerator will not allow the reversed cycle $-\mathscr{C}$ to occur. On the part of $-\mathscr{C}$ that corresponds to the part of $\mathscr{C}$ where $Q > 0$ the working body at the temperature $\theta(t)$ would be in contact with the furnace at the higher temperature θ^+, so heat would not flow out of the working body into the furnace. Bodies subsumed by the Doctrine of Latent and Specific Heats, as we know, are susceptible of *reversible processes only*. Here the reversed process, while possible indeed for the body, is *incompatible* with the surroundings. The experiences of the whole assembly—working body, furnace, and refrigerator—are not reversible. In this sense, then, the Carnot cycle, which is reversible not only by itself but also with its surroundings adjoined provides a reversible system, is compared by CARNOT with the experiences of an irreversible system. *Irreversibility reduces efficiency*, this argument seems to mean.

Second, we may say that while "spontaneous" flow of heat can do work, we can by using up work induce phenomena that are *not spontaneous*. In $-\mathscr{C}$ the heat is driven into the furnace and extracted from the refrigerator. There is then no conflict with the idea of reversibility, not only in the working body but also in the surroundings. In $-\mathscr{C}$ the body acts as a mechanically driven cooler. We may resort here to CARNOT's own mechanical analogy, the waterfall. The water descends spontaneously and does work. The water will not reascend spontaneously to the height from which it fell, but we by doing work upon it may force it back up the hill.

Both choices are matters of interpretation, not mathematics. In both cases the assertion about work done is one which can be phrased entirely in terms of the *working body alone*, with no reference to its effects upon the surroundings: Does a Carnot cycle do more work per unit heat absorbed than any other cycle with the same extremes of temperature? Within CARNOT's theory as it stands this mathematical question has an answer: yes or no. CARNOT does not attack this question at all. In §5M we shall provide a mathematical proof that the answer is yes. The proof will employ CARNOT's basic idea: The motive power of a quantity of heat is greater at higher temperatures, which idea we shall quantify also in §5M.

5F. CARNOT's Claim that the Efficiency of Carnot Cycles is Universal

Having convinced himself that a Carnot cycle for a given body is the most efficient, CARNOT is ready to compare the motive powers of Carnot cycles in different bodies. To this end he invents the famous argument about driving one engine backward to negate the work of another (pp. 20–22):

> But, if there were means of employing heat preferable to those we have used, that is, if it were possible by any method whatever to make caloric produce a quantity of motive power larger than we have made by our first series of operations, it would suffice to draw off a portion of this power in order to cause the caloric, by the method just indicated, to go up again...from the refrigerator to the furnace and re-establish things in their original state, thereby making it possible to recommence an operation altogether like the first, and so on. That would be not only perpetual motion but also the creation of boundless motive force with no consumption of caloric or of any other agent whatever. Creation of this kind is entirely contrary to the ideas received up to now, to the laws of mechanics and sound physics; it is inadmissible. Thus we ought to conclude that *the maximum motive power obtained from use of steam is also the maximum motive power realizable by any means whatever.*

A footnote explains that CARNOT does not really object to perpetual motion in the ordinary sense, exemplified by the ideal pendulum, but to a construction "capable of creating motive power in unlimited quantity".

The "second, more rigorous proof" on pp. 29–36 is just the same. CARNOT then infers (p. 38), "*The motive power of heat is independent of the agents used to realize it; its value is determined solely by the temperatures of the bodies between which is effected, finally, the transport of the caloric.*" The bodies to which he here refers are the furnace and the refrigerator. The "agents" are the possible working bodies: masses of steam, air, liquid, or even solid materials[1].

5G. Formal Statement and Elucidation of CARNOT's Claim of Universal Efficiency

Although CARNOT's expressions are vague, it is clear that by "the means... we have used" he refers to Carnot cycles. Thus he makes

Claim II (Universal Efficiency)[1]. *All Carnot cycles that absorb the same amount of heat and have the same operating temperatures have also the same motive power.*

---

[1] *Cf.* pp. 37–38: "any other body susceptible of changing its temperature by successive contractions and expansions, which includes all bodies in nature, or at least all those fit to realise the motive power of heat." On pp. 89–93 CARNOT explains why solids and liquids are not good working bodies. First, "they are susceptible of little change of temperature through change of volume". That is, the adiabats are too nearly coincident with the isotherms, so a Carnot cycle would not be possible except for small differences of temperature. In addition there are practical reasons: enormous forces would have to be applied, *etc.*

---

[1] The reader of CARNOT's book may be confused by a different and altogether independent claim of universality: The pressure function ϖ is (to within choice of R) "the same for all gases" (p. 46), namely, that given by the ideal gas law (2A.1), the main cases of which CARNOT calls (p. 46) "the law of Messrs. Gay-Lussac and Dalton" and (p. 51) "the law of Mariotte". CARNOT writes the equation on p. 74, footnote. CARNOT's successors will echo this claim of his for some decades. For example, no less a physicist than H. C. OERSTED, writing two years after CARNOT's treatise was published, claimed that his experiments on air, sulphurous acid gas, cyanogen, and "liquid bodies reducible to drops" in an apparatus capable of effecting pressures as great as 110.5 atm. justified the title of his paper: "Experiments proving that Mariotte's Law is applicable to all kinds of gases; and to all degrees of pressure under which the gases retain their aëriform state", *Philosophical Magazine* **68** (1826), 102–111. OERSTED mentions that JACOB [DANIEL?] BERNOULLI and EULER had held a contrary opinion, but nevertheless he regarded his experiments as serving merely to confirm what everyone already believed.

CARNOT's reasoning, for the most part, is essentially independent of this fancied particular universality. Because of CARNOT's leaning to arithmetic, a simple rule for recognizing use of this special assumption is to look for the number 267. The rule is not without exceptions, for by comic chance p. 267 of a book is cited on p. 59. Also, alas (pp. 81 and 82), the specific heat of 1 kg of air is 0.267. The other occurrences of 267 are on pp. 44–45, 60, 74–75, 79, 80, 81.

On p. 51, footnote, without using the number 267, CARNOT remarks deviations from the equation of state for ideal gases and asserts, "The theorems we shall derive here perhaps would not be exact if they were to be applied beyond certain limits of density or temperature; they ought not be regarded as true except within the limits of the laws of Mariotte and of Messrs. Gay-Lussac and Dalton themselves are confirmed."

The reader of this essay will have no trouble in recognizing the particular results that depend upon use of (2A.1), the equation of state of an ideal gas. Among these are $(5K.1)_2$, $(5K.5)_2$, (5L.6), (5O.5), and all the contents of §§5Q and 5R.

We have seen in §5E that his Claim I implies Claim II in the special case when only one body is considered. CARNOT's argument to extend that statement to two different bodies seems to apply, if it applies to anything, to every cycle. However, by bringing the assumptions out into the open we can reduce the passage to sense in terms of the concepts provided by the Doctrine of Latent and Specific Heats. We write $C^+_B(\mathscr{C})$ and $C^-_B(\mathscr{C})$ for the heat absorbed and the heat emitted by a body B in undergoing a cycle $\mathscr{C}$, and we write $L_B(\mathscr{C})$ for the work done by B in that cycle.

α. *Construction.* Let the body B_1 undergo some cycle $\mathscr{C}_1$, and let the body B_2 undergo the reverse $-\mathscr{C}_2$ of a cycle $\mathscr{C}_2$ so adjusted that

$$C^+_{B_1}(\mathscr{C}_1) = C^+_{B_2}(\mathscr{C}_2) \ . \qquad (5G.1)$$

Because

$$C^+(-\mathscr{P}) = C^-(\mathscr{P}) \ , \qquad C^-(-\mathscr{P}) = C^+(\mathscr{P}) \ , \qquad (2C.24)_r$$

we conclude that

$$C^+_{B_1}(\mathscr{C}_1) = C^-_{B_2}(-\mathscr{C}_2) \ ; \qquad (5G.2)$$

because

$$L(-\mathscr{P}) = -L(\mathscr{P}) \ , \qquad (2C.21)_r$$

we conclude that

$$L_{B_1}(\mathscr{C}_1) - L_{B_2}(\mathscr{C}_2) = L_{B_1}(\mathscr{C}_1) + L_{B_2}(-\mathscr{C}_2) \qquad (5G.3)$$

This much, indeed, rests upon *nothing more than reversibility* and the assumption that B_2 in undergoing $\mathscr{C}_2$ can absorb exactly as much heat as does B_1 in undergoing $\mathscr{C}_1$.

β. *Application.* Although CARNOT speaks of "any method whatever", his discussion refers only to Carnot cycles $\mathscr{C}_1$ and $\mathscr{C}_2$ that both *absorb the same amount of heat* and have *the same operating temperatures.* I cannot see that any definite conclusion results unless we restrict attention to these as being the competing "means of employing heat". Then we may conceive B_1 and B_2 as absorbing and emitting heat to *exactly two other bodies*: the furnace, whose temperature is θ^+, and the refrigerator, whose temperature is θ^-. In undergoing $\mathscr{C}_1$, the body B_1 absorbs heat from the furnace; in undergoing $-\mathscr{C}_2$ the body B_2 emits heat to the furnace. From (2) we see that after the two cycles have been completed *the furnace has neither lost nor gained heat.* If $L_{B_1}(\mathscr{C}_1) > L_{B_2}(\mathscr{C}_2)$, it follows from (3) that the *overall result of making* B_1 *traverse* $\mathscr{C}_1$ *and* B_2 *traverse* $-\mathscr{C}_2$ *is to do positive work while the net gain of heat by the furnace is null.*

γ. "*Sound physics*". Because $C^+_B(\mathscr{C}) = C^-_B(\mathscr{C})$ for every cycle $\mathscr{C}$ that B may undergo, the cycles $\mathscr{C}_1$ and $\mathscr{C}_2$ together result in *null net gain of heat not only for the furnace but also for the refrigerator*. Thus $\mathscr{C}_1$ and $-\mathscr{C}_2$ together serve to do positive work yet "re-establish things in their original state". This CARNOT considers "contrary to the laws of mechanics and sound physics".

δ. *Conclusion.* It is contrary to "sound physics" that in the cycles as above constructed we should obtain $L_{B_1}(\mathscr{C}_1) \neq L_{B_2}(\mathscr{C}_2)$. As the possibility that B_1 and B_2 are one and the same body is not excluded, we obtain the following conclusion:

Any two Carnot cycles, if they have the same operating temperatures and absorb the same amount of heat, have also the same motive power.

An equivalent statement in terms of the function G_B in the earlier conclusion

$$L(\mathscr{C}) = G_B(\theta^+, \theta^-, C^+(\mathscr{C})) \qquad (5E.5)_r$$

is as follows:

Let B_1 *and* B_2 *be any two fluid bodies, and let* G_{B_1} *and* G_{B_2} *be the corresponding functions* G_B. *Then on the intersection of the domains of* G_{B_1} *and* G_{B_2}

$$G_{B_1} = G_{B_2} \, . \qquad (5G.4)$$

5H. Critique of CARNOT's Argument to Support Universal Efficiency

CARNOT's argument in support of his Claim II refers to a system of four bodies: two working bodies, the furnace, and the refrigerator. The work done, however, is defined by (2C.20), which refers to the working bodies alone. CARNOT calls upon *properties of the environment of bodies in order to infer properties of bodies exposed to that environment.* Work cannot be done by two bodies of certain kinds, because if it were, *their environment would be unchanged.* This kind of argument does not provide a proof unless properties of the environment are specified along with the properties of the bodies on which it acts. Here the environment is not described by the Doctrine of Latent and Specific Heats, so there is no place in the formal structure where a proof using CARNOT's ideas could start.

Arguments of this kind are common in presentations of classical thermodynamics even today. Here is where they began. Mathematicians instinctively reject such arguments, because they stand above logic. Earlier theories of physics, having been created principally by mathematicians, made no appeal to properties of systems larger than the one being treated. In classical mechanics, for example, nobody ever suggested that something could not happen to a body because otherwise the environment of that body might suffer! This is the point in history where mathematics and physics, which had come together in the sixteenth century, began to part company.

Then there is the appeal to "sound physics". First of all, how can CARNOT object to "the creation of boundless motive force with no consumption of caloric or any other agent whatever"? Earlier (pp. 10–11) he has written that motive power is produced without "any real consumption of caloric...." That consequence of the Caloric Theory of heat is not acceptable today, but certainly it is no more incompatible with the laws of mechanics than is the theory of the ideal pendulum. After having long perpended this matter, I incline to think that the apparent contradiction merely reflects CARNOT's vagueness in expressing his ideas. In both cases, by "consumption" he means heat added. In the earlier instance it is heat added *to the working body*, which he takes as being null; in the latter instance it is heat added *to the furnace*. That all the heat taken from the furnace to effect positive work should have been restored to it, he regards as contrary to sound physics. Because he adopts the Caloric Theory, the heat added to the refrigerator is also null, so "things" are re-established "in their original state", not only the two working bodies but also the furnace and the refrigerator.

CARNOT's Claim I leads to the statement

$$L(\mathscr{C}) = G_{B}(\theta^{+}, \theta^{-}, C^{+}(\mathscr{C})) \ , \qquad (5E.5)_{r}$$

which refers to Carnot cycles alone. It would seem to be weaker than Claim I. Is it? As CARNOT himself showed in part (see §§5K–5R, below), (5E.5) gives rise to a mathematical structure, on the basis of which we may face a definite mathematical problem: *Is Claim I itself true or false*? This problem has two parts:

1. In the framework of the Caloric Theory.
2. In the more general framework of the Doctrine of Latent and Specific Heats.

While there is nothing in the early literature to suggest anyone ever sought a *mathematical* analysis of this question, the question itself is purely mathematical and should be so approached. The answers are as follows:

1. *Caloric Theory*. In §5M and again in §7H we shall prove that (5E.5) when specialized to the Caloric Theory implies Claim I. Therefore, *the Caloric Theory makes the statement* (5E.5) *equivalent to Claim I.*
2. *General Theory*, based on the Doctrine of Latent and Specific Heats. In Chapter 13 of *Concepts and Logic* the reader may see proof that Claim I does *not* generally follow from (5E.5). Thus CARNOT's arguments, which seem to invoke no specific theory of heat, do *not* ensure that (5E.5) shall be anything more than a necessary condition for the truth of Claim I.

In both cases the demonstrations rest upon the mathematical structure provided by (5E.5) and the theory of calorimetry, nothing more.

Nevertheless, by invoking "sound physics" we can still find a place for Claim I: Any thermodynamic theory that does not make Carnot cycles the most efficient should be rejected. It would not be "sound".

Claim II is another matter. To justify it, CARNOT can only appeal to the effect of the working body upon its surroundings, or, conversely, the "perpetual motion" the surroundings would be able to make it effect. Denial of one or another perpetual motion is to become a standard foot of clay for the colossus of thermodynamics as author after author tries to re-erect it. As BRIDGMAN[1] wrote,

> The guiding motif is strange to most of physics: namely, a capitalizing of the universal failure of human beings to construct perpetual motion machines of either the first or the second kind. Why should we expect nature to be interested either positively or negatively in the purposes of human beings, particularly purposes of such an unblushingly economic tinge? Or why should we expect that a formulation of regularities which we observe when we try to achieve these purposes should have a significance wider than the reach of the purposes themselves? The whole thing strikes one rather as a verbal *tour de force*, as an attempt to take the citadel by surprise.
>
> We usually do not proceed like this in other fields.

No, say the thermodynamicists. When reason and experiment fail, appeal to perpetual motion, or scoff at the credulity of those who would have it that "something" can arise from "nothing"[2].

Rather, CARNOT in effect lays down an ***assumption***:

It is impossible for a heat engine to have done positive work yet have restored to the furnace all the heat it previously absorbed from it and have withdrawn from the refrigerator all the heat it previously emitted to it.

This is the earliest of the many different statements physicists call *the Second Law of Thermodynamics*. We shall encounter other misty "Second Laws" below. As BRIDGMAN[3] put it

---

[1] BRIDGMAN [1941, p. 4 of the 1961 edition].

[2] Claims that it is "absurd" to think "something" can be created from "nothing" are common in the history of physics and even more common in the physics classroom. For example, to some Aristotelians it was absurd that a body could go on moving with "nothing to push it", and to some paedagogues now it is absurd that man ever could have believed that a vacuum (nothing) could have pulled up the cylinder of a pump (something). It is equally absurd that electromagnetic waves (something) can exist in a vacuum (nothing).

Even more, physicists today are prone to consider absurd the underlying idea about heat and work in CARNOT's theory, namely that a compressible body by merely transferring heat from a furnace to a refrigerator, without consumption of any, can thereby do positive work. As CARNOT's theory shows, there is nothing absurd about such an idea. That which is untrue is by no means always absurd. Alas, nowadays it is the proclaimed truths of science and society that are more likely to be absurd than the false superstitions of our ancestors.

[3] BRIDGMAN [1941, p. 116 of the reprint of 1961].

> There have been nearly as many formulations of the second law as there have been discussions of it. Although many of these formulations are doubtless roughly equivalent, and the proof that they are equivalent has been considered to be one of the tasks of a thermodynamic analysis, I question whether any really rigorous examination has been attempted from the postulational point of view and I question whether such an examination would be of great physical interest. It does seem obvious, however, that not all these formulations can be exactly equivalent, but it is possible to distinguish stronger and weaker forms.

The list of concepts extraneous to the formal structure that all these "Second Laws" employ would fill a page.

The vagueness and vacillation of CARNOT's concepts and assumptions—sufficiently witnessed by the numerous quotations we have presented and analysed—are typical of a theory not subjected to the discipline of mathematical statement, and perhaps unavoidable in such a theory.

The rest of this chapter concerns the mathematical theory that CARNOT began to build upon the basis of (5E.5) and his Claim II. It can be completed with perfect logic. Therefore what CARNOT did and did not do in regard to it can be subjected to the ordinary criteria of mathematical analysis.

5I. CARNOT's General and Special Axioms

According to CARNOT (see §5D), any difference of temperature suffices to provide motive power. Recalling the consequences of CARNOT's Claims I and II, respectively

$$L(\mathscr{C}) = G_{\mathrm{B}}(\theta^+, \theta^-, C^+(\mathscr{C})) \qquad (5\mathrm{E}.5)_{\mathrm{r}}$$

and

$$G_{\mathrm{B}_1} = G_{\mathrm{B}_2} \ , \qquad (5\mathrm{G}.4)_{\mathrm{r}}$$

we arrive at the basic principle upon which all the specific calculations of CARNOT's treatise are based. I shall call this principle

CARNOT's General Axiom

I. *Let the operating temperatures of a Carnot cycle be θ^+ and θ^-, and let the heat absorbed be $C^+(\mathscr{C})$. Then*

$$L(\mathscr{C}) = G(\theta^+, \theta^-, C^+(\mathscr{C})) \ , \qquad (5\mathrm{I}.1)$$

and

$$G(x, y, z) > 0 \quad \text{if} \quad x > y > 0 \quad \text{and} \quad z > 0 \ , \qquad (5\mathrm{I}.2)$$

II. *The function G is universal, the same for all bodies.*

[Statement II is traditional in its vagueness; in the sentence surrounding (5G.4) we have given it a precise meaning. We do not need to assume that every body is capable of undergoing a Carnot cycle having arbitrary operating temperatures and absorbing an arbitrary amount of heat. So simple an example as the Van der Waals fluid in the liquid region shows that such is not generally the case. Rather, we assume that the function $G(x, y, z)$ is defined for all positive z and all pairs x, y such that $x > y > 0$. The restriction of that G to arguments compatible with the constitutive domain and constitutive functions of some particular body determines through (1) the value $L(\mathscr{C})$ for every Carnot cycle $\mathscr{C}$ that that particular body may undergo. One function G does for all bodies.

[The arguments $x, y, 0$ and x, x, z do not correspond to Carnot cycles as we have defined them, but it is convenient to extend G to those arguments in such a way as to make it continuous at them:

$$G(x, y, 0) = 0 \ , \qquad G(x, x, z) = 0 \ .] \tag{5I.3}$$

CARNOT does not specify the function G. In his first enunciation of his General Axiom (p. 28–29) he states,

> ...the motive power of heat likewise depends upon the quantity of caloric employed and on what we can and shall indeed call the height of its fall, that is to say, the difference of temperatures of the bodies between which is effected the exchange of caloric. In the fall of water, the motive power is rigorously proportional to the difference of level between the upper and lower reservoirs. In the fall of caloric, the motive power doubtless increases with the difference of temperature between the hot body and the cold one, but we do not know whether it be proportional to this difference.

[These two statements together would seem to mean

$$G(x, y, z) = E(x - y, z) \ , \tag{5I.4}$$

$$\frac{\partial E}{\partial x}(x, z) > 0 \ ,]$$

but when CARNOT comes to calculate anything connected with finite differences of temperature, he always supposes that[1]

$$\begin{aligned} G(x, y, z) &= [F(x) - F(y)]z \ , \\ F(x) &> F(y) \quad \text{if} \quad x > y > 0 \ . \end{aligned} \tag{5I.5}$$

[1] It is difficult to locate in his treatise any explicit statement of (5). To see that CARNOT does in fact use it, we need only verify that a number of his specific conclusions are true if (5) holds but *false* otherwise: (5K.5) (as CARNOT himself stated it, namely with F' for μ), (5Q.1) and its consequences, and the considerations of §5S. If we were to deny that CARNOT used (5), we should thus have to maintain that he derived the aforementioned results by incorrect mathematics.

What little mathematical theory CARNOT presents, both that in the long footnote on pp. 73–79 and that given here and there in words, rests upon (5) rather than (1). [I shall call (5) ***CARNOT's Special Axiom.*** Usually we shall write $(5)_1$ in the form

$$L(\mathscr{C}) = [F(\theta^+) - F(\theta^-)]C^+(\mathscr{C}) \ , \tag{5I.6}$$

which follows by substituting $(5)_1$ into (1). The quantity $G(\theta^+, \theta^-, C^+)/C^+$, which has the dimensions of work ÷ heat, may be called the *efficiency* of the cycle. According to CARNOT's Special Axiom, the efficiency is $F(\theta^+) - F(\theta^-)$. If we select any constant J_0 bearing the dimensions of work ÷ heat, we may obtain a dimensionless efficiency $[F(\theta^+) - F(\theta^-)]/J_0$, but in CARNOT's theory there is nothing in favor of doing so, because one such constant is no better than any other.]

From $(5)_2$, which merely expresses a condition that F must satisfy in order that "whenever a difference of temperature exists, motive power can be produced" (p. 16), CARNOT much later (p. 94) draws the following conclusions:

> (1) The temperature of the fluid should be raised at once to the highest degree possible, so as to obtain a great fall of caloric, and consequently a large production of motive power.
> (2) For the same reason the cooling should be carried as far as possible.

In all his numerical calculations as well as in part of his mathematical theory, CARNOT is too cautious to use even his Special Axiom except when θ^+ and θ^- are taken as being nearly equal to each other. [In §5P, below, his caution will be proved justified.]

5J. Critique of CARNOT's General and Special Axioms. Scholia I–III. "Carnot's Function".

CARNOT makes three major assumptions:

1. The Caloric Theory of heat holds[1].
2. The motive power of a Carnot cycle $\mathscr{C}$ is determined by the heat $C^+(\mathscr{C})$ that it absorbs and by its operating temperatures θ^+ and θ^-.

[1] As (5C.3) shows, only a *change* of the amount of caloric in a body, not the absolute amount, plays a part in CARNOT's General and Special Axioms. Thus RUMFORD's observation that heat could be developed indefinitely by doing continual work in deforming a body did not bear in any way on CARNOT's theory or, later, on thermodynamics in general. In this context the reader may consult pp. 95–97 of S. C. BROWN's *Count Rumford, Physicist Extraordinary*, Garden City, Anchor Books, 1962.

3. Any two bodies which undergo Carnot cycles corresponding to the same data θ^+, θ^-, $C^+(\mathscr{C})$ have the same motive power.

These three assumptions are logically independent. In particular, the consequences of the second one can be developed with no reference to the other two.

CARNOT's General Axiom is not to be questioned during the period covered by this history. Even CLAUSIUS is to adopt it and use it (§8Bδ, below). On the contrary, REECH is to show later that CARNOT's Special Axiom (5I.5)—the only case of his General Axiom that CARNOT himself chooses to use—cannot be separated from the Caloric Theory of heat. We shall follow REECH's obscure analysis below in §9C, but here I give a simple proof that CARNOT's Special Axiom requires every Carnot cycle to emit all the heat it absorbs.

We consider a given Carnot cycle $\mathscr{C}$ as shown in Figure 4, and we extend its adiabats downward[2] until they intersect the isotherm at a temperature θ_0 slightly less than θ^-. In this way we obtain two new Carnot cycles: one with operating temperatures θ^- and θ_0, which we shall label $\mathscr{C}'$, and one with temperatures θ^+ and θ_0, which we shall label $\mathscr{C}''$. By use of the reversal theorems (2C.21) we see that

$$L(\mathscr{C}'') = L(\mathscr{C}) + L(\mathscr{C}') . \tag{5J.1}$$

The reversal theorem (2C.24) shows that

$$C^+(\mathscr{C}'') = C^+(\mathscr{C}) , \quad C^+(\mathscr{C}') = C^-(\mathscr{C}) . \tag{5J.2}$$

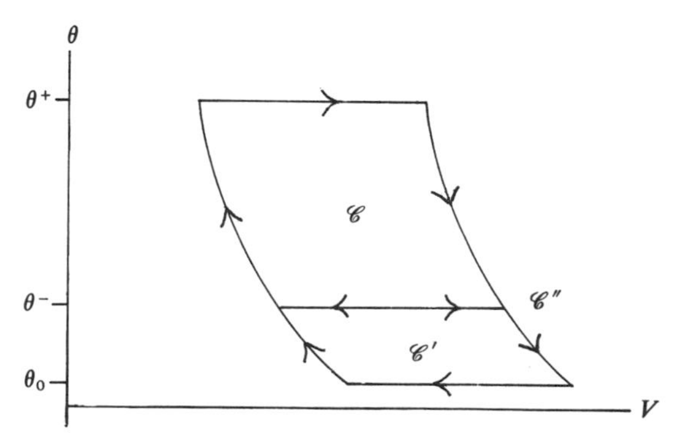

Figure 4. Construction to prove that Carnot's Special Axiom requires heat to be conserved in Carnot cycles

[2] The modern student knows that a construction of this kind is not possible without some assumptions regarding the constitutive domain. As analytical precision was very uncommon in works on physics in the nineteenth century, I think it would be out of place to attempt it in a history, so I pass over details of this kind in silence.

If we now apply to each term in (1) the equation in CARNOT's Special Axiom, namely

$$L(\mathscr{C}) = [F(\theta^+) - F(\theta^-)]C^+(\mathscr{C}) \;, \tag{5I.6$_r$}$$

and thereafter use (2), we find that

$$\begin{aligned}[F(\theta^+) - F(\theta_0)]C^+(\mathscr{C}) = {} & [F(\theta^+) - F(\theta^-)]C^+(\mathscr{C}) \\ & + [F(\theta^-) - F(\theta_0)]C^-(\mathscr{C}) \;, \end{aligned} \tag{5J.3}$$

so

$$[F(\theta^-) - F(\theta_0)]C^+(\mathscr{C}) = [F(\theta^-) - F(\theta_0)]C^-(\mathscr{C}) \;. \tag{5J.4}$$

Because of the inequality in CARNOT's Special Axiom, namely

$$F(x) > F(y) \quad \text{if} \quad x > y > 0 \;, \tag{5I.5$_{2r}$}$$

we conclude that $C^+(\mathscr{C}) = C^-(\mathscr{C})$ for every Carnot cycle $\mathscr{C}$. We have established

SCHOLION I: *Carnot's Special Axiom is compatible with the reversal theorems only if heat is neither lost nor gained in any Carnot cycle.*

CAUTION. Scholion I refers to *finite differences* of temperature. Of course, if $\theta^+ - \theta^-$ is very small, so is $C^+(\mathscr{C}) - C^-(\mathscr{C})$, irrespective of any assumption about heat and work.

Physicists are wont to say that CARNOT "knew the Second Law of Thermodynamics without knowing the First." The Second Law in this context is regarded as a statement about the motive power of Carnot cycles. We have just proved that *such a claim is false*. The assumption about motive power CARNOT actually used, his Special Axiom, *by itself implies that Carnot cycles conserve heat*. CLAUSIUS' "First Law", as we shall see in §8A, implies that a cycle $\mathscr{C}$ in which $C^+(\mathscr{C}) = C^-(\mathscr{C})$ does no work at all. Thus CARNOT's "Second Law", whatever may be meant by the term, is *incompatible* with CLAUSIUS' "First Law". Since CLAUSIUS' "Second Law" is *compatible* with his "First Law", it cannot be equivalent to CARNOT's "Second Law". So much we get by pure logic, not having to commit ourselves to statement of what all these "Laws" mean. Of course we retain compatibility for infinitesimal differences of temperature, in view of the caution stated just after Scholion I.

We shall see in Act III that no such thing can be said of CARNOT's General Axiom. On the other hand, that axiom is not the "Second Law" but a far less restrictive statement, as may be learned from the program and results in *Concepts and Logic*. In §5H we have analysed the statement (§5F) CARNOT used in order to infer his General Axiom (§5I). That statement is weaker than anything today called the "Second Law". In §8A we shall see that CLAUSIUS will impose a stricter prohibition in order to support CARNOT's claims after deleting their one apparent reference to the Caloric Theory, namely that the quantities of heat in both refrigerator and furnace are left unchanged.

Had CARNOT put in mathematical form his often repeated assertions of reversibility, he could easily have seen the simple argument based on Figure 4. His failure to do so is typical of the theorist who tries to get along without mathematics. That CARNOT, whenever he treated finite differences of temperature, did employ the one and only case of his loosely stated General Axiom compatible both with the theory of calorimetry and with his own further assumption (5C.1), is one more example of his astonishing ability to guess right.

CARNOT was wise in not elevating to a general principle his suggestion that the motive power be a function of the difference $\theta^+ - \theta^-$. Of course such a relation is possible if E in (5I.4) is a linear function of $\theta^+ - \theta^-$, but from the reversal theorem it can be shown that the linear dependence is the only admissible one within the Caloric Theory of heat. I leave the proof to the reader. For a linear function, the theorem we have just proved above shows that the Caloric Theory is implied.

The distinction we have just taken pains to express disappears if we limit attention, as CARNOT nearly always did, to infinitesimal differences of temperature. REECH (§9C, below) was to prove that the function G that figures in the first statement in CARNOT's General Axiom,

$$L(\mathscr{C}) = G(\theta^+, \theta^-, C^+(\mathscr{C})) \ , \qquad (5\text{I}.1)_r$$

must be linear in its third argument:

$$L(\mathscr{C}) = K(\theta^+, \theta^-)C^+(\mathscr{C}) \ , \quad K(x, y) > 0 \ \text{ if } \ x > y > 0 \ , \quad K(x, x) = 0 \ . \qquad (5\text{J}.5)$$

Other than REECH, no early author considered any other possibility; perhaps it is "obvious" that for any given operating temperatures a Carnot cycle which absorbs twice as much heat effects twice as much work. Except in regard to REECH's researches, whenever in this tragicomedy we shall refer in any quantitative way to CARNOT's General Axiom we shall employ (5).

Doing so, we replace K by its linear approximation and thereby reduce the very statements of CARNOT's General and Special Axioms to a *common form* when θ^+ and θ^- are both infinitesimally near to θ and differ by the amount $\Delta\theta$:

$$L(\mathscr{C}) \approx \mu(\theta)\Delta\theta C^+(\mathscr{C}) \ . \qquad (5\text{J}.6)$$

The only difference is that

$$\mu(\theta) = \begin{cases} \left.\dfrac{\partial K}{\partial x}(x, \theta)\right|_{x=\theta} & \text{according to CARNOT's } \textit{General} \text{ Axiom,} \\ F'(\theta) & \text{according to CARNOT's } \textit{Special} \text{ Axiom.} \end{cases} \qquad (5\text{J}.7)$$

We summarize this fact as[3]

SCHOLION II: *When applied to Carnot cycles corresponding to infinitesimal differences of temperature, CARNOT's General and Special Axioms are indistinguishable. For finite differences of temperature, they are not.*

The distinction made in Scholion II is delicate. We shall see in §8C that a passage in CLAUSIUS' paper of 1850 comes near to it. It was certainly grasped by KELVIN, whose expression of it we shall present in §9B. Perhaps because he did not state it so broadly, it seems not to be widely understood[4]. Here, perhaps, lies a reason for some of the confusion and obscurity which still, after a century of debate, surround CARNOT's legacy.

For purposes of interpretation and criticism we may combine Scholia I and II to yield

SCHOLION III: *Insofar as CARNOT treats only infinitesimal differences of temperature, his results derive from his* General *Axiom and do* not *require that heat be conserved in Carnot cycles. Insofar as CARNOT brings to bear his* Special *Axiom for finite differences of temperature, his results generally are* false *if heat is not conserved in Carnot cycles.*

Nevertheless, the qualification "generally" before "false" in Scholion III cannot be omitted, for some of CARNOT's statements about finite differences of temperature remain true even if the Caloric Theory is abandoned. Indeed, if two statements are incompatible in general, it is nevertheless possible to draw from both together conclusions that do not require the truth of either.

In this section we have stated our conclusions in terms of Carnot cycles alone. They can be strengthened to refer to all simple cycles. Indeed, the mathematically competent reader will show easily that if each point in the common domain of Λ_V, K_V, and ϖ may be inclosed by an arbitrarily small Carnot cycle, then the statement that $C^+(\mathscr{C}) = C^-(\mathscr{C})$ for all simple Carnot cycles $\mathscr{C}$ suffices that $C^+(\mathscr{C}) = C^-(\mathscr{C})$ for all simple cycles, or for all cycles

[3] Some historians have made much of the extent to which CARNOT's theory and CLAUSIUS' agree in higher approximations, but such agreement merely obscures the conceptual issues of the theory.

[4] In a review written just after CLAUSIUS' paper appeared HELMHOLTZ [1855, p. 576] showed in his way that "all consequences CLAPEYRON drew from this equation [(5J.6)] without integrating it remain valid." Also "It is another matter when laws for finite differences of temperature are derived by integrating this equation...." This review like others, has been regarded as ephemeral. It is considerably clearer than the works of discovery.

HELMHOLTZ [1859, *2*], perhaps implying that CLAUSIUS was skimpy in acknowledging his debt to his predecessors, in his review of CLAUSIUS [1856, *1*] repeated the observation quoted just above.

in a simply connected domain[5]. Thus Scholia I and II may be extended: Within the framework of the Doctrine of Latent and Specific Heats, we may say that (6) *is a local equivalent to CARNOT's General Axiom if the Caloric Theory of heat is assumed; otherwise, it is only a very particular consequence of that axiom.*

We shall reiterate this conclusion in a different way in §5L.

The function μ appearing in (6) is to play a central role in the development of thermodynamics. KELVIN[6] is to call it "Carnot's function". Because of the inequality asserted in CARNOT's General Axiom, namely

$$G(x, y, z) > 0 \quad \text{if} \quad x > y > 0 \quad \text{and} \quad z > 0 \;, \tag{5I.2$_r$}$$

early authors seem to have concluded that

$$\mu > 0 \;. \tag{5J.8}$$

In fact, however, the limit process used to derive (6) leads only to the weaker condition

$$\mu \geqq 0 \;. \tag{5J.9}$$

Careful examination of the matter shows that (9) cannot be replaced by (8). Theorem 7 in Chapter 9 of *Concept and Logic* shows that there are an increasing function g and a positive function h such that $\mu = g'/h$. Thus (9) follows, but as g' may vanish on a set with empty interior, so may μ. An example is provided in Remark 6 after Corollary 11.2 in Chapter 10 of *Concepts and Logic.*

The distinction has bearing upon the theory of absolute temperature, which we shall discuss below in §11H.

5K. CARNOT's Treatment of his Cycle

[To clarify the course of thought[1], we shall first use a general equation of state (2A.2), although CARNOT himself always uses the ideal gas law $pV =$

---

[5] If $\mathscr{A}$ is the region inclosed by the simple cycle $\mathscr{C}$, AMPÈRE's transformation yields

$$\begin{aligned} C^{+}(\mathscr{C}) - C^{-}(\mathscr{C}) &= \int_{\mathscr{C}} (\Lambda_V dV + \mathrm{K}_V d\theta) \;, \\ &= \int_{\mathscr{A}} \left(\frac{\partial \Lambda_V}{\partial \theta} - \frac{\partial \mathrm{K}_V}{\partial V}\right) dV d\theta \;. \end{aligned} \tag{A}$$

If $\mathscr{C}$ is a Carnot cycle which can be shrunk down to a point, (5C.1) implies that (5M.3) holds at that point. Then the right-hand side of (A) vanishes, no matter what be $\mathscr{A}$. Hence (5C.1) holds for all simple cycles $\mathscr{C}$. Likewise, (5M.3) implies the existence of $\mathrm{H_C}$ in a simply connected domain.

[6] THOMSON [1851, §2]. Earlier THOMSON [1849, §30] had called it "Carnot's coefficient".

[1] I am indebted to Dr. KEITH HUTCHISON for pointing out an error in my first estimate of this part of CARNOT's work, written in 1970, and for explaining to me some of CARNOT's arithmetic.

$R\theta$. For the most part in this section, I rely upon the long footnote (pp. 73–79) which CARNOT himself assures the reader is no more than "the analytical translation" of "some of the propositions above" into "algebraic language".

[As is shown by his crude numerical calculations of the motive power of 1 kg of carbon (above, §5A), CARNOT is unable to analyse a cycle corresponding to a finite difference of temperature.] To find the motive power of heat, he begins from an expression for the work L done by the fluid body in traversing an isothermal path $\mathscr{P}_\theta$ at temperature θ from V_a to V_b:

$$L(\mathscr{P}_\theta) = \int_{V_b}^{V_a} \varpi(V, \theta) dV = R\theta \log \frac{V_b}{V_a} , \tag{5K.1}$$

the second form being valid only for an ideal gas. For this elementary result he cites a paper of 1818 by PETIT[2]. [Perhaps so as to evade the problem of calculating the work done on an adiabat,] CARNOT considers a Carnot cycle with infinitesimal difference of temperature, for then the adiabats, being infinitesimally short, contribute nothing to the work done. Although he employs $(1)_2$, valid only for an ideal gas, his reasoning, [which is general, is only obscured by this useless specialization; if applied to $(1)_1$, it] yields[3]

$$\begin{aligned} L(\mathscr{C}) \approx L(\mathscr{P}_{\theta+\Delta\theta}) + L(-\mathscr{P}_\theta) &\approx \left[\frac{\partial}{\partial\theta}\int_{V_a}^{V_b} \varpi(V, \theta) dV\right] \Delta\theta, \\ &= \left[\int_{V_a}^{V_b} \frac{\partial p}{\partial \theta}(V, \theta) dV\right] \Delta\theta . \end{aligned} \tag{5K.2}$$

CARNOT's General Axiom leads to the statement

$$L(\mathscr{C}) \approx \mu(\theta)\Delta\theta C^+(\mathscr{C}) , \tag{5J.6$_r$}$$

so

$$L(\mathscr{C}) \approx \mu(\theta) C(\mathscr{P}_\theta)\Delta\theta . \tag{5K.3}$$

Comparison of (2) and (3) yields

$$C(\mathscr{P}_\theta) = \frac{1}{\mu(\theta)}\int_{V_a}^{V_b} \frac{\partial p}{\partial\theta}(V, \theta) dV . \tag{5K.4}$$

---

[2] PETIT [1818, p. 292]. CARNOT is generous, for PETIT, who had not even mentioned the fact that θ must be constant for this result to hold, had claimed to calculate the effect of an engine by equating the power of the working body to the rate of increase of the kinetic energy of the piston. *Cf.* CARNOT's later remark on p. 86 of *Réflexions*. From PETIT's paper we may judge how scantly the physicists of the day understood the science of mechanics as it was practised at that very time by geometers such as LAPLACE, POISSON, and CAUCHY.

[3] CARNOT in effect states that on the isothermal paths $-\mathscr{P}_\theta$ and $\mathscr{P}_{\theta+\Delta\theta}$

$$L(\mathscr{P}_{\theta+\Delta\theta}) = \int_{V_a}^{V_b} \varpi(V, \theta + \Delta\theta) dV ,$$

$$L(-\mathscr{P}_\theta) = \int_{V_c}^{V_d} \varpi(V, \theta) dV \approx -\int_{V_a}^{V_b} \varpi(V, \theta) dV$$

when $\Delta\theta$ is small.

For a body of ideal gas the quadrature on the right-hand side may be effected (p. 76):

$$C(\mathscr{P}_\theta) = \frac{R}{\mu(\theta)}\int_{V_a}^{V_b}\frac{dV}{V} = \frac{R}{\mu(\theta)}\log\frac{V_b}{V_a}\ . \tag{5K.5}$$

This is ***CARNOT's main theorem***, which he has already expressed in words (p. 52–53): "*When a gas changes in volume without change of temperature, the quantities of heat absorbed or emitted by that gas are in arithmetic progression if the increments or decrements of volume are found to be in geometric progression.*"

5L. Critique of CARNOT's Treatment of his Cycle. Scholion IV.

Anyone familiar with the theory of line integrals sees that the general relation

$$L = \int_{\mathscr{P}} \varpi(V, \theta)dV \qquad (2C.20)_{3r}$$

when applied to a simple cycle $\mathscr{C}$ yields[1]

$$L(\mathscr{C}) = \iint_{\bar{\mathscr{A}}} dpdV = \int_{\mathscr{A}} \frac{\partial p}{\partial\theta}dVd\theta\ ; \tag{5L.1}$$

as is shown in Figure 2 in §5C, $\mathscr{A}$ denotes the region of the V–θ quadrant inclosed by $\mathscr{C}$, while $\bar{\mathscr{A}}$ denotes the region inclosed by the corresponding curve in the p–V quadrant. The convention of sign here is that the region $\mathscr{A}$ must lie on the right-hand side as $\mathscr{C}$ progresses. CARNOT's result

$$L(\mathscr{C}) \approx \left[\int_{V_a}^{V_b}\frac{\partial p}{\partial\theta}(V, \theta)dV\right]\Delta\theta \qquad (5K.2)_{3r}$$

is a special case of (1). Restriction to this special case merely obscures the reasoning. Also, to calculate the heat absorbed on a simple isothermal path $\mathscr{P}_\theta$, we appeal directly to the Doctrine of Latent and Specific Heats:

$$Q = \Lambda_V(V, \theta)\dot{V} + \mathrm{K}_V(V, \theta)\dot{\theta}\ , \qquad (2C.4)_r$$

$$\Lambda_V > 0\ , \qquad \mathrm{K}_V > 0\ . \qquad (2C.5)_r$$

A glance shows that

$$C(\mathscr{P}_\theta) = \int_{V_a}^{V_b}\Lambda_V(V, \theta)dV\ , \qquad V_b > V_a\ . \qquad (5L.2)_r$$

---

[1] CLAPEYRON [1834, §II] stated the differential form of (1) in words so casual as to give the impression that it was common knowledge.

From CARNOT's General Axiom in its reduced form

$$L(\mathscr{C}) = K(\theta^+, \theta^-)C^+(\mathscr{C}) \ , \quad K(x, y) > 0 \ \text{ if } \ x > y > 0 \ , \quad K(x, x) = 0 \ , \tag{5J.5)$_r$}$$

by use of (1) and (2) we see that

$$\int_{\mathscr{A}} \frac{\partial p}{\partial \theta} dV d\theta = K(\theta^+, \theta^-) \int_{V_b}^{V_a} \Lambda_V(V, \theta^+) dV \ . \tag{5L.3}$$

Dividing this equation by $(V_b - V_a)(\theta^+ - \theta^-)$ and then passing to the limit as $V_b \to V_a$ and $\theta^+ \to \theta^-$, we show that

$$\mu \Lambda_V = \frac{\partial p}{\partial \theta} \ , \tag{5L.4}$$

"Carnot's function" μ being given as follows:

$$\mu(\theta) = \begin{cases} \left. \dfrac{\partial K}{\partial x}(x, \theta) \right|_{x=\theta} & \text{according to CARNOT's } \textit{General} \text{ Axiom,} \\ F'(\theta) & \text{according to CARNOT's } \textit{Special} \text{ Axiom.} \end{cases} \tag{5J.7)$_r$}$$

CARNOT's argument, which was to be extended to a general equation of state by CLAPEYRON (§6A, below), is a loose version of just this. In fact, if we substitute (2) into the relation

$$C(\mathscr{P}_\theta) = \frac{1}{\mu(\theta)} \int_{V_a}^{V_b} \frac{\partial p}{\partial \theta}(V, \theta) dV \tag{5K.4)$_r$}$$

and then differentiate with respect to V_b, we obtain (4). Thus (4) deserves to be called the ***General CARNOT–CLAPEYRON Theorem.***

The spectators, recalling the distinctions made in §5J, will see that the protagonist while seeming to smelt lead has cast a gold ingot. The reasoning rests upon CARNOT's General Axiom alone, and μ is given in terms of that axiom by $(5J.7)_1$. However, CARNOT and his popularizer CLAPEYRON appealed only to CARNOT's Special Axiom. For them, μ necessarily has the form $(5J.7)_2$, and (4) is neither more nor less than

$$F' \Lambda_V = \frac{\partial p}{\partial \theta} \ . \tag{5L.5}$$

This is the ***Special CARNOT–CLAPEYRON Theorem***[2].

---

[2] The notation F is CARNOT's; CLAPEYRON wrote C for $1/F'$. CARNOT wrote T for R/F', that is, RC, R being the constitutive constant of the gas. F and C are universal functions, but CARNOT's T is not. The notation μ is KELVIN's (§7H, below).

As is stated in §§2A and 2C, we follow the custom of all early writers on thermodynamics in assuming that $\Lambda_V > 0$ and $\partial p/\partial\theta > 0$. Not only is (4) consistent with

For an ideal gas the General and Special CARNOT-CLAPEYRON Theorems reduce, respectively, to

$$\mu\Lambda_V = \frac{p}{\theta} = \frac{R}{V}\,, \qquad F'\Lambda_V = \frac{p}{\theta} = \frac{R}{V}\,. \tag{5L.6}$$

CARNOT's theorem (5K.5) is neither more nor less than an integrated form of the last of these formulae. Only the awkwardness of CARNOT's mathematics made this theorem seem anything but an immediate consequence of his assumptions.

Here begins another spiral of misunderstanding and confusion. The distinction between (4) and its special case (5) is formally plain enough, once stated; both CLAUSIUS (below, §8C) and KELVIN (below, §9B) were to grasp it, the latter very firmly; the first author to make it clear, and the last until 1975, will be REECH in 1853 (below, §9C), but nobody was to pay attention to him. The General CARNOT-CLAPEYRON Theorem (4) expresses μ through quantities at least in principle accessible to experiment: $\mu = \partial p/\partial\theta \div \Lambda_V$. If CARNOT's *Special* Axiom is adopted (and hence, Scholion I tells us, so is the Caloric Theory), then $(5J.7)_2$ applies, $\mu = F'$, and so[3]

$$F = \int \mu d\theta\,, \qquad L(\mathscr{C}) = \left(\int_{\theta-}^{\theta+} \mu d\theta\right) C^+(\mathscr{C})\,. \tag{5L.7}$$

The latter of these formulae expresses neither more nor less than CARNOT's Special Axiom. If, on the contrary, we do not specialize CARNOT's General Axiom but leave it as it is, the function μ as given by $(5J.7)_1$ is only the restriction of one partial derivative of the function K to one line in the plane of its two arguments. Thus μ does *not* determine G. In summary we have

SCHOLION IV: *Let CARNOT's General Axiom be accepted, and let μ have been determined. If the General Axiom is specialized to the Special Axiom (thereby, as Scholion I states, implicitly requiring that heat be conserved in Carnot cycles), then the motive power of heat associated with finite differences of temperature* is *determined. Without that specialization or some other, μ does* not *determine that motive power.*

CARNOT's Special Axiom, which requires that F be an increasing function, but also (5) requires that F' shall not vanish, not even at isolated points. This limitation is satisfied by all examples of F that CARNOT gave.

However, *a priori* restriction of the signs of Λ_V and $\partial p/\partial\theta$ is not necessary to the mathematical development of CARNOT's ideas, and no such restriction is laid down in *Concepts and Logic*. The reader interested in the complications that are encountered when there may be curves on which $\partial p/\partial\theta = 0$ may look at Figures 9 and 10 in §7, Lemma 1 in §8, and Definition 17 and Theorem 7 in §9 of that work. That theorem delivers (4) with no *a priori* restriction on the signs of Λ_V and $\partial p/\partial\theta$, and it allows μ to vanish on a set of temperatures with empty interior.

[3] W. THOMSON [1849, Equation 7].

The importance of the CARNOT-CLAPEYRON Theorem is central. Until a theory relating heat to work is laid down, the constitutive functions Λ_V and ϖ remain independent. The relation (4) shows that neither is arbitrary, once the other be known. That is, the caloric and thermal properties of the fluid body impose a condition upon each other, a condition determined by a function of temperature which relates the work done by the body in an infinitesimal Carnot cycle to the heat absorbed in that cycle. CARNOT was the first to see that a *theory connecting heat and work imposes restrictions upon the constitutive function of bodies*. The discovery of such restrictions has been the *essence of thermodynamics* from his day to ours.

CARNOT's expressions in words are often less specific than the consequences that do follow from his theory. An example is his famous theorem about arithmetic and geometric progressions quoted at the end of the preceding section. The law really derived by CARNOT is

$$C(\mathscr{P}_\theta) = \frac{R}{\mu(\theta)} \int_{V_a}^{V_b} \frac{dV}{V} = \frac{R}{\mu(\theta)} \log \frac{V_b}{V_a} \qquad (5K.5)_r$$

with μ specialized to F'. His statement in words makes no reference to the origin of μ. His reasoning, as we have shown, suffices to obtain from his General Axiom the result (5K.5) with general μ. Thus again CARNOT's theorem on progressions follows although the Special Axiom, which he invoked so as to derive it, need not hold.

This much should suffice to foreshadow the influence CARNOT's reasoning will exert, long after his Special Axiom has been rejected.

5M. Critique: Interconvertibility of Heat and Work as Implied by CARNOT's Theory. Proof that CARNOT's Cycles are Indeed the Most Efficient.

The Special CARNOT-CLAPEYRON Theorem, namely

$$F'\Lambda_V = \frac{\partial p}{\partial \theta}, \qquad (5L.5)_r$$

implies the whole panoply of thermodynamic relations, both local and integral, of CARNOT's theory. Few of these were noticed by the pioneers who came after CARNOT; even fewer by CARNOT himself. To obtain some of them, we first recall that CARNOT assumed the existence of the heat function $\mathrm{H_C}$, so

$$C(\mathscr{P}) = \mathrm{H_C}(V_2, \theta_2) - \mathrm{H_C}(V_1, \theta_1) . \qquad (5B.1)_r$$

Then

$$Q = \dot{\mathrm{H}}_\mathrm{C} = \frac{\partial \mathrm{H_C}}{\partial V} \dot{V} + \frac{\partial \mathrm{H_C}}{\partial \theta} \dot{\theta} . \qquad (5M.1)$$

Comparison with the Doctrine of Latent and Specific Heats, expressed in part by the relation

$$Q = \Lambda_V(V, \theta)\dot{V} + \mathrm{K}_V(V, \theta)\dot{\theta} \ , \qquad (2C.4)_r$$

yields precise determination of Λ_V and K_V:

$$\Lambda_V = \frac{\partial \mathrm{H_C}}{\partial V} \ , \qquad \mathrm{K}_V = \frac{\partial \mathrm{H_C}}{\partial \theta} \ , \qquad (5M.2)$$

and the corresponding condition of integrability

$$\frac{\partial \Lambda_V}{\partial \theta} = \frac{\partial \mathrm{K}_V}{\partial V} \ . \qquad (5M.3)$$

Of course, the formulae (1), (2), and (3) are together equivalent to the formulae

$$\mathrm{K}_p = -M\left(\frac{\partial p}{\partial \theta}\bigg/\frac{\partial p}{\partial \rho}\right)\frac{\partial \mathrm{H_L}}{\partial \rho} \ , \qquad \mathrm{K}_V = M\frac{\partial p}{\partial \theta}\frac{\partial \mathrm{H_L}}{\partial p} \ , \qquad (3F.1)_r$$

$$\frac{\partial}{\partial \rho}\left(\frac{\mathrm{K}_V}{\dfrac{\partial p}{\partial \theta}}\right) + \frac{\partial}{\partial p}\left(\frac{\mathrm{K}_p}{\dfrac{\partial p}{\partial \theta}\bigg/\dfrac{\partial p}{\partial \rho}}\right) = 0 \ , \qquad (3F.4)_r$$

formulae implied by LAPLACE's theory but not written out by him. The tragicomic muse showed LAPLACE and CARNOT, each in his own way, how to make these trivial relations seem profound: LAPLACE, the mathematician, by choosing the awkward variables p and ρ and at the same time plunging into intricate molecular calculations, and CARNOT, the engineer, who chose the natural variables V and θ, by stringing out the argument in vague words and arithmetic. We have already noticed CARNOT's reluctance to introduce Λ_V in the few equations he gives, although he has declared at the outset that he expects his readers to be thoroughly familiar with the concept of latent heat. He certainly uses $(2)_2$ frequently, but I cannot find in his book anything equivalent to its companion, $(2)_1$. It is not clear whether he knows about conditions of integrability, though they were already old in his day[1]. KELVIN,

[1] I see no reason to follow the custom of those writers on thermodynamics who by calling every statement of some part of a condition of integrability a "Maxwell relation" seem to imply that EULER's theorem

$$\frac{\partial^2 f}{\partial x \partial y} = \frac{\partial^2 f}{\partial y \partial x} \ ,$$

under appropriate (and by thermodynamicists never stated) hypotheses on f, becomes a great discovery of physics as soon as thermodynamic interpretations are attached to x, y, and f. The term derives from a footnote by MAXWELL [1871, p. 167]. MAXWELL's third "thermodynamic relation" is neither more nor less than the General CARNOT-CLAPEYRON Theorem in its final form $(7I.3)_3$ but with Λ_V expressed in terms of the entropy H through the relation $\Lambda_V = \theta\partial\mathrm{H}/\partial V$, which is an immediate consequence of (11E.6), below.

at the very sunset of the Caloric Theory, will be the first to publish (3) and to recognize it as being the essence of that theory: From it, "other remarkable conclusions...might have been drawn," so that "experimental tests might have been suggested" (see §9B, below).

Indeed, we can use (3) to find right away the full connection between heat and work that CARNOT's theory implies. We know that

$$L(\mathscr{C}) = \iint_{\mathscr{A}} dp dV = \iint_{\mathscr{A}} \frac{\partial p}{\partial \theta} dV d\theta \ ; \qquad (5L.1)_r$$

we have the Special CARNOT-CLAPEYRON Theorem:

$$F' \Lambda_V = \frac{\partial p}{\partial \theta} \ . \qquad (5L.5)_r$$

Hence

$$\begin{aligned} L(\mathscr{C}) &= \iint_{\mathscr{A}} F' \Lambda_V dV d\theta \ , \\ &= \iint_{\mathscr{A}} \left[F' \Lambda_V + F \left(\frac{\partial \Lambda_V}{\partial \theta} - \frac{\partial \mathrm{K}_V}{\partial V} \right) \right] dV d\theta \ , \qquad (5M.4) \\ &= \iint_{\mathscr{A}} \left[\frac{\partial}{\partial \theta} (F \Lambda_V) - \frac{\partial}{\partial V} (F \mathrm{K}_V) \right] dV d\theta \ ; \end{aligned}$$

the second line is a consequence of (3). By applying AMPÈRE's transformation of a line integral around a simple circuit into an integral over the region that it bounds, we conclude from (4) that for any simple cycle[2]

$$\begin{aligned} L(\mathscr{C}) &= \int_{\mathscr{C}} F(\Lambda_V dV + \mathrm{K}_V d\theta) \ , \\ &= \int_{t_1}^{t_2} F(\theta) Q dt \ , \end{aligned} \qquad (5M.5)$$

retaining the convention of sign laid down in §5L. We see at a glance that this formula allows us to recover the equation in CARNOT's Special Axiom:

$$L(\mathscr{C}) = [F(\theta^+) - F(\theta^-)] C^+(\mathscr{C}) \ . \qquad (5I.6)_r$$

We may regard (5) as the extension of that equation to an arbitrary simple cycle. More important than that is the general interpretation of (5):

In a cyclic process, according to CARNOT's theory a unit of heat absorbed at the temperature θ is equivalent to $F(\theta)$ units of positive work, while a unit of heat emitted at the temperature θ is equivalent to $F(\theta)$ units of negative work.

---

[2] This formula seems not to be generally known even today. It is an immediate consequence of specializing REECH's formula $(9C.1)_2$ to the Caloric Theory. However, as the proof in the text shows, it need not be obtained in that way. Use of AMPÈRE's transformation is elegant but not necessary. CARNOT could have obtained (5) by hacking differential elements in the common style of physicists in his period.

We note that F may not be a constant function, for that would contradict $(5I.5)_2$, which expresses CARNOT's often-repeated statement that any difference of temperature suffices to effect work if a suitable engine is employed. We may express this result as follows:

CARNOT's theory requires that heat and work be interconvertible, but it forbids them from being uniformly so[3].

While today we reject (5) with any F except a constant one, there would have been no reason for CARNOT to be alarmed by that relation with an F that increases with θ. It expresses one possible interpretation of his grand idea: In its power to do work, heat at high temperature is different from heat at low temperature. Today we still accept that grand idea, though we give it a different specific interpretation, for we do not allow a body that does cyclic work through "thermal agency" to emit all the heat it absorbs in a cycle. We apply the grand idea to the heat absorbed *alone*, not to all the heat exchanged in a cycle.

The formula (5) provides a way to calculate the motive power of any cycle. Thus it puts in our hands a tool for calculating the greatest motive power that a unit of heat can provide in a cycle $\mathscr{C}$ whose maximum and minimum temperatures are $\theta_{\max}$ and $\theta_{\min}$. We may write (5) in the form

$$L(\mathscr{C}) = \int_{\mathscr{T}^+} F(\theta)Q\,dt - \int_{\mathscr{T}^-} F(\theta)(-Q)\,dt \ , \tag{5M.6}$$

$\mathscr{T}^+$ and $\mathscr{T}^-$ being the parts of the interval $]t_1, t_2[$ on which $Q > 0$ and $Q < 0$, respectively. Since $Q > 0$ in the first integrand, while $-Q > 0$ in

[3] Since CARNOT's considerations refer mainly to cycles, (5) suffices for the purpose here, but it is easy to extend the idea to arbitrary processes, as is shown by Theorem 8 in §9 of *Concepts and Logic*. CARNOT's theory implies the existence of an internal pro-energy $\mathrm{E_C}(V, \theta)$ such that in any process

$$\dot{\mathrm{E}}_\mathrm{C} = F(\theta)Q - p\dot{\mathrm{V}} \ .$$

In fact the special CARNOT-CLAPEYRON Theorem (5L.5) may be interpreted as a condition of integrability sufficient that $\mathrm{E_C}$ exist locally, and

$$\mathrm{E_C} = F\mathrm{H_C} - \int \varpi dV + f \ ,$$

f being a suitably chosen function of θ alone; for an ideal gas

$$\mathrm{E_C} = R\left(\frac{F}{F'} - \theta\right) \log V + f \ .$$

The reader who desires to check these formulae must remember to use not only (5L.5) but also (2). With the choice $a\theta$ for F (*cf.* (5S.6)) the internal pro-energy of an ideal gas thus reduces to a function of θ alone, just as in the later thermodynamics. With the prophetic choice (5N.7), on the contrary, $\mathrm{E_C}$ depends on V as well as θ.

the second, the fact that F is an increasing function allows us to conclude that

$$\begin{aligned} L(\mathscr{C}) &\leqq F(\theta_{\max})C^+(\mathscr{C}) - F(\theta_{\min})C^-(\mathscr{C}) \ , \\ &= [F(\theta_{\max}) - F(\theta_{\min})]C^+(\mathscr{C}) \ , \end{aligned} \tag{5M.7}$$

the second step being a consequence of the fact that $C^+(\mathscr{C}) = C^-(\mathscr{C})$ in every cycle $\mathscr{C}$. Equality holds in $(7)_1$ if $C^+(\mathscr{C}) = 0$. Once this trivial possibility be set aside, the reasoning that delivered $(7)_1$ shows that equality holds if and only if $\theta = \theta_{\max}$ on $\mathscr{T}^+$ and $\theta = \theta_{\min}$ on $\mathscr{T}^-$. This is precisely CARNOT's Claim I, as stated and criticised in §5E. *Among all cycles that absorb any heat, Carnot cycles*[4] *and they alone are the most efficient for given extremes of temperature.* In §5H we have remarked that the specific conclusion

$$L(\mathscr{C}) = G_B(\theta^+, \theta^-, C^+(\mathscr{C})) \tag{5E.5$_r$}$$

which we have drawn from CARNOT's Claim I makes that claim either false or else demonstrable and hence superfluous as an axiom. Here, by a few lines of calculus, we have shown that Claim I when specialized to the Caloric Theory is true, a consequence of (5E.5). In the literature of thermodynamics from the beginnings until the 1970s I have found no trace of the elementary analysis just given. In fact, I have not found (5) or any verbal statement equivalent to it.

Hindsight!, the Historians of Science will say. It is no credit to CARNOT that he expressed himself so obscurely that a path for bringing to bear the hindsight provided by seventeenth-century calculus was not cleared through his book until the late twentieth century.

---

[4] Because of the constitutive inequality $(2C.5)_1$ all Carnot cycles look like Figure 3 in §5C. If, as is necessary in the later thermodynamics, we drop $(2C.5)_1$, other kinds of Carnot cycles become possible. The following sketch shows one for water near 4°C at nearly atmospheric pressure:

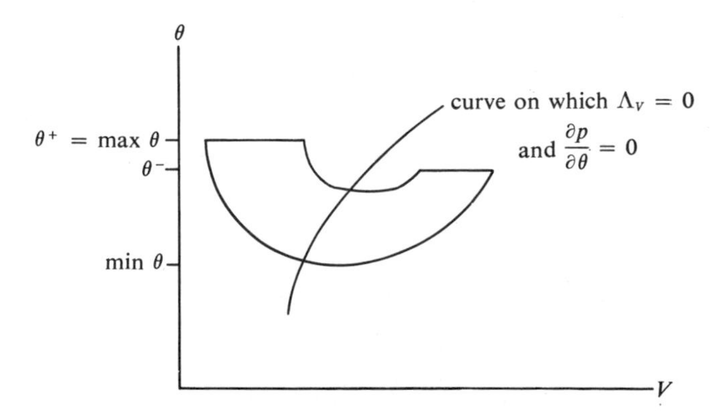

The assertion and proof given in the text above are still valid, provided $\theta_{\max}$ be made to mean the supremum of θ on $\mathscr{T}^+$, while θ^- is made to mean the infimum of θ on $\mathscr{T}^-$. In the case drawn $\theta^- \neq \min \theta$, but $\theta^- = \theta_{\min}$ as redefined.

5N. Critique: Dimensional Invariance of CARNOT's Theory

Hindsight again! It is strange that no one has ever subjected CARNOT's assumptions to a dimensional analysis. First of all, CARNOT's General Axiom (5I.1) tells us that if we choose some particular Carnot cycle $\mathscr{C}_0$, on which the heat absorbed is $C^+(\mathscr{C}_0)$, we shall obtain a definite amount of work $L(\mathscr{C}_0)$. Thus $L(\mathscr{C}_0)$ and $C^+(\mathscr{C}_0)$ are certain constants, so $L(\mathscr{C}_0)/C^+(\mathscr{C}_0)$ is a constant bearing the dimensions of work ÷ heat. Every Carnot cycle furnishes us such a constant. We choose one of these and call it J_0. Then the equation in CARNOT's Special Axiom, namely

$$L(\mathscr{C}) = [F(\theta^+) - F(\theta^-)]C^+(\mathscr{C}) \ . \tag{5I.6}_r$$

may be written in the form

$$L(\mathscr{C}) = J_0[f(\theta^+) - f(\theta^-)]C^+(\mathscr{C}) \ , \tag{5N.1}$$

the function within the brackets being dimensionless. A unit of heat produces a certain number of units of work; the factor of proportionality is the product of a constant J_0, which merely converts units from one arbitrary scale to another, and a *dimensionless* function of two temperatures.

We now turn to the function f. Since the units of temperature are independent of the units of heat and work, f must satisfy the relation

$$f(A\theta^+) - f(A\theta^-) = f(\theta^+) - f(\theta^-) \ , \tag{5N.2}$$

at least for all A in a small interval about 1. Solving this functional equation, we see that f is proportional to $\log \theta$. Thus *CARNOT's Special Axiom is dimensionally invariant if and only if CARNOT's function is of the form*

$$F(\theta) = J \log \theta + \text{const.} \tag{5N.3}$$

and the motive power of a Carnot cycle $\mathscr{C}$ is then given by

$$L(\mathscr{C}) = J\left(\log \frac{\theta^+}{\theta^-}\right) C^+(\mathscr{C}) \ . \tag{5N.4}$$

J is an arbitrary constant having the dimensions of work ÷ heat. Because of $(5I.5)_2$, $J > 0$. As we shall see in §5S, this particular choice of F is one to which CARNOT inclined, but he did not exclude others.

Of course it does not follow that other choices of F are inadmissible, provided we allow F to vary from one body to another. If F is to differ from (3), it must have as a further argument some particular constant temperature θ_0. Written in full, (5I.6) must then be

$$L(\mathscr{C}) = J\left[g\left(\frac{\theta^+}{\theta_0}\right) - g\left(\frac{\theta^-}{\theta_0}\right)\right] C^+(\mathscr{C}) \ , \tag{5N.5}$$

g being a dimensionless function of its dimensionless arguments, and indeed all of CARNOT's own hypothetical examples are of this form (*cf.* §5S, below).

If in (5) $g(x) = A \log x + B$, (4) results; otherwise the ratio $L/(JC^+)$ cannot be determined by θ^+ and θ^- alone. The third temperature θ_0 in (5) could have one of two meanings. First, it could be a universal constant bearing the dimensions of temperature. There is neither evidence in nature of any such particular universal temperature nor any indication that CARNOT believed there was such, so this possibility must be set aside. The second and only remaining possibility is that the particular temperature θ_0 is proper to the body in question, for example, a boiling point or a melting point. If so, the right-hand side of (5) is not a universal function, common to all bodies. On the contrary, CARNOT always claimed the motive power of heat to be independent of the agents used to realize it. Therefore, *CARNOT's Special Axiom is inconsistent with his claim of universality unless his function F has the form* (3).

No such objection can be raised against CARNOT's General Axiom

$$L(\mathscr{C}) = G(\theta^+, \theta^-, C^+(\mathscr{C})) \ , \tag{5I.1$_r$}$$

$$G(x, y, z) > 0 \quad \text{if} \quad x > y > 0 \quad \text{and} \quad z > 0 \ . \tag{5I.2$_r$}$$

Dimensional considerations, though they appear in primitive form in FOURIER's work (above, §4E), were not familiar in the early nineteenth century. They would have sufficed to determine at once CARNOT's function F. In §5T we shall show that the result, namely (3), leads to conclusions so objectionable as to have caused CARNOT's theory to be rejected, had they been perceived.

We see that (3) is equivalent to

$$F'(\theta) = J/\theta \ . \tag{5N.6}$$

Because of $(5J.7)_2$, we can write this result in the prophetic form

$$\mu = J/\theta \ . \tag{5N.7}$$

It will be important to recall that this determination of μ is compatible with CARNOT's theory.

5O. CARNOT's Numerical Evaluation of the Motive Power of Heat

As we have stated in §5A, CARNOT concluded from his theory that 1000 units of heat passing from a body maintained at the temperature of 1° to another body maintained at 0° would produce, in acting upon air, 1.395 units of motive power. To obtain this number, [which is one of his most striking conclusions,] CARNOT presented only strings of [opaque, ugly, and] approximate arithmetic (pp. 79–82).

[Nevertheless, his idea is a simple one. To grasp it, the student's easiest course is to turn to CLAPEYRON's calculation[1] of the same quantity. Although CLAPEYRON's analysis is just as loose, it is presented in "the usual dialect of mathematicians" rather than arithmetic and hence can be understood easily. Once we have traced the argument, we can recognize it as being just what CARNOT had expressed in his own way. Moreover, we can see that there is nothing that need be loose in it, and we can make it still easier to understand by referring it to a general equation of state rather than to the example of the ideal gas, to which both CARNOT and CLAPEYRON confined their presentations. In addition, we can see that the reasoning itself makes no direct use of the existence of a heat function, which CLAPEYRON brought in only so as to cancel it out again.]

To begin with,

$$Q = \Lambda_p(V, \theta)\dot{p} + \mathrm{K}_p(V, \theta)\dot{\theta} \ . \qquad (2\mathrm{C}.8)_\mathrm{r}$$

Hence

$$Q = \mathrm{K}_p\dot{\theta} + \Lambda_p\left(\frac{\partial p}{\partial \theta}\dot{\theta} + \frac{\partial p}{\partial V}\dot{V}\right) . \qquad (5\mathrm{O}.1)$$

In an adiabatic process $V = f(\theta)$; writing $dV/d\theta$ for the derivative of f, we obtain from (1)

$$\frac{1}{\Lambda_p} = -\frac{\dfrac{\partial p}{\partial \theta} + \dfrac{\partial p}{\partial V}\dfrac{dV}{d\theta}}{K_p} . \qquad (5\mathrm{O}.2)$$

Now we invoke the General CARNOT-CLAPEYRON Theorem:

$$\mu\Lambda_V = \frac{\partial p}{\partial \theta} . \qquad (5\mathrm{L}.4)_\mathrm{r}$$

Because

$$\Lambda_p = \Lambda_V \bigg/ \frac{\partial p}{\partial V} , \qquad (2\mathrm{C}.9)_{1\mathrm{r}}$$

we conclude that

$$\mu = \frac{1}{\Lambda_p}\cdot\frac{\dfrac{\partial p}{\partial \theta}}{\dfrac{\partial p}{\partial V}} . \qquad (5\mathrm{O}.3)$$

Substitution of (2) into (3) yields

$$\mu = -\frac{\dfrac{\partial p}{\partial \theta}}{\mathrm{K}_p}\left[\frac{\dfrac{\partial p}{\partial \theta}}{\dfrac{\partial p}{\partial V}} + \frac{dV}{d\theta}\right] . \qquad (5\mathrm{O}.4)$$

---

[1] CLAPEYRON [1834, §VI].

For an ideal gas this statement becomes

$$\mu = R \frac{1 - \dfrac{\theta}{V}\dfrac{dV}{d\theta}}{\theta K_p} \tag{5O.5}$$

which is what CLAPEYRON obtained, following in CARNOT's footsteps.

CARNOT (p. 43) attributed to POISSON the experimental datum[2] that in an adiabatic compression of air at 0°C

$$\frac{\theta dV}{V d\theta} = -\frac{267}{116} = -2.30 \ . \tag{5O.6}$$

He accepted also (p. 81) the value of K_p for 1 kg of air at 0°C obtained by DELAROCHE & BÉRARD:

$$K_p = 0.267 \text{ Kcal/°C} \ . \tag{5O.7}$$

Putting these numbers into (5) yields

$$\mu(0°\text{C}) \quad \text{for air} = 1.395 \ . \tag{5O.8}$$

This is the first of CARNOT's two celebrated determinations of the motive power of heat. We shall not consider the second, for it employs statements relating steam to water.

5P. Critique of CARNOT's Numerical Evaluation of the Motive Power of Heat

CARNOT's analysis rests upon the General CARNOT-CLAPEYRON Theorem alone; *it makes no use of the Caloric Theory.* Indeed, it derives from CARNOT's General Axiom; in Act IV we shall see that CLAUSIUS will make an assumption essentially equivalent to that axiom while rejecting the Caloric Theory. Thus the traditional amazement over CARNOT's having got a good numerical answer from a false theory is groundless, for the simple reason that the theory he really used at this point was not false. True, CARNOT wrote F' where we have written μ, but the reasoning by which he concluded that

$$\mu = R \frac{1 - \dfrac{\theta}{V}\dfrac{dV}{d\theta}}{\theta K_p} \tag{5O.5$_r$}$$

made no use of his Special Axiom except for infinitesimal differences of temperature, and there it is indistinguishable from his General Axiom.

[2] We have explained this datum above in §3A.

CARNOT's *evaluation of* μ *at* 0°C *would have led to its modern value*[1] *there, had he used modern data for* K_p *and for* $(\theta/V)dV/d\theta$ *in an adiabatic process.*

CARNOT's method has a capital advantage over later ones. Namely, beyond the Doctrine of Latent and Specific Heats it rests upon CARNOT's General Axiom *alone*. It does not assume any relation whatever between the specific heats of ideal gases[2]. The function μ as determined by it need not be assumed the same for all bodies. It requires neither the Caloric Theory nor any other specializing assumption regarding the work done by absorption of heat. Within the broad limits imposed by CARNOT's General Axiom, it is purely *experimental* and thus allows an experimental check on the Caloric Theory, the First Law, and the specific heats of gases.

For all this it pays a price. It is not a general determination of the motive power of heat. To know μ is not enough for that. As we have shown in §5L, the motive power of infinitesimal Carnot cycles does not generally determine the motive power of finite ones.

5Q. CARNOT's Theory of Specific Heats

[Unlike his analysis of the Carnot cycle, CARNOT's theory of specific heats really refers only to ideal gases.] It rests upon CARNOT's main theorem:

$$C(\mathscr{P}_\theta) = \frac{\mathrm{R}}{\mu(\theta)}\int_{V_a}^{V_b}\frac{dV}{V} = \frac{R}{\mu(\theta)}\log\frac{V_b}{V_a}\,. \qquad (5\mathrm{K}.5)_r$$

---

[1] I read the remark made by THOMSON [1853, *1*, §86] about CARNOT's determination of μ as showing that he finally reached this conclusion.

As will be seen in §8B, the "modern value" must agree with the value of J/θ at 0°C, J being the mechanical equivalent of a unit of heat. The statement in the text does not refer at all to that agreement. Rather, μ is defined by $(5\mathrm{J}.7)_1$ for *any theory compatible with CARNOT's General Axiom*. Therefore, both CARNOT's theory and CLAUSIUS' are included. Through (5J.6) the function μ for *both* theories determines the work done in Carnot cycles with infinitesimal difference of operating temperatures. CARNOT's calculation of μ through (5O.5), resting, as it does, directly on quantities accessible to experiment, is "modern" because it holds *in common* for CLAUSIUS' theory and CARNOT's.

[2] If γ = const., we may invoke the LAPLACE-POISSON Law (3D.5) and so conclude that in an adiabatic process $(V/\theta)d\theta/dV = -(\gamma - 1)$. Then (5O.5) reduces to

$$\mu = \frac{R}{(1-\gamma^{-1})\mathrm{K}_p\theta}\,.$$

If *both* specific heats are constant, this formula implies (5N.7) and evaluates the constant J therein just as MAYER is to do; *cf.* (7B.3), below. If μ is the same for all fluid bodies, then so is J, so to evaluate J once and for all it suffices to assume that one ideal gas has constant specific heats. *Cf.* Axiom V of *Concepts and Logic*.

This appeal to modernity blunts the edge of CARNOT's method, which is independent of such specializing assumptions. Besides, as we shall see below in §5T, CARNOT does not accept the LAPLACE-POISSON law.

Because CARNOT assumes that

$$C(\mathscr{P}) = \mathrm{H_C}(V_2, \theta_2) - \mathrm{H_C}(V_1, \theta_1) \ , \qquad (5\text{B}.1)_\text{r}$$

$C(\mathscr{P}_\theta)$ may be expressed also in terms of the heat function $\mathrm{H_C}$. Equating the two expressions yields

$$\mathrm{H_C}(V, \theta) - \mathrm{H_C}(V_a, \theta) = \frac{R}{F'(\theta)} \log \frac{V}{V_a} \ . \qquad (5\text{Q}.1)$$

Defining U as follows:

$$U(\theta) \equiv \mathrm{H_C}(1, \theta), \qquad (5\text{Q}.2)$$

by comparison with (1) CARNOT obtains his *heat function for an ideal gas* (p. 77):

$$\mathrm{H_C}(V, \theta) = \frac{R}{F'(\theta)} \log V + U(\theta) \ . \qquad (5\text{Q}.3)$$

[With the aid of (3) the relation between temperature and volume along the adiabat connecting (V_0, θ_0) to (V, θ) becomes obvious:

$$\frac{R}{F'(\theta)} \log V + U(\theta) = \frac{R}{F'(\theta_0)} \log V_0 + U(\theta_0) \ . \qquad (5\text{Q}.4)$$

However, CARNOT apparently does not see it;] instead, he gives a somewhat more complicated method[1] (pp. 62–64) which would lead to the same result, but he works it out only in a special case, which we shall consider in §5S.

CARNOT knows that

$$\mathrm{K}_V = \frac{\partial \mathrm{H_C}}{\partial \theta} \ . \qquad (5\text{M}.2)_{2\text{r}}$$

Thus (3) yields[2]

$$\mathrm{K}_V = R \left(\frac{1}{F'}\right)' \log V + U' \ . \qquad (5\text{Q}.5)$$

By a difficult verbal argument[3], CARNOT concludes also (p. 59) that "The

---

[1] First he uses (5K.5) to find the heat emitted in an isothermal compression from (V_0, θ_0) to (V, θ_0). Then he uses (2C.4) to calculate the heat absorbed in an isochoric process connecting (V, θ_0) to (V, θ), on the supposition that K_V is a function of V only and hence is constant in this particular process. In the Caloric Theory these two quantities of heat are equal. To obtain (4) by this reasoning of CARNOT's, we need only use his general formula (5), below, rather than take K_V as being a function of V alone.

[2] The long verbal argument on pp. 56–58 regards the temperature as constant and leads to a corresponding statement of (5) in words about progressions; a complicated and obscure way of taking account of variation of temperature appears on pp. 53–64; (5) itself is obtained in the footnote on p. 77, where it is numbered (5).

[3] On pp. 43–46 a very obscure argument, full of incomprehensible arithmetic, leads to the conclusion that "The difference between the specific heat at constant pressure

difference between the specific heat at constant pressure and the specific heat at constant volume is always the same, whatever be the density of the gas, provided the weight remain the same." The reader is expected to remember that the whole argument, starting some pages earlier (p. 55), presumes that the gases "be taken and maintained at a certain invariable temperature. But these theorems furnish no means of comparing the quantities of heat emitted or absorbed by elastic fluids which change in volume at different temperatures." To determine those quantities we should require (p. 56) "some other data that physics today will not supply us." [This is one way of stating that F is an unknown function of temperature. The formula that expresses CARNOT's conclusions succinctly is

$$\mathrm{K}_p - \mathrm{K}_V = \frac{R}{\theta F'} \; ; \qquad (5Q.6)$$

it is an immediate consequence of the relations

$$\mathrm{K}_p - \mathrm{K}_V = \frac{V\Lambda_V}{\theta} \quad \text{and} \quad F'\Lambda_V = \frac{R}{V} \, . \qquad (2C.14)_{3r}, (5L.6)_{4r}$$

More generally, we may use the General CARNOT-CLAPEYRON Theorem for an ideal gas

$$\mu\Lambda_V = \frac{R}{V} \, , \qquad (5L.6)_{2r}$$

and so show that for an ideal gas

$$\mathrm{K}_p - \mathrm{K}_V = \frac{R}{\theta\mu} \; ; \qquad (5Q.7)$$

CARNOT's General Axiom implies that the difference of specific heats of an ideal gas is a function of θ alone.]

5R. Critique of CARNOT's Theory of Specific Heats

Here CARNOT's mathematics is clear and sound.

One of the "other remarkable conclusions" from (5M.3), conclusions

---

and the specific heat at constant volume is the same for all gases." CARNOT explains that the pressure [and the temperature] are held constant and "the specific heats are measured with respect to the volumes." The argument leading to the statement quoted above appears on pp. 58–59. Of course I do not imply that those arguments are wrong. The explicit relation

$$\frac{\mathrm{K}_p - \mathrm{K}_V}{V} = \frac{p}{\theta^2 F'}$$

was stated by CLAPEYRON [1834, end of §3] and THOMSON [1849, Eq. (12)]; it is an immediate consequence of (6).

from which, as KELVIN is to write in 1851, "experimental tests might have been suggested", we have seen already in the form

$$K_p - K_V + p\frac{\partial K_p}{\partial p} + \rho\frac{\partial K_V}{\partial \rho} = 0 \,, \tag{3F.5$_r$}$$

which LAPLACE could easily have derived from his formulation of the Caloric Theory. With CARNOT's choice of variables an equivalent statement follows by solving $(2C.14)_3$ for Λ_V, calculating $\partial\Lambda_V/\partial\theta$, and putting the result into (5M.3):

$$V\frac{\partial K_V}{\partial V} = \frac{\partial}{\partial\theta}[\theta(K_p - K_V)] \,. \tag{5R.1}$$

The possibility that $K_p = K_V$ is excluded by (2C.17) (*cf.* also Footnote 3 to §5C). Therefore (1), like (3F.5), shows that *the Caloric Theory of heat does not allow both specific heats of an ideal gas to be constant.*

For CARNOT's purpose this fact, some consequences of which we have foreseen in §3F, is especially harsh.

It is CARNOT's misfortune that he weaves his original and sound views on the motive power of fire into a theory of heat that while logically sound is untenable in physics. As we shall see in the two following sections, this very matter of the specific heats defeats CARNOT's theory as an explanation of facts of nature.

5S. CARNOT's Attempts to Determine his Function F

[CARNOT perceives the double purpose of a theory of specific heats: first, to support the general results of thermodynamics by exhibiting some rules of proportion or of dependence or independence which accord with known experimental facts, and second, to calculate the motive power of heat in numerical examples.

[In CARNOT's theory the key formula is

$$K_V = R\left(\frac{1}{F'}\right)' \log V + U' \,. \tag{5Q.5$_r$}$$

First of all, he must show that its general form squares with what is known about the dependence of K_V upon V and θ. Having done so, he may use a few specific values of K_V and so determine F, more or less, to within two arbitrary constants. The motive power of heat then follows by use of the equation in his Special Axiom: For a Carnot cycle $\mathscr{C}$,

$$L(\mathscr{C}) = [F(\theta^+) - F(\theta^-)]C^+(\mathscr{C}) \,. \tag{5I.6$_r$}$$

Above all, CARNOT desires to find some numbers that can be used in the design of steam engines. To this end, he must find the particular F that conforms best with the phenomena of nature.]

In the first place (pp. 55, 60), DELAROCHE & BÉRARD concluded from their experiments that the specific heat of a gas [at constant pressure] was a [slowly] increasing function of its volume [when the temperature was held constant]. In CARNOT's theory, likewise, (5Q.5) and its companion

$$K_p - K_V = \frac{R}{\theta F'} \tag{5Q.6$_r$}$$

show that (p. 59) "...both specific heats increase as the density of the gas diminishes [provided $F' \neq$ const.], but their difference does not change" [if the temperature is kept constant]. A footnote mentions that Messrs. GAY-LUSSAC & WELTER have found by direct experiments, cited by LAPLACE, that the ratio of specific heats, γ, "varies very little with the density of the gas. From what we have just seen, it is the difference that should remain constant, not the ratio. Because, however, the specific heat[s] of gases for a given weight vary very little with the density, it is obvious that the ratio itself experiences only small changes." Thus CARNOT's theory seems to fit the data at hand, provided $F' \neq$ const. [Doubtless CARNOT has perceived that in his theory it is impossible for both K_p and K_V to be constants, as we have seen in the preceding section.]

When we try to determine F (p. 69), "To start with, it would seem natural enough to suppose that for equal differences of temperature, the quantities of motive power produced are equal.... Doubtless, such a law would be very remarkable, but there are no reasons sufficient to assume it *a priori*. We shall discuss its reality by rigorous reasoning." [CARNOT proceeds to explore some possible embodiments of his central conviction: A quantity of heat can do more work at higher temperature than at lower. He gives a simple [and indeed rigorous] argument[1] (pp. 71–72) by which, again after appeal to the results of

[1] The argument is made clear by a diagram:

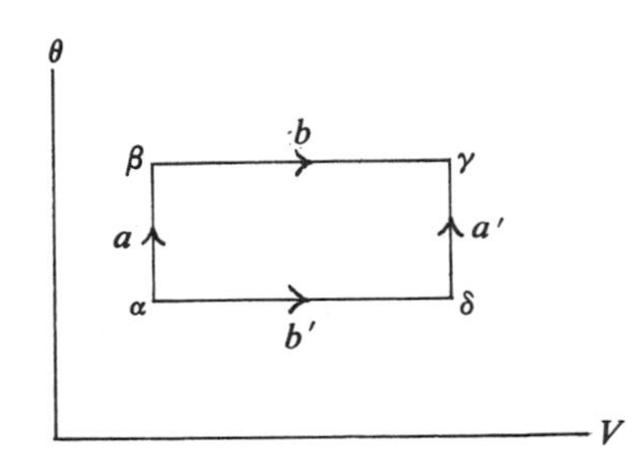

on which a, b, a', and b' stand for the amounts of heat gained on the respective paths $\alpha\beta$, $\beta\gamma$, $\delta\gamma$, and $\alpha\delta$. According to the Caloric Theory, $a + b = a' + b'$. For any given

DELAROCHE & BÉRARD, he concludes that (p. 72)

> *The quantity of heat due to the change of volume of a gas is the more considerable, the higher is the temperature....*
>
> These unequal quantities of heat would produce...equal quantities of motive power for equal falls of caloric, taken at different heights on the thermometric scale, whence the following conclusion may be drawn: *The fall of caloric produces more motive power in the lower ranges [of temperature] than in the higher ones.*

But (p. 73) this difference

> should be very small; it would be zero if the air's capacity for heat remained constant, despite changes of density. According to the experiments of Messrs. Delaroche and Bérard, this capacity varies little, so little indeed that the differences noted could strictly be attributed to some errors of observation or to some circumstances not taken into account.

Earlier (pp. 28–29), we recall (§5I), he had written, "the motive power doubtless increases with the difference of temperature between the hot body and

---

θ, the experiments of DELAROCHE & BÉRARD show that $K_p(V', \theta) > K_p(V'', \theta)$ if $V' > V''$. By (5Q.7) it follows that $K_V(V', \theta) > K_V(V'', \theta)$. Thus the quantities of heat gained on the isochoric paths $\alpha\beta$ and $\delta\gamma$ satisfy the relation $a' > a$. Hence $b > b'$.

CARNOT's statement amounts to

$$\frac{\partial \Lambda_V}{\partial \theta} > 0 .$$

Indeed, by use of (5M.3) and (5Q.7) we see that according to CARNOT's theory

$$\frac{\partial \Lambda_V}{\partial \theta} = \frac{\partial K_V}{\partial V} = \frac{\partial K_p}{\partial V} ,$$

so Λ_V is an increasing function of θ if and only if K_p is an increasing function of V. This conclusion fails if we relinquish the Caloric Theory. Falling back then upon the General CARNOT-CLAPEYRON Theorem $(5L.6)_2$, we see that for an ideal gas

$$\frac{\partial \Lambda_V}{\partial \theta} = -\frac{R\mu'}{V\mu^2}$$

on the assumption that $\mu > 0$, and so

$$\frac{\partial \Lambda_V}{\partial \theta} > 0 \quad \Leftrightarrow \quad \mu' < 0 .$$

The thermodynamics of CLAUSIUS incorporates the determination $\mu = J/\theta$, so the right-hand statement holds. It is a tribute to the insight of CARNOT that his expectation here is confirmed, although both the theory and the experimental evidence he used to support it were to be rejected later.

the cold one...." In terms of CARNOT's Special Axiom his two statements about motive power are, respectively[2]:

1. F increases
2. F' decreases

} as θ increases.

In a long footnote (pp. 73–79) CARNOT goes over the same ground in a somewhat more precise fashion, using some algebraic notation. He considers two special cases, both of them hypothetical, or largely so.

Special Case 1. In order that K_V "be constant at all temperatures" [but not independent of V] (p. 77), inspection of (5Q.5) yields the necessary and sufficient conditions

$$1/F'(\theta) = \theta/J + d \ , \qquad J = \text{const.} \neq 0 \ , \qquad [\text{and } U/R = e\theta + \text{const.}] \ ; \tag{5S.1}$$

thus

$$F(\theta) = J \log(\theta + Jd) + \text{const.} \ , \tag{5S.2}$$

[If $J > 0$ and $d \geqq 0$, this example satisfies both of CARNOT's conditions listed above. We note that if $d = 0$ and if we replace F' by the more general μ, $(1)_1$ reduces to the prophetic relation

$$\mu = J/\theta!] \qquad (5\text{N}.7)_r$$

The variation of temperature and volume along an adiabat is given by

$$\theta - \theta_0 = \frac{(\theta_0 + Jd) \log \dfrac{V_0}{V}}{\log V + Je} \ , \tag{5S.3}$$

[as CARNOT had shown earlier (p. 66), and as we easily conclude from (5Q.4)]. According to CARNOT, this formula approximates very well the results of

[2] CARNOT's requirements may be stated also in terms of the function K that occurs in (5J.5), which expresses CARNOT's General Axiom as reduced by REECH. In that way they become free of the Caloric Theory:

1. $K(\theta + h, \theta)$ is an increasing function of the positive variable h for each fixed θ.
2. $K(\theta + h, \theta)$ is a decreasing function of θ for each fixed, positive h.

It is a further tribute to CARNOT's insight that both these statements of his are verified by the thermodynamics of CLAUSIUS. Indeed, from (9A.15), below, we see that

$$\frac{\partial}{\partial h} K(\theta + h, \theta) = \frac{\theta}{(\theta + h)^2} > 0 \ ,$$

$$\frac{\partial}{\partial \theta} K(\theta + h, \theta) = -\frac{h}{(\theta + h)^2} < 0 \ .$$

experiment. In virtue of CARNOT's Special Axiom (5I.5), the corresponding motive power of a Carnot cycle is (p. 78)

$$L(\mathscr{C}) = J \log \left(\frac{\theta^+ + Jd}{\theta^- + Jd} \right) C^+(\mathscr{C}) \; . \tag{5S.4}$$

CARNOT reminds us that the hypothesis $K_V = f(V)$ has not yet been sufficiently verified by experiment, so (4) should for the time being be accepted only for a moderate range of temperatures.

[Also, if $\theta^+ - \theta^- = h$,

$$\frac{L(\mathscr{C})}{C^+(\mathscr{C})} = J \log \left(1 + \frac{h}{\theta^- + Jd} \right) \, , \tag{5S.5}$$

which indeed diminishes slowly if h is fixed and $\theta^- \to \infty$.]

Special Case 2. If we suppose that K_V is a function of θ alone (p. 78), then (5Q.5) requires that $F' =$ const., so

$$F = a\theta + \text{const.} \; , \qquad a = \text{const.} > 0 \; , \tag{5S.6}$$

and the motive power of a Carnot cycle $\mathscr{C}$ is

$$L(\mathscr{C}) = a(\theta^+ - \theta^-)C^+(\mathscr{C}) \; ; \tag{5S.7}$$

that is, "the motive power produced would be found exactly proportional to the fall of caloric."

The first special case, expressed by (2), is that which results if CARNOT's statement about "more considerable" quantities of heat absorbed in isothermal expansion is interpreted as a strict proportion. The second special case is the one CARNOT considered first but rejected later because it failed to agree with the results of DELAROCHE & BÉRARD. It satisfies the first condition listed above but does not satisfy the second.

Although CARNOT inclines toward the first special case, in the end he adopts neither and takes no final position regarding the motive power associated with a finite fall of caloric[3]. [CARNOT's numerical calculation has been

---

[3] Here, it seems, is the place to return to the mistaken idea, dismissed in Footnote 5 to §5A, that CARNOT really meant to distinguish "chaleur" from "calorique". Its origin has been traced to LIPPMANN and CALLENDAR, in both cases unjustly.

G. LIPPMANN on p. 79 of his *Cours de Thermodynamique*, Paris, Carré, 1889, after his analysis of a Carnot cycle by use of the relation $L(\mathscr{C})/J = (\theta^+ - \theta^-)\, \Delta H$ remarked, "One can express this result by saying that the production of work is due to the fall of an invariable quantity of entropy H, which passes from θ^+ to θ^-." There is nothing wrong with this statement, but it is parallel to CARNOT's results *only* in the special case when (5S.6) holds, and CARNOT drew back from endorsing this special case. CARNOT himself, as we have seen above in §5K, did not analyse a Carnot cycle in this way. ZEUNER [1860, §10] attributed this approach to REECH [1856], but below in §9A we shall see that in fact it is due to RANKINE (1851).

H. L. CALLENDAR in §17 of the article "Heat", *Encyclopaedia Brittanica*, 11th to 14th editions, made CLAPEYRON responsible for all sins then commonly laid upon CARNOT:

analysed in §5O. As has been explained in §5A, his final numerical estimates are little more than guesses.]

5T. Critique: CARNOT's Dilemma

We have seen in §5R that the Caloric Theory forbade the specific heats of an ideal gas to be both constant. In CARNOT's special case, governed by the relations

$$K_V = R\left(\frac{1}{F'}\right)' \log V + U' \quad \text{and} \quad K_p - K_V = \frac{R}{\theta F'}, \qquad (5Q.5)_r, (5Q.6)_r$$

this dilemma is acute.

By use of those formulae we easily verify that in CARNOT's theory each of the following conditions is equivalent to the rest:

1. γ is a function of θ alone.
2. K_p is a function of θ alone.
3. K_V is a function of θ alone.
4. $F = a\theta + \text{const.}$, $a = \text{const.} > 0$.
5. $L(\mathscr{C}) = a(\theta^+ - \theta^-)C^+(\mathscr{C})$.

"Unfortunately, in describing Carnot's cycle, he assumed the caloric theory of heat, and made some unnecessary mistakes, which Carnot (who, we now know, was a believer in the mechanical theory) had been very careful to avoid." In manufacturing his estimate of CARNOT's degree of care CALLENDAR would have us believe that CARNOT when he was writing the *Réflexions* was already a secret "believer in the mechanical theory", despite CARNOT's own explicit statement to the contrary, which we have quoted above in the text of §5A.

Misconceptions aside, it is astonishing how many of CARNOT's conclusions remain acceptable in the later thermodynamics if *we* choose to regard his "chaleur" as "heat" and his "calorique" as "entropy". Among them are everything CARNOT writes about isothermal processes and many things he writes about cycles. A counterexample is provided by the statement we have numbered (5Q.5); that statement does *not* remain correct if H_C is regarded as entropy, for K_V is then $\theta\partial H_C/\partial\theta$, not $\partial H_C/\partial\theta$. The later final choice (5N.7) for F' does *not* make (5Q.5) reduce to the final form of K_V for an ideal gas, namely $K_V = U'$.

I have relegated these remarks to a footnote because I think that fortuities of this kind mislead rather than enlighten. A theory of physics grows neither from augury nor from haruspication, much less from hunches or stabs in the dark. The spectator of this tragicomedy, if he has the capacity to learn from what he has seen, will recognize CARNOT as the man who established and illustrated the program of thermodynamics, the great creator of concepts and principles.

Any one of them implies also that

$$\gamma - 1 = \frac{R}{a\theta \mathrm{K}_V}\,, \tag{5T.1}$$

so

$$\mathrm{K}_V = \text{const.} \quad \Rightarrow \quad \gamma - 1 \propto \frac{1}{\theta}\,, \tag{5T.2}$$

and

$$\gamma = \text{const.} \quad \Rightarrow \quad \mathrm{K}_V \propto \frac{1}{\theta}\,, \qquad \mathrm{K}_p \propto \frac{1}{\theta}\,. \tag{5T.3}$$

Also

$$\mathrm{K}_p = \text{const.} \quad \Rightarrow \quad \mathrm{K}_V = -\frac{R}{a\theta} + \text{const.} \tag{5T.4}$$

We recognize these results as being special cases of the theorem we have noticed in §3F; as we remarked there, they stand in glaring conflict with common expectation of the behavior of gases and are most implausible; they show that *if either* K_V *or* γ *is constant,* K_p *must decrease as the temperature increases.* In the limit as $\theta \to \infty$, (2) implies that $\gamma \to 1$, while (3) implies that $\mathrm{K}_V \to 0$. Worst of all, (4) implies that $\mathrm{K}_V \to -\infty$ as $\theta \to 0$. But a basic assumption of the Doctrine of Latent and Specific Heats is

$$\mathrm{K}_V > 0\,. \tag{2C.5)$_{2r}$}$$

Therefore, we must either *reject altogether the possibility that* K_p = const. or else interpret it as admissible only in a limited range of temperatures.

These facts may explain why CARNOT refuses to consider the case in which γ = const.

Yet in his call for new experiments so as to determine K_V as a function of V and θ, CARNOT turns his back on one of the most reliable and complete bodies of experimental data at his time, namely, those regarding the speed of sound in air. These data, going back more than a century, fully supported two qualitative relations satisfied by the speed c:

$$c^2 \propto \theta\,, \qquad c^2 > \frac{p}{\rho}\,. \tag{5T.5}$$

Now if we accept, as CARNOT does, LAPLACE's contention that sound is an adiabatic motion of an ideal gas, we are led inescapably to LAPLACE's results: In general

$$c^2 = \gamma \frac{\partial p}{\partial \rho}\,, \tag{3C.1)$_r$}$$

and for an ideal gas

$$c^2 = \gamma p/\rho = \gamma r\theta\,. \tag{3C.2)$_{1,2r}$}$$

As we have shown in §3F, these results do not require γ to be constant. However, the theory as expressed by (3C.2) is consistent with experiment as summarized by (5) *if and only if*

$$\gamma = \text{const.} > 1 \; , \qquad (5T.6)$$

the gas, as always, being supposed ideal. Therefore, in order to square his theory with these two facts, one theoretical and the other experimental, CARNOT would have been forced to reject his first special case:

$$1/F'(\theta) = \theta/J + d \; , \qquad J \neq 0 \; , \qquad (5S.1)_{1,2r}$$

adopt his second special case:

$$F = a\theta + \text{const.} \; , \qquad a = \text{const.} > 0 \; , \qquad (5S.6)_r$$

and stomach the disagreeable conclusion (3). *This is not a matter that called for new experiments. CARNOT's theory, because it was inextricably entwined with the Caloric Theory, was doomed from the start by a basic flaw*, a flaw we have demonstrated in §3F.

CARNOT was not one to reject good experimental data. I think his failure here—and failure is the only word for it—reflects his clumsiness in mathematics. Since LAPLACE's work was snarled in long calculations, we may not exclude the possibility that CARNOT could not follow it, or if he could, that he was unable to disentwine the solid parts of it from LAPLACE's fantasies about molecules of caloric, atmospheres, and attractions. Perhaps he thought[1] that LAPLACE had *needed* to assume γ constant in order to get (6) at all, so that no test of the hypothesis was provided by the result, and some other theory of adiabatic motion in an ideal gas might have squared equally well with (5). The distinction here is rather fine for a person not accustomed to mathematical criticism.

CARNOT nowhere mentions the LAPLACE-POISSON law (3D.5) of adiabatic change:

$$p\rho^{-\gamma} = \text{const.} \; , \qquad (3D.5)_{1r}$$

which follows from assuming that $\gamma = \text{const.}$ It is possible that he did not know of it[2]. We have seen above in §5Q that CARNOT was able to treat

[1] CARNOT cites LAPLACE twice. While his note on p. 30 might refer only to LAPLACE [1816], on p. 59 he cites the *Mécanique Céleste* for the experiments of GAY-LUSSAC & WELTER, which are mentioned only in Book 12 of Volume 5 (LAPLACE [1823]), in the course of LAPLACE's exposition of his own theory of specific heats and adiabatic change.

[2] I have reached this conclusion after having long regarded it as untenable. I justify it as follows.

The work of LAPLACE, published in 1822 and 1823, we have described and analysed above in §§3C and 3F. While anyone who followed it in detail would have been able to read off the LAPLACE-POISSON law from a glance at (3C.11), LAPLACE himself did not record it, and only a person adept in long formal calculations, patient and determined beyond the

adiabatic changes in a special case. He could easily have done so also on the assumption that γ = const. As we have seen, he would then have been forced to adopt his second special choice of F, namely (5S.6), using which in his own formulae (5Q.3) and $(5M.2)_2$ would have shown him that

$$U = \frac{R}{(\gamma - 1)a} \log \theta + \text{const.} \ , \tag{5T.7}$$

Putting (4) and (5S.6) into (5Q.4)—the formula for the adiabats which follows by inspection from CARNOT's own determination of his heat function—we obtain the LAPLACE-POISSON law.

Certainly CARNOT's theory is compatible with the LAPLACE-POISSON law, but had CARNOT preferred that appropriate special case rather than the one he regarded as conformable with the experiments of DELAROCHE & BÉRARD, it would not have saved his theory. In the two special cases in question the motive power of a Carnot cycle is, respectively,

$$L(\mathscr{C}) = J \log \left(\frac{\theta^+ + Jd}{\theta^- + Jd} \right) C^+(\mathscr{C}) \ , \qquad (5S.4)_r$$

and

$$L(\mathscr{C}) = a(\theta^+ - \theta^-) C^+(\mathscr{C}) \ . \qquad (5S.7)_r$$

If we hold $C^+(\mathscr{C})$ and θ^- fixed, then according to either of these formulae

$$L(\mathscr{C})/C^+(\mathscr{C}) \to \infty \quad \text{as} \quad \theta^+ \to \infty \ . \tag{5T.8}$$

Both formulae imply that a given quantity of heat can produce arbitrarily great motive power in a Carnot cycle with the furnace at sufficiently high temperature.

---

ordinary, and of strong stomach could have penetrated (or can now penetrate) the dense thicket, found (3C.11), and interpreted it in adiabatic processes.

On p. 43 CARNOT attributes to POISSON the statement that air rises in temperature by one degree when its volume is reduced by the 116th part. CARNOT uses this number as if it were a datum of experiment; he does not cite the publication where he found it. It occurs in the first memoir of POISSON [1808, p. 363] (POISSON himself [1823, *1*, p. 7] was to cite the passage incorrectly as being on p. 334). There POISSON obtained it as a consequence of the *theoretical assumption* (3A.1), forced into agreement with the measured speed of sound by giving k in (3A.5) its measured value, namely 0.4254 at 6°C. As we have seen in §3B, POISSON's treatment of 1808, tentative as it is, does not require the motion to be adiabatic or γ to be constant. CARNOT, following LAPLACE, assumes the motion adiabatic and thus can accept POISSON's figure 1/116 as being appropriate to adiabatic changes; as we have seen above in §5O, this is just what CARNOT does.

POISSON's final treatment [1823, *1* and *2*], in which he assumes that γ = const., obtains and publishes the LAPLACE-POISSON law, and makes its status fairly clear, is simple enough for any good graduate of the Ecole Polytechnique to read, but it appeared in 1823, and CARNOT's treatise was published on June 12, 1824. Any claim that CARNOT could have seen POISSON's papers of 1823 in time to use their contents in his manuscript would have to be substantiated by precise specification of their dates of publication and of CARNOT's return of proofsheets.

In modern terms, the efficiency of Carnot cycles is not bounded above. The abundant "Second Laws" of modern thermodynamics are interpreted, according to the divinations of our contemporary experts, as denying truth to such a conclusion. Of course it contradicts also the "First Law".

In refraining from trying to apply his results to specific cases except when the difference $\theta^+ - \theta^-$ is small, CARNOT shows his practical good sense. For finite differences of temperature, both choices of F he considers—the one he regards as supported by the experiments of DELAROCHE & BÉRARD and the one that he should have to accept, were he to take seriously the results of experiment on the speed of sound—both, I say, lead to ridiculous conclusions, refuted by the simplest experience with heat engines. To both cases[3] we may apply his own words (p. 21): "Creation of this kind is entirely contrary to the ideas presently received, to the laws of...sound physics. It is inadmissible."

In CARNOT's work there is no sign of the scrupulous analysis of concepts, the inexorable rejection of every postulate or axiom not necessary to the end desired, that is the essence of mathematics[4] in general, of the rational mechanics of the seventeenth and eighteenth centuries in particular.

It was not experiment that was wanting; it was mathematics.

5U. CARNOT's Bequest

It is now obvious to the spectator that CARNOT founded thermodynamics, founded it not only in physical concept but also in program and in schema. Of more lasting importance than any specific calculation is his having perceived that a theory of the work effected by heat necessarily *restricts the class of constitutive relations allowed*, and it was he himself who obtained in principle the central restriction, the General CARNOT-CLAPEYRON Theorem:

$$\mu\Lambda_V = \frac{\partial p}{\partial \theta} \,. \qquad (5L.4)_r$$

---

[3] CARNOT's theory does not imply (8) for all choices of F. For example, if $F(\theta) = J(1 - e^{-k\theta})$, then $L(\mathscr{C})/C^+(\mathscr{C}) < J$ for all values of θ^+ and θ^-. Of course we know today that no choice of CARNOT's F can square even roughly with experimental fact except when $(\theta^+ - \theta^-)/\theta^+$ is small.

[4] Some historians and physicists have suggested that CARNOT may have worked out all his theory by the aid of mathematics and then expressed it in everyday language for the presumed benefit such treatment would confer upon engineers. Although there is no evidence whatever on which to base this charitable imagination, it may be true nevertheless, but it has nothing to do with the mathematical standard of CARNOT's work. Such reasoning as he does present, whether in words or in symbols, is almost always right as far as it goes. Both kinds are equally disorderly and equally uncritical, equally inferior to the standard that had been set by the rational mechanics of the preceding century and was being maintained in CARNOT's day by CAUCHY and others.

CARNOT treated the principles of thermodynamics as restrictions upon *materials*, not upon *processes* in a body of given material. His own result $(5L.6)_2$ we may interpret as the requirement that *for all ideal gases*,

$$\frac{V\Lambda_V}{R} = \text{one and the same function of } \theta \text{, namely } \frac{1}{F'(\theta)} \ . \tag{5U.1}$$

CARNOT's mastery of the concept expressed by (1) is reflected in his line of thought which we have quoted above in §5S, although he does not mention (1) itself there. His first claim regards isothermal expansion; it is equivalent to $\partial\Lambda_V/\partial\theta > 0$. Upon its heels comes his second claim, which refers to motive power; it is equivalent to $(F')' < 0$. He writes as if he took for granted that the former implied the latter. Certainly no such thing is true without a theory of the motive power of heat. CARNOT's theory delivers the constitutive restriction (1), which makes his first and second claims equivalent.

A less obvious but equally important debt we owe CARNOT was pointed out by LIPPMANN[1]. He observed that while the physicists and chemists of the early nineteenth century might now seem to have stood but a short step from the creation of thermodynamics,

> In fact they were far away from it; they lacked one idea. They would have had to introduce into physics in place of the conception of molecular motion *mechanical work*, which is a quantity measurable by experiment. Only the geometers[2] used the notion of work: Sadi Carnot devised how to introduce it into physics; that was the start of modern physics.

Few men have done so much to found any science as has CARNOT for thermodynamics. Our analysis of his work has revealed magnificent success in part, though failure in the end. CARNOT was by no means alone in failing much while doing much. The curse of thermodynamics has been, not that, as happened in every other branch of physics, the great creators occasionally erred or failed, but that their successors have treasured the errors and the deficiencies while neglecting to seize, purify, and exploit the successes. CARNOT's splendid general conception, embodied in his great General Axiom:

$$L(\mathscr{C}) = G(\theta^+, \theta^-, C^+(\mathscr{C})) \ , \tag{5I.1$_r$}$$

$$G(x, y, z) > 0 \quad \text{if} \quad x > y > 0 \quad \text{and} \quad z > 0 \ , \tag{5I.2$_2$}$$

---

[1] G. LIPPMANNN, *Cours de Thermodynamique*, Paris, Carré, 1889. See pp. 3–4. While PETIT [1818] may have been the first to apply the concept of work in the theory of heat, CARNOT was certainly the first to do so effectively.

[2] The geometers to whom LIPPMANN referred must have been DANIEL BERNOULLI and EULER, perhaps also CAUCHY and LAZARE CARNOT. I have found nothing about work in the writings of LAPLACE and FOURIER. FOURIER applied to the theory of heat the kind of mathematics the geometers had invented for use in mechanics, but he did not broaden the domain of mechanics itself.

and the function G is universal, the same for all bodies; his specific reasoning leading to the General CARNOT-CLAPEYRON Theorem:

$$\mu\Lambda_V = \frac{\partial p}{\partial \theta} \; ; \tag{5L.4$_\text{r}$}$$

and his perception that the nature of the specific heats provides the key to thermodynamics, were to be little understood. Instead, his preference for undisciplined, unmathematical arguments; his primitive use of the infinitesimal calculus; his tendency to sweeping claims about maxima without specifying what is held fixed and what is allowed to vary, or even the variables upon which the thing maximized depends; his predilection for steam and coal; his appeal to irrelevant or at best merely ancillary experimental details; his reluctance to face the test of comparing his own results with the successful theories of others; and, finally and above all, his confusion of the constitutive properties of special substances with the general relations between heat, work, and change of temperature—all these were cherished, enshrined, and magnified by his successors. Such became the tradition of the subject. An eruption of paper covered with symbols and the data of experiments could have spared, had CARNOT stated his assumptions clearly in mathematical terms and given explicit mathematical proofs of his deductions from them.

In no way rejecting or disregarding LIPPMANN's penetrating estimate, in full admiration for CARNOT's amazing grasp of the physics of heat and uncanny sense of what was right despite his being hobbled by a basically unsound concept of heat itself, I must adjoin a counterbalanced judgment. Among physicists of the first rank, CARNOT is the first who was not in at least equal measure a mathematician. Thermodynamics is the first mathematical science to have been invented without the control afforded by patient, merciless, mathematical criticism. It has suffered from this congenital defect from 1824 until now.

On the physical side, most unfortunate is CARNOT's failure to give any position to irreversible processes, despite their everyday familiarity and despite FOURIER's work, already famous though only recently published. Here, too, CARNOT's limitation is to become characteristic of thermodynamics: Rather than extend the existing successful theories so as to embrace new ranges of phenomena, thermodynamics will rule them out from the start. Having put on the stage as protagonist a pygmy, the ideal gas, CARNOT appoints as director a Mephistopheles who tells him it makes no difference which way he goes. The "reversible" process, a prototype of Liberal Philosophy, is to keep thermodynamics turning in ineluctable circles for over a century. For such processes the time makes no difference. Thus the letter t is free, and CARNOT, unlike FOURIER and LAPLACE, uses it for temperature. As all thermodynamicists were to follow CARNOT, it came to seem impossible that thermodynamics could ever mention the time. The very letter for it was already used up! Even KELVIN, a virtuoso in heat conduction and mechanics, in his papers on thermodynamics refrained from using t for anything but temperature and from introducing any letter at all to denote the time. Thus begins that quality of classical thermodynamics that to the modern student is most striking: its *timelessness*.

CARNOT set one great stone in the foundation of a general thermodynamics. FOURIER had set another, at an opposite corner. Cornerstones these were, not a framework or even a substructure. The successors of FOURIER and CARNOT, blindly eager to perpetuate their failures rather than promote their successes, until the very last three lustra built outward and away from each other in clusters of little chapels, with no thought to finish the foundation on which the cathedral was to rise.

6. Distracting Interlude: CLAPEYRON and DUHAMEL

ché a tutti un fil di ferro i cigli fóra
e cusce....
DANTE, *Purgatorio* XIII, 70–71.

6A. Confusion by Awkward Variables: CLAPEYRON

The theory of CARNOT was taken up by CLAPEYRON in a memoir[1] published in 1834. While CLAPEYRON (§I) hails "the idea which serves as a basis of [CARNOT's] researches" as being "both fertile and beyond question", he deplores CARNOT's preference for "avoiding the use of mathematical analysis" in favor of "a chain of difficult and elusive arguments" so as to arrive at "results which can be deduced easily from a more general law."

[It has become customary to claim that CLAPEYRON's translation of CARNOT's ideas into mathematics was unfaithful. On the contrary, my analysis of CARNOT's work with great pains and pain has led me to opine that CLAPEYRON did indeed present CARNOT's theory faithfully—faithfully, I say, though not well.]

In §II CLAPEYRON states that "...a quantity of mechanical action and a quantity of heat which can pass from a hot body to a cold body are quantities of the same kind, and...it is possible to replace the one by the other...." [To those who do not distinguish verbal science from rational science, this statement might be confused[2] with what is now called "The First Law of Thermodynamics". It is nothing of the kind. As his subsequent mathematical analysis shows, CLAPEYRON is merely paraphrasing what CARNOT had written to the same effect. In the theory of CARNOT and CLAPEYRON heat and work are indeed interconvertible, but not uniformly: The factor of conversion depends upon the temperature. *Cf.* §5M, above.

[1] CLAPEYRON [1834].

[2] In annotating this passage in his edition of CARNOT's treatise, MENDOZA calls it "an unambiguous statement of the First Law of Thermodynamics".

[CLAPEYRON's work lies at a lower level of concept than CARNOT's[3], the product of an expounder, not a creator. In one respect, however, CLAPEYRON is more careful than CARNOT, for] in referring to the Carnot cycle he always writes "maximum effect", "maximum quantity of action", *etc.*, [while CARNOT's expression often leaves the reader in doubt whether he refers to all engines, to engines whose working substance is ideal gas, or only to those alleged to be the best possible ones.]

Although the scope of CLAPEYRON's concrete analysis is the same as CARNOT's, there are two differences.

[First, CLAPEYRON's mathematics, although loose, is clear and easy[4]] besides referring to a general gas rather than an ideal one[5]. Second, while CLAPEYRON introduces the diagrams which have since become standard and which we have used already in our discussion of CARNOT's treatise, [the tragicomic genius of thermodynamics tells him to obscure the results, as LAPLACE had done before him, by using the V–p quadrant instead of the V–θ quadrant. To estimate the resulting complexity we need only compare (3F.2) with (5M.2), or (3F.4) with (5M.3).] On the other hand, CLAPEYRON brings into the open that essential quantity, the latent heat Λ_V, which CARNOT had avoided treating directly. [Despite] CLAPEYRON's claim to discover "some new relations" (§III), [he does not present a single new idea of his own and merely says in his own way what CARNOT had said, if somewhat more specially and obscurely, before him; thus our analysis of his work can

---

[3] *E.g.*, CLAPEYRON's explanation (§III) of CARNOT's result (5K.5) is, "equal volumes of all elastic fluids, taken at the same temperature and subject to the same pressure, when compressed or expanded by the same fraction of their volumes, emit or absorb the same absolute quantity of heat." The phrase "and subject to the same pressure" is taken to refer to initial pressure.

CLAPEYRON's wording of CARNOT's argument about the two cycles that cancel each other's transfer of heat from the furnace to the refrigerator (end of §II) is so sloppy that I cannot say whether his failure to mention the refrigerator is by intention or by oversight. He refers to "a quantity of action ... created out of nothing and without consumption of heat, an absurd result which would lead to the possibility of creating force or heat at no cost and without limit." It seems to him that denial of any such possibility "can be accepted as a fundamental axiom of mechanics; no one has ever dreamt of objecting to Lagrange's demonstration of the principle of virtual velocities using pulleys, and this seems to me to depend on something similar." What CLAPEYRON means by "consumption" is not clear, and I doubt that any serious student of the foundations of mechanics has ever taken LAGRANGE's pulleys (!) as being anything more than rhetoric, so for thermodynamics they provide a sorry precedent. Earlier (§I) CLAPEYRON has written that CARNOT's "demonstrations are founded on *the absurdity of supposing the possibility of creating motive power or heat out of nothing.*"

[4] CLAPEYRON, like CARNOT before him, uses differentials along paths. I present his arguments in terms of derivatives with respect to time, since these are easier for modern students to follow securely.

[5] It is only in this regard that the theory of CLAPEYRON could be claimed to rest on "a more general law". CLAPEYRON carries out all the analysis first for a perfect gas and then does it over again for a general equation of state.

be brief and can confine itself to the effects of his unfortunate choice of independent variables.]

We have seen how CARNOT obtained his main result (5K.5) by considering Carnot cycles with infinitesimal difference of temperature. CLAPEYRON takes advantage of this fact from the outset by using a Carnot cycle with all four of its parts infinitesimal. Since for CLAPEYRON $Q = \dot{\mathrm{H}}_{\mathrm{Cl}}(V, p)$, say, he finds (§V) that

$$Q = \dot{\mathrm{H}}_{\mathrm{Cl}} = \frac{\partial \mathrm{H}_{\mathrm{Cl}}}{\partial V}\dot{V} + \frac{\partial \mathrm{H}_{\mathrm{Cl}}}{\partial p}\dot{p} \tag{6A.1}$$

in any process (*cf.* (3C.7)). Since $\theta = \theta(V, p)$, in an isothermal process

$$0 = \dot{\theta} = \frac{\partial \theta}{\partial V}\dot{V} + \frac{\partial \theta}{\partial p}\dot{p} \ , \tag{6A.2}$$

so

$$\Lambda_V \dot{V} = Q|_{\theta=\mathrm{const.}} = \left(\frac{\partial \mathrm{H}_{\mathrm{Cl}}}{\partial V} - \frac{\frac{\partial \theta}{\partial V}}{\frac{\partial \theta}{\partial p}} \frac{\partial \mathrm{H}_{\mathrm{Cl}}}{\partial p} \right) \dot{V} \ . \tag{6A.3}$$

In this notation the Special CARNOT-CLAPEYRON Theorem, namely

$$F'\Lambda_V = \frac{\partial p}{\partial \theta} \ , \tag{5L.5$_r$}$$

takes the form (§V)

$$\frac{\partial(\mathrm{H}_{\mathrm{Cl}}, \theta)}{\partial(V, p)} = C \ ; \tag{6A.4}$$

in terms of the function F in CARNOT's Special Axiom (5I.5), CLAPEYRON's function $C = 1/F' = 1/\mu$, so C is "a function of temperature which is the same for all [bodies]." CLAPEYRON considers the result as a differential equation for H_{Cl}, integrates it in the special case of an ideal gas, and discusses its solution for a general equation of state. After some manipulations he obtains the formula (end of §V)

$$\Lambda_p = -C\frac{dV}{d\theta} \ , \tag{6A.5}$$

in which the ordinary derivative refers to an isobaric process. While CLAPEYRON regards this formula as "the most general consequence we can get from this axiom: It is absurd to suppose that force or heat can be created from nothing and at no cost", [in fact it is a trivial consequence of $(2C.9)_1$ and (5L.5), interpreted by referring (2C.1) to a process in which $\dot{p} = 0$].

CLAPEYRON remarks (§I) that LAPLACE and POISSON had based their work on the hypothesis that $\gamma = $ const., but he refrains from committing himself to that hypothesis and from remarking that it would require, according to the theory he presents, that $C = \text{const.}$, while all the data he adduces show that C is not a constant. As we have remarked in §5Q, CARNOT himself seemed not to notice the general equation for adiabatic change his theory implied, namely,

$$\frac{R}{F'(\theta)} \log V + U(\theta) = \frac{R}{F'(\theta_0)} \log V_0 + U(\theta_0) \ . \qquad (5Q.4)_r$$

[Surely we should expect that CLAPEYRON in his mathematical presentation would find and discuss this relation,] but he does not. He goes so far as to state (§II) that the adiabatic changes in a Carnot cycle follow "an unknown law". [Although CARNOT's failure in 1824 to compare his new theory with the then recent LAPLACE-POISSON theory of adiabatic change can be explained or at least condoned, in CLAPEYRON's hands ten years later the same failure is a case of plain negligence.]

As has been described in §5O, CLAPEYRON presented CARNOT's numerical evaluation of the motive power of heat in more comprehensible form. The slight difference in his result, 1.41 instead of CARNOT's 1.395, results only from his use of DULONG's value for the adiabatic compressibility, namely $(-0.421)^{-1}$, instead of POISSON's value -2.30.

[Slight as was CLAPEYRON's originality, his influence upon classical thermodynamics has been heavy and lasting. Even the notations in an ordinary thermodynamics book today are essentially those he introduced. In matters of physical principle his style of argument also became standard. Lacking altogether CARNOT's grasp, insight, and genius, there CLAPEYRON merely copied and magnified CARNOT's weaknesses. His exposition is a haze of words about maxima and all possible processes in every conceivable material, followed by some simple mathematics dealing only with the most special of substances, namely, a gas, under the most degenerate conditions, namely, uniform fields of density and temperature. The confusion of the general principles of thermodynamics with constitutive properties of a material whose response to deformation is a hydrostatic pressure determined by the volume, temperature, and perhaps a few further scalar parameters became and has remained inherent to thermodynamic ritual. Indeed, nowadays some fakirs of thermodynamics claim that the first problem of thermodynamics is to decide what the "state" is, while in truth the illusion that there is such a thing as a "state" has been abundantly refuted by the kinetic theory of gases for over a century and has not the least to do with the fundamental theory of heat, temperature, and work.

[How] CLAPEYRON could claim (§V), [of course echoing CARNOT,] that his results hold "for all bodies in nature—solids, liquids, or gases," [it is hard to see.] The year before, he and LAMÉ had published a joint paper on the theory of elastic solids, in which the very concept of scalar pressure does not generally exist. The scientific paths of these two men, like the two aspects of a true thermodynamics, seem not to have crossed again. LAMÉ became an expert on the field theories of elasticity and the conduction of heat. His *Lectures on the Analytical Theory of Heat*, published a quarter of a century later, follow straight out of the work of FOURIER and DUHAMEL and do not so much as mention the thermodynamics of CARNOT and CLAUSIUS. [Between the "mechanical theory of heat" and the "analytical theory of heat", created separately at about the same time, had been erected an adiabatic wall. One was a mathematical field theory, clearly stated, conceptually meager, and abounding in initial-value problems and boundary-value problems. The other was a physical theory of lumped parameters, given to extravagant and altogether unjustified claims of generality, pregnant but abortive[6].]

6B. Confusion by Linearizing Everything: DUHAMEL

DUHAMEL it was to whom fell the next entry upon the stage. In the second of his two attempts he succeeded in formulating the theory of thermo-elasticity for isotropic materials subject to infinitesimal strains and infinitesimal differences of temperature. [Anyone approaching this subject today lays down as one of his axioms some expression of the principle of conservation of energy. DUHAMEL, on the contrary, gives no evidence of grasping any such idea, and his work does not in the least foreshadow that principle, which was not to be formulated until the 1850s.] Rather, adopting the attachment to linearity of "the illustrious author of the mathematical theory of heat", he blindly superimposes one effect upon another.

In his first memoir[1] DUHAMEL follows "the same course as Mr. Poisson" so as to calculate the overall forces exerted upon parts of a body composed of stationary molecules, except that he allows the intermolecular repulsive forces to depend upon the temperature. [The details of the approximate calculation based upon this speculative model need not concern us.] DUHAMEL's main result is the following constitutive relation for an isotropic material (his Equation (5))[2]:

$$-5\delta'\mathbf{T} = (\operatorname{tr}\mathbf{E})\mathbf{1} + 2\mathbf{E} - 5\delta(\theta - \theta_0)\mathbf{1} \; ; \qquad (6B.1)$$

[6] The wall could even separate two parts of one and the same man's mind. The great universal mathematician and physicist POINCARÉ lectured on thermodynamics in 1888/9 and on the propagation of heat in 1893/4. If we may trust the published texts of the two courses, neither alluded even once to the subject of the other, and the modes of thought and levels of mathematical precision in the two would seem, were it not for the common name on the two title pages, to belong to two different persons.

[1] DUHAMEL [1838, pp. 445–464]. In reporting the results I use the notations for coefficients DUHAMEL introduced on p. 462 and adopted in his second memoir.

[2] In the notation now current, the constitutive relation for the stress in an isotropic thermo-elastic material in infinitesimal strain is

$$\mathbf{T} = \lambda(\operatorname{tr}\mathbf{E})\mathbf{1} + 2\mu\mathbf{E} + m(\theta - \theta_0)\mathbf{1} \; .$$

DUHAMEL's theory, like POISSON's and all other early static-molecular theories, required

T is the stress tensor, **E** is the infinitesimal strain tensor, θ_0 is the temperature at which the stress vanishes if the strain does, and δ and δ' are constitutive coefficients: the linear dilatation produced by unit increase of temperature [when the stress vanishes], and the linear dilatation corresponding to unit normal tension on the entire boundary [when $\theta = \theta_0$]. Thus, whatever the molecular properties used, DUHAMEL's result asserts that *the stress in a thermo-elastic material is obtained by superimposing upon the elastic stress a hydrostatic pressure proportional to the increase of temperature.* In this first attempt DUHAMEL avoids commitment to any particular differential equation for the temperature, but the reader is left with the presumption that FOURIER's equation, namely

$$\rho C \frac{\partial \theta}{\partial t} = K \Delta \theta \ , \qquad (4E.1)_r$$

remains valid[3].

Such a presumption would be false, as DUHAMEL was soon made to see. In a second memoir[4], which was published before the first one, he writes:

> It is generally assumed that all bodies release heat when they are compressed, and absorb it when they are expanded, whence it follows that there is a noticeable difference between the specific heats at constant volume and at constant pressure. This is the principle which serves as the base of my theory, and I assume that the quantity of heat released is proportional to the increase suffered by the density, provided that increase be very small.

Thus

> the equations of Mr. Navier require a modification, which that learned geometer has indeed foreseen and of which he has spoken in his report on the theory I have given.... This modification refers to the heat developed or absorbed in the changes of density which can accompany the vibrations; it can be calculated only by the theory I have just recalled.

---

that $\lambda = \mu$. Then DUHAMEL's coefficients are related as follows to the now current ones:

$$\delta' = -\frac{1}{5\mu} \ , \qquad \delta = -\frac{m}{5\mu} \ .$$

The coefficient δ is nowadays called the *coefficient of thermal expansion,* while m is the *stress-temperature modulus.*

[3] F. E. NEUMANN [1843, §10, Equation (B)] arrives at "general equations" which may be abbreviated as $\mathscr{P}$ grad $\theta = -\text{div}\,\mathbf{T}$. In his long memoir I find no counterpart or generalization of FOURIER's (4E.1).

[4] DUHAMEL [1837, pp. 2, 4–5].

That theory is the LAPLACE-POISSON theory of adiabatic change:

> I assume that each infinitely small particle takes instantly all the heat that the compression it experiences can give it. This is the hypothesis assumed by the physicists who have treated this question in the case of gases, and one should recognize that it is much better founded in the case of solids. For if we were to suppose a molecule to take an appreciable time to receive the increase of temperature that a compression should confer upon it, the slowness of the operation of diffusion of heat in solids would allow us to neglect its effect during this interval of time, which cannot be anything else than extremely small.

By use of the sort of cycle imagined by POISSON (above, §3D), DUHAMEL now concludes[5] that, in effect, the increase in temperature "which would result in general from a small increase . . . of the density" would be given by

$$\dot{\theta} = -\frac{\gamma - 1}{3\delta} \operatorname{tr} \dot{\mathbf{E}} \ . \tag{6B.2}$$

"This general relation between the corresponding variations of the density and the temperature [in an adiabatic process] has been known for a long time. . . ." DUHAMEL then simply adds the right-hand side of (2) to that of FOURIER's equation (4E.8), so obtaining the following field equation for the temperature:

$$\frac{\partial \theta}{\partial t} = \frac{1}{\rho C} \operatorname{div}(K \operatorname{grad} \theta) - \frac{\gamma - 1}{3\delta} \operatorname{tr} \dot{\mathbf{E}} \ . \tag{6B.3}$$

[Indeed, (2) does hold for infinitesimal adiabatic dilatational motion of a body for which DUHAMEL's constitutive relation (1) is valid. To see this, we need only suppose that $\mathbf{E} = e\mathbf{1}$; of course $e = \frac{1}{3}(1 - \rho/\rho_0)$, and so from (1) we obtain $\mathbf{T} = -p\mathbf{1}$, and

$$\begin{aligned} \delta' p &= e - \delta(\theta - \theta_0) \ , \\ &= \frac{1}{3}\left(1 - \frac{\rho}{\rho_0}\right) - \delta(\theta - \theta_0) \ . \end{aligned} \tag{6B.4}$$

Although this "thermal equation of state" holds only in dilatations, use of those special deformations suffices to deliver connections among coefficients which are, by hypothesis, independent of the deformation. We begin from the calorimetric relation

$$Q = -(\mathrm{K}_p - \mathrm{K}_V) \frac{\partial p/\partial \rho}{\partial p/\partial \theta} \dot{\rho} + \mathrm{K}_V \dot{\theta} \ , \tag{2C.13$_{1r}$}$$

[5] DUHAMEL [1837, p. 9].

assume that the infinitesimal dilatation is adiabatic, and then use the fact that $\dot{\rho} + \rho_0 \operatorname{tr} \dot{\mathbf{E}} = 0$ to obtain (2).

[While DUHAMEL does not explain his steps further, they are clear. He does not assume that the motion of the thermo-elastic body is adiabatic. Rather, his tacit axiom is one of superposition: *For small deformations and changes of temperature, the rate of increase of temperature is the sum of two*:

1. *That which would arise from the conduction of heat if the body were rigid.*
2. *That which would arise from adiabatic dilatational motion with the same condensation $\dot{\rho}/\rho$ as in the actual motion, if the body did not conduct heat.*

[Thus the magic of linearity enabled DUHAMEL to formulate the theory of infinitesimal thermo-elasticity for isotropic bodies without facing any of the basic physical problems of such a theory. It is hard not to consider his method of deriving (3) a stroke of luck—if luck it can be called to rest content with picking up a gold piece which lies upon the doorsill of a caché of diamonds.

[The passion to linearize before thinking, displayed again and again in nineteenth-century physics, brought its gains and its losses. Its gains, most brilliant in the work of FOURIER himself, reflect its easy freedom from having to face difficult conceptual problems. The losses reflect the same cause: There are many ways to get an approximate linear theory, and these need not suggest the true physical principles that underlie it. DUHAMEL's thermo-elasticity is a case in point. There are no grounds at all to support DUHAMEL's superposition of effects and in particular his choice of an adiabatic process in the solid under circumstances when it behaves like a fluid, and his expression for the coefficient of thermal expansion in (3) means nothing[6]. Moreover, DUHAMEL's equation (3), while anyone today who looks at it will see that it must somehow reflect the principle of conservation of energy, did not foreshadow or suggest that principle[7] to him or to anyone else, so far as the printed record witnesses[8].]

---

[6] What DUHAMEL writes as $-\frac{1}{3}(\gamma - 1)/\delta$ is just a further empirical coefficient like δ'. Indeed, by substituting into $(2C.9)_2$ the values of $\partial p/\partial \rho$ and $\partial p/\partial \theta$ calculated from (4) we easily show that

$$-\frac{\gamma - 1}{3\delta} = -\frac{M\rho_0 \Lambda_V}{\rho^2 K_V},$$

so DUHAMEL's coefficient is, as it ought to be, proportional to the latent heat with respect to volume. Glancing at (3), we might think that the term representing the effect of dilatation upon temperature vanished if $\gamma = 1$, but this is not so, since p is a function of ρ only, so by (4) we see that $\delta = 0$ also. (More generally, in Footnote 13 to §2C we have seen that if $\gamma = 1$, then $\Lambda_V \partial p/\partial \theta = 0$.) All one can say is that DUHAMEL's expression for the coefficient of thermal expansion becomes meaningless if $\gamma = 1$.

[7] Indeed, its relation to that principle is somewhat subtle, as may be seen from the derivation presented by D. E. CARLSON in §7 of "Linear thermoelasticity", *Handbuch der Physik* VIa/2, ed. C. TRUESDELL, Berlin *etc.*, Springer-Verlag, 1972.

[8] In the beautiful paper on elasticity he wrote as a boy of 19, not yet matriculated at Cambridge, MAXWELL [1850, Eq. (13) and Cases IX and X] considered thermo-elasticity

DUHAMEL, following the example of FOURIER, turned to the solution of important special problems and the proof of general theorems in his linear theory. [On that theory he has left his mark as the founder and a major discoverer[9]. For thermodynamics, however, DUHAMEL's action was a lasting defeat. Not only was there to be a lapse of 126 years until thermo-elasticity should be incorporated into a general scheme of thermodynamics, but also thermo-elasticity was to serve as the pilot case for the creation of that scheme.]

---

briefly. He rediscovered only a part of DUHAMEL's theory, not reaching even any special case of DUHAMEL's basic equation (3).

Apparently THOMSON [1855] was the first to attempt a treatment of thermo-elasticity on the basis of general ideas concerning heat and work. Following the lead of GREEN for ordinary elasticity, he assumed the existence of an elastic potential, which he allowed to depend upon θ. He did not reach a set of complete and general equations.

F. NEUMANN [1885, §59] in his lectures of 1857/1858, some years after the principle of equivalence of heat and work had been established, did not refer to it in connection with thermo-elasticity. Although he had been himself one of the pioneers of thermodynamics, he derived (3) by much the same obscure process as DUHAMEL had used to discover it, with no remark that it might be related to general thermodynamic concepts and principles.

[9] More than that, DUHAMEL [1837, p. 3] was the first to learn, and this despite his superposition of terms to get his differential equations, that deformation and change of temperature could not be determined separately: "The theory of the propagation of heat thus becomes dependent upon the mechanical theory that determines the changes of position required by the interior equilibrium of a body unequally heated, and the second theory depends in turn on the first, so that neither can be treated separately.... Thus the two great physical theories that for some years past have most occupied the attention of geometers are found to be intimately connected." The "two great physical theories" both leave CARNOT's ideas in oblivion. Unfortunately, those ideas, probably because of their obscure presentation, did not attract the geometers.

7. Act III. Equivalence, Conservation, Interconvertibility: When and of What?

> ... noi ci mettemmo per un bosco
> che da neun sentiero era segnato.
> Non fronda verde, ma di color fosco;
> non rami schietti, ma nodosi e 'nvolti....
>
> DANTE, *Inferno* XIII, 2–5.

7A. Critique: What Did Janus See in 1842?

That heat could sometimes cause mechanical effect, and much of it, had been known since the disaster that befel STREPSIADES while he was cooking the haggis for the feast of Zeus, but apparently it was the sooty proliferation of the steam engine in the early nineteenth century that first roused physicists to pay much attention to the phenomenon. As CARNOT had seen, and as CLAPEYRON had made widely known, by absorbing and emitting heat a given body undergoing a cyclic process may do a definite amount of work, and by doing work cyclically a body may absorb and emit definite amounts of heat. Certain ideal bodies, described by the theory of calorimetry, give out in undergoing the reverse of a given process the heat they would gain and the work they would do in the given process. In this sense, then, it was known that

1. *Heat and work are* interconvertible *in cyclic processes.*

CARNOT's theory provides an instance: As we have seen in §5M, for a cycle $\mathscr{C}$ CARNOT's theory implies that

$$L(\mathscr{C}) = \int_{t_1}^{t_2} F(\theta) Q dt \, . \qquad (5M.5)_{2r}$$

A unit of heat absorbed at the temperature θ in a cyclic process effects $F(\theta)$ units of work overall. A unit of heat emitted at the temperature θ in a cyclic process uses up $F(\theta)$ units of work overall. The term "overall" reminds us that we may add to the integrand $F(\theta)Q$ any function of t whose integral over cyclic processes is null. That addend may give rise to additional work done or used up in portions of the process, but such additional work is cancelled by an equal amount used up or done, respectively, in the remaining portions of the process.

In (5M.5) the function F might vary from one body to another: It might be a constitutive function, just as ϖ, Λ_V, and K_V are constitutive. For CARNOT such would be "contrary to sound physics". In CARNOT's theory F is a *universal* function, the same for all bodies:

2. *Heat and work are* universally *interconvertible in cyclic processes.*

In §§5N and 5S we have seen examples of universal interconvertibility in CARNOT's sense. In his theory we may take the ratio $L(\mathscr{C})/C^+(\mathscr{C})$ for some given Carnot cycle $\mathscr{C}$ and so obtain a constant J having the dimensions of work ÷ heat. The J in the relation

$$L(\mathscr{C}) = J\left(\log\frac{\theta^+}{\theta^-}\right)C^+(\mathscr{C}) \tag{5N.4)$_r$}$$

provides an example: J is the ratio of work done to heat absorbed in a Carnot cycle whose operating temperatures are $e\theta$ and θ for any θ, provided CARNOT's F has the special form

$$F(\theta) = J\log\theta + \text{const.} \qquad (5N.3)_r$$

A somewhat more general possibility is provided by

$$L(\mathscr{C}) = J\log\left(\frac{\theta^+ + Jd}{\theta^- + Jd}\right)C^+(\mathscr{C})\ . \qquad (5S.4)_r$$

But such a J is not a universal equivalent. For example, according to (5N.3) the ratio of work done to heat absorbed in a Carnot cycle whose operating temperatures are $e^2\theta$ and θ is $2J$.

The later thermodynamics, some form of which we are taught in school today, assumes something more: Conversion of units of heat into units of work in cyclic processes is *independent of the temperature* at which the conversion is effected. That is, the relation

$$L(\mathscr{C}) = \int_{t_1}^{t_2} F(\theta)Q\,dt \qquad (5M.5)_{2r}$$

holds with F replaced by a *universal constant* J:

3. *Heat and work are* uniformly *interconvertible in cyclic processes.*

In mathematical statement,

$$L(\mathscr{C}) = JC(\mathscr{C}) = J[C^+(\mathscr{C}) - C^-(\mathscr{C})]\ . \qquad (7A.1)$$

As far as this assumption goes, the constant J might be constitutive, but Assumption 2 forbids that and makes *J a universal constant*, which allows us to express *a unit of heat as being for all bodies and all temperatures equivalent*[1], in the sense just specified, *to J units of work*.

To CARNOT Assumption 3 was excluded. The Caloric Theory, which he adopted and employed, made $C^+(\mathscr{C})$ and $C^-(\mathscr{C})$ equal in all cycles, so (1) would then imply that $L(\mathscr{C}) = 0$ in *all* cycles $\mathscr{C}$. No heat engine would then be possible, and the inequality

$$F(x) > F(y) \quad \text{if} \quad x > y > 0 \,, \tag{5I.5$_{2r}$}$$

which is a part of CARNOT's Special Axiom, would be contradicted. Thus the constant J cannot be regarded[2] as a special choice of CARNOT's F. Uniform interconvertibility as expressed by (1) *requires that the Caloric Theory be rejected.* Of course the Caloric Theory and uniform interconvertibility while incompatible with each other are compatible, each by itself, with CARNOT's General Axiom:

$$L(\mathscr{C}) = G(\theta^+, \theta^-, C^+(\mathscr{C})) \,, \tag{5I.1$_r$}$$

$$G(x, y, z) > 0 \quad \text{if} \quad x > y > 0 \quad \text{and} \quad z > 0 \,, \tag{5I.2$_r$}$$

and the function G is universal.

Everyone has read that CARNOT at some time between 1824 and 1831, the year in which he died, abandoned the Caloric Theory, and that in his notes, not published until 1878, he entered a numerical value for the universal and uniform mechanical equivalent of a unit of heat[3]. The example provided by

---

[1] CLAUSIUS [1850, §I]: "the principle of the equivalence of Heat and Work", clearly in reference to cyclic processes only.

[2] The reader confused by this statement should recall the position of CARNOT's Special Axiom, established in Scholion I of §5J. The matter is made fully clear by Theorem 7 in §9 of *Concepts and Logic*. That theorem evaluates CARNOT's function μ in the General CARNOT-CLAPEYRON Theorem (5L.4):

$$\mu = g'/h \,, \qquad h > 0, \, g' > 0 \text{ for almost all } \theta \,.$$

CARNOT's theory corresponds to the special case in which $h = \text{const}$. Uniform interconvertibility as expressed by (1) corresponds to the special case in which

$$g'/h = J \,.$$

These statements were first published in 1975.

[3] Anyone who has studied §5O of this tragicomedy will recall that CARNOT's method of determining the numerical value of μ was independent of his special assumptions beyond his General Axiom. Thus the merely numerical side was already taken care of. The calculation neither required that μ so obtained should equal J/θ for some constant J nor implied that if such were the case J would turn out to satisfy (1). We know today that with correct experimental data CARNOT's method of calculation, if applied to a sufficiently broad range of temperatures, would have shown that $\mu = J/\theta$ very nearly, θ being measured by an air thermometer, but the calculation itself and the physical ideas on which it rests would never have suggested that.

CLAPEYRON, which we have disposed of in §6A, should teach us to distrust the early attributions assiduously seined from the unmathematical literature. Certainly the claims of uniform interconvertibility were rejected at first by leading physicists and chemists. We who have been brought up to take it for granted may have trouble seeing just what was the difficulty of first grasping it. There should be no difficulty at all. We must not forget that every scientist is, like ourselves, brought up with a set of beliefs he has not been encouraged to question. Only the exceptional man knows how to ask an important question. Still more exceptional is the man who can answer one.

This tragicomical history, since it concerns mathematical theory and its experimental basis or confirmation or contradiction, is not the place to recount or contrast beliefs on scientific questions. Indeed, that is become the special province of Historians of Science. Here I will barely mention some particular beliefs about heat, just in case some spectator is not already familiar with them at least roughly.

A. *Heat is merely a manifestation of intestine motion.*

That is, what our senses perceive as hot and cold could by finer scrutiny be expressed in terms of mass and the change of place as time goes on. Structural or epistemological, according to taste, this belief is irrelevant to thermodynamics, though it was and still is important for motivation. It is a very old idea, which had been favored by many philosophers and physicists[4] of earlier periods, and toward the end of the eighteenth century it came to be known as "the *vis viva* theory of heat". While the concepts "heating" and "heat added", which, following the pioneers of thermodynamics, we have used and shall continue to use in this history, refer to the result of a *process* undergone by the body, the *vis viva* theory reflects a concept of the "total heat" *residing* in a body. It is not a single model or theory but an approach to theories and models. Some effective calculations on the basis of particular models of this kind had been made by EULER and DANIEL BERNOULLI[5], using different hypotheses about the intestine motion whose energy was to be identified with total heat, but by the 1800s physics was become a profession, so the work of

---

[4] *Cf.* the remarks of BRIDGMAN [1941, pp. 9–10 of the edition of 1961]:

> An understanding of the attitude of physicists toward thermodynamics and kinetic theory is, I think, to be sought only in the realm of psychology. Ever since the days of the Greek philosophers or of Lucretius human speculation has run straight to the atomic. At first there was absolutely no experimental justification for this, or logical justification either, for that matter. From our present point of vantage we must not draw the conclusion that because atoms have now been found in the laboratory our primitive urge to analyze into atoms was therefore justified. It just seems to be a fact about our thinking machinery that we must have our atoms....

[5] I have written the history of quantitative aspects of the subject through 1865: "Early kinetic theories of gases", *Archive for History of Exact Sciences* **15**, 1–66 (1975).

long dead thinkers was not regarded. The historical record[6] shows that some[7] physicists and chemists of the first quarter of the nineteenth century saw no conflict between using one or another *vis viva* theory for physics and chemistry (*i.e.* popular explanation of how things really are) and some aspect of the Caloric Theory for mathematics (*i.e.* calculation of the phenomena, for comparison with experimental data). If addition of caloric was the agent by which the parts were made to move, then for the mathematical theory it was defter to discuss the transfer of that agent rather than the fine structure of the intestine motion which it reflected. Thus, so long as nothing specific was required as the outcome, it was not hard to believe in theories of both kinds at once.

In the early 1800s the *vis viva* theories were thought ill adapted to calculate anything, as WATERSTON was made aware, to his scathe, even as late as 1845, on the very eve of the new departures in thermodynamics that the spectators of this tragicomedy are just about to see. Indeed, had the *vis viva* theories been developed to the level of quantitative science, they might have served as structural models for the Caloric Theory. To the modern student it may seem strange that the *vis viva* theories, while attempting to reduce heat to invisible motion, seem to have been unable to include any effect of visible motion. In some way all the *vis viva* was assumed to be heat, leaving none over for the body as a whole, so ordinary or gross kinetic energy was always tacitly assumed absent in early calculations. Classical thermodynamics never outgrew this curious limitation, which it later, much later, came to include within the mystic idea "quasistatic".

B. *Heat is only a kind of "force" or "energy"; hence heat and work are universally and uniformly interconvertible in all circumstances.*

This idea, one of the many sometimes called the "First Law of Thermodynamics", would seem to include as a special case the universal and uniform Interconvertibility of Heat and Work in cyclic reversible processes (Statements 2 and 3, above). Like the *vis viva* theories it refers to the "total heat" resident in a body. As Janus could have seen, *Claim B is unsound.* Only with the discovery of *internal energy* in 1850 was it to be reduced to something admissible.

The spectator may envision now a steady progress toward the principle that heat and work are uniformly and universally interconvertible in cyclic

---

[6] The fullest account is given by Fox in his *Caloric Theory*, cited above in Footnote 2 to §2C. *Cf.* also MENDOZA's introduction to his edition of CARNOT [1824] and Chapter 2 of S. C. BROWN, *Benjamin Thompson, Count Rumford*, Oxford, *etc.*, Pergamon Press, 1967.

[7] There were also some who refused to adopt the *vis viva* theory in any of its many forms. FOURIER [1822, p. xvi of the edition in his *Œuvres*] wrote, "Whatever may be the generality of mechanical theories, they do not apply at all to the effects of heat."

processes. If so, he will be disappointed. He will not encounter the basic statement (1) until after five years of wrangling about other ideas. One of these is the vague, almost philosophical Claim B: Heat and work or "force" are "equivalent". The other is a claim that has often been confused with (1) but is in fact different:

C. *Heat and work are universally and uniformly interconvertible in* isothermal *processes.*

This statement is trivially true for a cyclic isothermal process, since for such a process both work and net gain of heat are null. Unlike Assertions 1, 2, and 3, and like Assertions A and B, it bears upon *processes that are not cyclic*; unlike Assertions A and B, it refers only to heating and heat added, not to total heat.

In the literature of physics and its history some works that deal only or mainly with this peculiar, questionable, and in fact untypical and misleading sort of interconvertibility are often by error regarded as having achieved something called "the First Law of Thermodynamics". We turn now to analysis of those works. As Janus could have seen, both this generally unsound idea about isothermal processes and also the vague belief in some overriding "equivalence" were to play their parts, for weal and woe, in preparing the ground for the resolution we shall witness in Act IV.

Now for Act III. Nobody will expect the actors in it to enjoy Janus' power to look into the future, but some may be astonished to see them unable to look backward, either! It is a new cast. The tragicomic muse will deny to all but one of them the power to understand CARNOT's ideas and achievement.

7B. MAYER's Assertion

[As the first to attempt any specific use of the idea that heat and work are interconvertible[1], the tragicomic muse of thermodynamics chose the muzziest of all her muzzy retinue: ROBERT MAYER, a gifted and thoughtful physician who knew no mathematics and whose mode of reasoning was emasculated by the school of *Naturphilosophie*[2], from which he was just then beginning to free himself. That in 1842 he did assert force, by which he may have meant something like what is now called "energy", to be indestructible, is certain, and many physicists sooner or later acknowledged that he had

[1] The name "interconvertibility of heat and work" derives from RANKINE's "convertibility of heat and mechanical power" [1850, ¶2] and "dynamical convertibility of heat" [1851, *3*, ¶40]. There is no doubt that RANKINE meant *universal* interconvertibility by his term and that he included internal energy as a kind of work.

[2] MAYER [1845, Introduction].

been the first to do so[3], but he supported his claim neither by experiment nor by mathematics but instead by a discourse[4] that recalls the lessons in natural philosophy which SOCRATES delivered from his aerial basket to STREPSIADES:]

> Forces are causes: Accordingly, we may in relation to them make full application of the principle: *Causa aequat effectum*.... In a chain of causes and effects, a term or a part of a term can never...become equal to nothing. This first property of all causes we call their indestructibility.
>
> ...Forces are therefore indestructible, convertible, imponderable objects....
>
> Without recognizing the causal connection between motion and heat, it is just as difficult to explain the production of heat by friction as it is to give any account of the motion that disappears.

[This is the sort of paper no scientist would look at twice unless he were in search of a reason to deny priority to someone else.]

Though MAYER's next paper begins in much the same tone, [it displays far better grasp of physics. Unfortunately,] this thoughtful paper was rejected by a professional organ of science. MAYER published it as a monograph, [which nobody read]. In it[5] MAYER states clearly, "Heat is a force; it may be transformed into mechanical effect," and in reference to a steam engine, "the work done [Leistung] by the machine is inseparably bound to a consumption [Konsumo] of heat." Though MAYER does not mention CARNOT and CLAPEYRON, [this sentence shows that he rejected outright their view, according to which work could be gained by merely letting down heat from a higher temperature to a lower one.

[In our analysis of the mathematical theory, MAYER's papers are treated only because of] the contents of the last paragraph of the first one:

> By applying the principles that have been set forth to the relations subsisting between the heat and the volume of gases, we find that the sinking of a mercury column by which a gas is compressed is equivalent to the quantity of heat set free by the compression, and hence it follows, if the ratio between the capacity for heat of air under constant pressure and its capacity under constant volume be taken as 1.421, that the warming of a given weight of water from 0° to 1°C corresponds to the fall of an equal weight from the height of about 365 metres.

---

[3] *Cf.* THOMSON [1851, §4] and the nobly expressed Note 5 HELMHOLTZ adjoined to the reprint of his paper [1847] in Volume 1 of his *Abhandlungen*, 1882.

[4] MAYER [1842].

[5] MAYER [1845, §3].

[While this passage is totally incomprehensible[6], anyone can see that] it provides a numerical value[7] for some interconversion of heat and work. *Cf.* the possibilities we have listed in §7A.

The second paper of MAYER[8] explains the calculation. When a gas expands under a piston which exerts a constant pressure p, it gives out work at the rate $p\dot{V}$, where $V(t)$ is the volume of the gas. At the same time, heat is imparted to the gas at the rate $K_p\dot{\theta}$. In effect, MAYER regards the heating that does work as being $(K_p - K_V)\dot{\theta}$. MAYER's general idea of "equivalence"

---

[6] Although physicists of the misty sort, like MACH, have proclaimed MAYER clear, my statement is borne out by two contemporaries of MAYER who were great pioneers of thermodynamics.

First, although the relation (2) is nowadays often, and justly, attributed to MAYER, such was MAYER's obscurity that CLAUSIUS [1850, Eq. (10a)] had to arrive at that relation by himself. His statements regarding it make it clear that he regarded it as his own discovery. He cited MAYER's first paper, but, as he himself was to write to MAYER on 15 June, 1862, he had not looked at the second paper when it first came out because its title gave him no idea that it might concern the mechanical theory of heat. This letter opens a short correspondence which is printed in MAYER's *Kleinere Schriften und Briefe*, Stuttgart, Cotta'sche Buchhandlung, 1893. CLAUSIUS' oversight is easy for me to understand. When, some thirty years ago, I began to study the origins of thermodynamics, the references I then found directed me only to MAYER's first paper. Those references attributed (2) to MAYER, but after puzzling and puzzling over his words I finally decided that it was not there. Now that the whole matter is clear, I reassert my old conclusion. The only difference is that MAYER's second paper makes it certain he had perceived and used (2) in the numerical calculation, the result of which he had published in his first paper.

Second, THOMSON [1851, §4], misunderstanding the same passage, spoke of MAYER's "false analogy between the approach of a weight to the earth and a diminution of the volume of a continuous substance, on which an attempt is founded to find numerically the mechanical equivalent of a given quantity of heat."

[7] A carelessly worded phrase in a paper by JOULE led some of the older general historians of science to give credit for the first calculation of this equivalent to RUMFORD. In fact, RUMFORD's writings reflect no idea of the interconvertibility of heat and work, and the calculation in question was by JOULE himself, so as to show that the new idea he was promulging squared roughly with the old and famous data of RUMFORD. *Cf.* S. C. BROWN, *Benjamin Thompson, Count Rumford*, Oxford *etc.*, Pergamon Press, 1967; see p. 15.

I remark also that an isolated calculation such as MAYER's or the one erroneously attributed to RUMFORD could not have carried conviction at the time, since, as we shall see in §7D, the existence of J, universal in the sense that $\mu = J/\theta$ with one and the same J for all bodies, is *not* inconsistent with the Caloric Theory. Also we have seen in §5M that CARNOT's theory presumed the existence of a universal factor for converting units of heat to units of work, but that factor had to depend upon θ.

WATERSTON in a paper submitted in 1845 to the Royal Society but not published until 1892 calculated the mechanical equivalent of a unit of heat on the basis of his kinetic-molecular theory of gases. The number he obtained is close to MAYER's. U. HOYER, "Über Waterstons mechanisches Wärmeäquivalent", *Archive for History of Exact Sciences* **19** (1978), 371–381, has analysed the calculation and has shown that the apparent agreement is due to three compensating errors.

[8] MAYER [1845, §3].

makes him expect that the mechanical power produced by this heating will be proportional to it. If the factor of proportionality is J, then the power corresponding to this heating is $J(\mathrm{K}_p - \mathrm{K}_V)\dot{\theta}$, so

$$J(\mathrm{K}_p - \mathrm{K}_V)\dot{\theta} = p\dot{V} , \qquad p = \text{const.} \tag{7B.1}$$

The equation of state of a body of ideal gas is $pV = R\theta$; hence $p\dot{V} = R\dot{\theta}$ when $p = $ const., so (1) becomes

$$J(\mathrm{K}_p - \mathrm{K}_V) = R , \tag{7B.2}$$

[a celebrated formula which we shall call ***MAYER's Assertion*** regarding ideal gases. Because of the constitutive inequality

$$\mathrm{K}_p > \mathrm{K}_V \qquad (2\mathrm{C}.10)_{2\mathrm{r}}$$

we may solve (2) for J]:

$$J = \frac{R}{(1 - \gamma^{-1})\mathrm{K}_p} . \tag{7B.3}$$

All three quantities on the right-hand side are accessible to measurement, and substitution of three particular values available to MAYER yields the number he obtained for J.

"The same result is obtained if instead of atmospheric air another simple or compound kind of gas is used for the calculation. 'Heat = mechanical effect' is independent of the nature of an elastic fluid, which serves only as a tool for effecting the transformation of the one force into the other." As an illustration MAYER carries out the numerical calculation for carbon dioxide and for olefiant gas, obtaining exactly the same numerical value for J.

7C. Preliminary Critique of MAYER's Assertion

MAYER is vague, and what he really means is difficult to discern in general. His specific calculation, however, rests upon a specific idea: *In a process at constant pressure, the heat used to produce expansion is universally and uniformly interconvertible with work.* That is the meaning of his basic assertion:

$$J(\mathrm{K}_p - \mathrm{K}_V)\dot{\theta} = p\dot{V} , \qquad p = \text{const.} \qquad (7\mathrm{B}.1)_\mathrm{r}$$

Apparently MAYER presumes that K_p and K_V are constants, but that is not necessary for his calculations. If K_p and K_V are functions of θ, we could test the uniformity of J further, should we evaluate the right-hand side of (7B.3) by use of data gathered at different temperatures, but MAYER says nothing of that.

No-one understood MAYER's paper when it appeared. It can be understood only by hindsight.

Even so, logical analysis is difficult, for we cannot tell how much MAYER knew of what was the common property of trained theorists of his day, or of that, how much he accepted. Indeed, MAYER seems to have been virginally innocent of all earlier thermodynamic theory[1]. While his ignorance may have helped him feel free in his originality, it prevented him from seeing the far-reaching consequences his ideas would have produced, had he framed them in terms of the mathematical structure that embodies the basic Doctrine of Latent and Specific Heats, the theory which all early theorists accepted tacitly if not always with full understanding, and with CARNOT's General Axiom. In those terms logical analysis becomes easy, but I defer it until §7E, after we shall have encountered another assertion, which the theory of calorimetry makes equivalent to MAYER's.

7D. HOLTZMANN's Assertion

In 1845, the year of MAYER's second paper, appeared an essay by HOLTZMANN, likewise published as a separate work. HOLTZMANN regards POISSON's assumption that the ratio of specific heats is constant as being "certainly incorrect, as is evident from CLAPEYRON's treatise...." He states[1], "The effect of the heat added to the gas is...either increase of temperature combined with increase of elasticity [pressure], or a mechanical action, or a combination of the two; and a mechanical action is the equivalent of the increase of temperature." Also, "I term *unit of heat the heat which on its addition to any gas is capable of producing the mechanical action a*, *i.e.*, to use a definite measure, that which can raise *a* kilogrammes 1 metre." HOLTZMANN's main application of these claims is to state in effect[2] that *an ideal gas undergoing an isothermal process interconverts heat and work uniformly*; *the factor of interconversion is universal for ideal gases.*

---

[1] Anyone who expects MAYER to have had some concrete ideas about the theory of heat and work can easily disabuse himself by reading MAYER's account, composed late in 1850 and hence *after* CLAUSIUS' first memoir on thermodynamics had appeared, of how he had observed and reasoned so as to arrive at his earlier ideas: *Bemerkungen über das mechanische Aequivalent der Wärme*, Heilbronn, Johann Ulrich Landherr, 1851 = pp. 235 ff. of MAYER's *Die Mechanik der Wärme*, Stuttgart, 1867 = pp. 243–302 of *ibid.*, 2nd ed., 1874 = pp. 235–276 of *ibid.*, 3rd ed., 1893. Transl. J. C. FOSTER, "The mechanical equivalent of heat", pp. 316–355 of *The Correlation and Conservation of Forces*, ed. E. L. YOUMANS, New York, Appleton, 1865. Transl. R. B. LINDSAY, "Comments on the mechanical equivalent of heat", pp. 198–231 of R. B. LINDSAY, *Julius Robert Mayer, Prophet of Energy*, Oxford *etc.*, Pergamon, 1973.

---

[1] HOLTZMANN [1845, Preface and ¶¶1–2]. I have seen this work only in the translation published in TAYLOR's *Scientific Memoirs*, from which, consequently, my quotations derive.

[2] HOLTZMANN [1845, ¶3]: $p \cdot dv/dq = a$. HOLTZMANN's explanation of this formula is pretty vague. Indeed, it is often hard to match what HOLTZMANN says with what he does.

According to the Doctrine of Latent and Specific Heats

$$Q = \Lambda_V(V, \theta)\dot{V} + \mathrm{K}_V(V, \theta)\dot{\theta} \ . \tag{2C.4$_r$}$$

The definition of work done by a fluid body is

$$L \equiv \int_{t_1}^{t_2} p(t)\dot{V}(t)dt \ . \tag{2C.19$_r$}$$

HOLTZMANN's claim thus assumes the form $p\dot{V} = J\Lambda_V(V, \theta)\dot{V}$. Hence

$$J\Lambda_V = p \ , \tag{7D.1}$$

in which J is a universal constant bearing the dimensions of work per unit heat. The assumed relation (1) between the constitutive functions Λ_V and ϖ, the latter being denoted here by the more usual symbol p, will play a great role in the development of thermodynamics. [We shall call it ***HOLTZMANN's Assertion.*** Henceforth we shall take it as referring only to an ideal gas, for so it is to be interpreted by CLAUSIUS and KELVIN.] HOLTZMANN himself (§5) promulged it more generally. [The limitations CARNOT's General Axiom imposes upon its possible validity we shall disclose below through Theorem 3 in §7I.]

In his calculations (¶4) HOLTZMANN assumes the existence of a heat function, essentially LAPLACE's, which we have denoted by $\mathrm{H_L}$ in §6A. [It is easier to use CARNOT's heat function $\mathrm{H_C}$ and the corresponding determination of Λ_V and K_V:

$$\Lambda_V = \frac{\partial \mathrm{H_C}}{\partial V} \ , \qquad \mathrm{K}_V = \frac{\partial \mathrm{H_C}}{\partial \theta} \ , \tag{5M.2$_r$}$$

Putting (1) into $(5\mathrm{M}.2)_1$ yields a differential equation, the integral of which for an ideal gas can be written as

$$\mathrm{H_C} = \frac{R\theta}{J} \log V + U(\theta) \ . \tag{7D.2}$$

By $(5\mathrm{M}.2)_2$ it follows that

$$\mathrm{K}_V = \frac{R}{J} \log V + U'(\theta) \ ,] \tag{7D.3}$$

results which HOLTZMANN obtains [by more complicated means] using $\mathrm{H_L}$, and by further calculation he proves [MAYER's Assertion]:

$$J(\mathrm{K}_p - \mathrm{K}_V) = R \ , \tag{7B.2$_r$}$$

on the basis of which he calculates (¶8) a numerical value for J [as had MAYER before him]. He regards experiment as showing that γ ought not depend upon θ, and so from (3) and (7B.2) he concludes (¶6) that $U' =$ const.

An equation for the adiabats (¶40) follows from (2) and HOLTZMANN's conclusion that $U' =$ const. From it he calculates the work done on an adiabat and hence after various approximations and guesses obtains a definite formula for the efficiency of an engine.

[HOLTZMANN's basic assumption really is that for an ideal gas traversing an isothermal path $\mathscr{P}_\theta$

$$L(\mathscr{P}_\theta) = JC(\mathscr{P}_\theta) \ , \tag{7D.4}$$

in which J is a constant universal for ideal gases. The restatements[3] by later authors say the same thing in words.]

7E. Preliminary Critique of HOLTZMANN's Assertion

We might think that HOLTZMANN's two assumptions, namely

$$J\Lambda_V = p \qquad (7D.1)_r$$

and the Caloric Theory of heat, were inconsistent. In fact, there is no logical contradiction: As HELMHOLTZ[1] is soon to remark, HOLTZMANN's heat function is only a special case of CARNOT's. Had HOLTZMANN chosen to use CARNOT's model of an engine, he could have used CARNOT's theory to evaluate its motive power.

CLAUSIUS[2] was just in dismissing HOLTZMANN's work for its adherence to the Caloric Theory. Nevertheless, as we shall see in §8B, CLAUSIUS will soon appropriate HOLTZMANN's Assertion.

While MAYER may have known little of the Doctrine of Latent and Specific Heats, HOLTZMANN by using CLAPEYRON's presentation of CARNOT's theory certainly gives us the right to criticize his work in terms of the concepts used by his forebeers. We look again at the formula for the difference between the specific heats of an ideal gas which follows from the theory of calorimetry, indifferently to hypothesis about the nature of heat or its connection with work:

$$\mathrm{K}_p - \mathrm{K}_V = \frac{R\Lambda_V}{p} \ . \qquad (2C.14)_{2r}$$

It shows at a glance that

$$J(\mathrm{K}_p - \mathrm{K}_V) = R \quad \Leftrightarrow \quad J\Lambda_V = p \ . \tag{7E.1}$$

Thus *the theory of calorimetry makes MAYER's Assertion equivalent to*

---

[3] CLAUSIUS [1850, §I]: "A permanent gas, when expanded at constant temperature, takes up [verschluckt] only so much heat as is consumed [verbraucht] in doing external work during the expansion."

HELMHOLTZ [1855, p. 567]: "[A]ll the heat taken on by a gas in expansion without change of temperature is transformed into mechanical work; let this idea be called the principle of Heat-Work in gases."

THOMSON [1852, *1*, §65]: "[T]he work spent in the [isothermal] compression of a[n ideal] gas is ... exactly the mechanical equivalent of the [heat evolved]."

[1] HELMHOLTZ [1847, §IV].

[2] CLAUSIUS [1850]. *Cf.* also HELMHOLTZ [1855, p. 589].

HOLTZMANN's[3], and J is the same constant in each. Indeed, KELVIN[4] was to call (7D.1) "MAYER's Hypothesis" and never to mention HOLTZMANN's name in connection with it. The analysis given in §§5R and 5T shows that if we graft HOLTZMANN's Assertion upon the Caloric Theory, γ cannot be a constant other than 1; MAYER's Assertion (7B.2), equivalent to HOLTZMANN's, shows that $\gamma \neq 1$. We may draw these same conclusions from HOLTZMANN's own formulae[5]. Thus HOLTZMANN's theory makes a constant value for γ altogether impossible.

7F. HELMHOLTZ's Weakest Work

In 1847 HELMHOLTZ published his paper, *On the conservation of force*[1]. For HELMHOLTZ the total energy was the sum of the kinetic and potential energies of a system of mass-points subject to pairwise equilibrated central mutual forces. The theorem he proved was no more than a special case of the energy theorem of analytical dynamics, [by no means new in 1847.

[The tragicomic daemon of thermodynamics exacted his toll again. First, the repetition or rediscovery of known] properties of dynamical systems gives rise to all sorts of [unsupported] claims about heat and electricity. [Second, in this early paper the great physicist and able mathematician HELMHOLTZ was later to become is led into several crude errors. As these have no bearing on thermodynamics, there is no point in recounting them here[2].] HELMHOLTZ argues against the Caloric Theory and in favor of a mechanical interpretation of heat as equivalent to "force".

---

[3] HOLTZMANN [1845, ¶20] noticed that $J(K_p - K_V) = R$ in consequence of (7D.1), but to do so he unnecessarily made use of his heat function.

[4] THOMSON [1852, *I*, §65].

[5] HOLTZMANN [1845, Equation (IV) in ¶7].

[1] HELMHOLTZ [1847].

[2] The published discussion begins with a criticism by CLAUSIUS [1853] and the reply to it by HELMHOLTZ [1854].

In annotating the reprint of the paper for his collected works in 1881, HELMHOLTZ remarked that LIPSCHITZ had pointed out an error, but he did not say what it was. He tacitly admitted it, however, by proceeding to derive an energy theorem for motions subject to fields of force that do not have a potential, accordingly cannot enter the equations of motion as he writes them in his paper, and hence furnish counterexamples to his assertion of 1847.

Concerning this paper F. KLEIN writes on p. 227 of his *Vorlesungen über die Entwicklung der Mathematik im 19. Jahrhundert*, Teil 1 (ed. R. COURANT & O. NEUGEBAUER), Berlin, Springer, 1926, "Finally the details are often fumbling and incomplete, corresponding to such fragmentary study of the literature as Helmholtz was able to effect in his seclusion in Potsdam." The American reader can scarcely help wondering if Potsdam in 1847 could have been any more secluded than New Haven in 1876.

In fact HELMHOLTZ does not state his assumptions clearly, but in each interpretation I can conjecture the theorem he asserts is at least partly false. He seems not yet to have mastered the concepts and structure of analytical dynamics.

The paper concludes with applications of HELMHOLTZ's general idea to various branches of physics and chemistry. As for thermodynamics[3], "we have yet to investigate what relation may hold between the attempts of Clapeyron and Holtzmann and ours to derive the force-equivalent of heat." HELMHOLTZ criticizes the early experiments of JOULE; he remarks that HOLTZMANN's results, "for which speak the numerous consequences consonant with experience," are a special case of CLAPEYRON's.

HELMHOLTZ[4] remarks further that by comparing one of HOLTZMANN's results with one of CLAPEYRON's we may infer that

$$\mu = J/\theta \ . \tag{5N.7$_r$}$$

[In §5S we have seen that CARNOT himself had inclined to an equivalent statement. Later CLAUSIUS[5] will attribute (5N.7) itself to HOLTZMANN.

I do not find the formula stated in HOLTZMANN's paper, but in his preface he writes that CLAPEYRON's formulae "contain an undetermined function, which in the more direct way I have taken is determined...." The displayed equation in his §11 reflects his determination of μ, which HELMHOLTZ recognized. Of course it is (5N.7).

In a small table HELMHOLTZ shows that (5N.7) agrees pretty well with CLAPEYRON's numerical values of μ at several temperatures. [Because both CLAPEYRON and HOLTZMANN had used the Caloric Theory of heat, this passage in HELMHOLTZ's work was to be set aside later. In fact it is perfectly sound. First, CLAPEYRON's method of calculating μ was CARNOT's, and we have seen in §5O that CARNOT's method rested upon the General CARNOT-CLAPEYRON Theorem alone, making no use of the Caloric Theory. Likewise, if we compare HOLTZMANN's Assertion with the General CARNOT-CLAPEYRON Theorem specialized to an ideal gas:

$$\mu \Lambda_V = \frac{p}{\theta} \ , \tag{5L.6$_{1r}$}$$

[we derive (5N.7) by inspection! Thus if we accept CARNOT's General Axiom, we may regard HELMHOLTZ's table as *giving experimental support to HOLTZMANN's Assertion*, the gas being supposed ideal. Of course, it would be too much to attribute to HELMHOLTZ at this time such intimate mastery of the logic of thermodynamics.]

---

[3] HELMHOLTZ [1847, §IV].

[4] HELMHOLTZ [1847, §IV] selects one of HOLTZMANN's intermediate steps, which we may write as

$$V\frac{\partial \mathrm{H_{Cl}}}{\partial V} - p\frac{\partial \mathrm{H_{Cl}}}{\partial p} = \frac{pV}{J} \ .$$

CLAPEYRON's (6A.4), specialized to an ideal gas, is

$$V\frac{\partial \mathrm{H_{Cl}}}{\partial V} - p\frac{\partial \mathrm{H_{Cl}}}{\partial p} = \frac{R}{\mu} \ .$$

Comparison yields (5N.7).

[5] CLAUSIUS [1856, *1*].

While HELMHOLTZ presents nothing whatever that would connect thermodynamics with the specific theorem of mechanics he has claimed to prove at the beginning of his paper, [his placing an energy theorem alongside his claim that "force" is equivalent to heat may have served thermodynamics indirectly. Mechanical energy may be stored as "potential" energy. If a body is capable of receiving heat as well as motion, its capacity to store energy should not be lost. The sweeping claims of MAYER and of HELMHOLTZ himself that heat and work are "equivalent" must be modified to allow storage as well as interconversion. We shall not encounter that idea explicitly until the next act.

[There is no doubt that HELMHOLTZ's paper was widely read. Though it effected no positive advance, it was right for the moment,] and it seems to have made converts to the new view of heat. Certainly it was admired by many of the leading savants. For example, MAXWELL[6], writing thirty years later, while he passed lightly over HELMHOLTZ's alleged theorem, went on to say that

> the scientific importance of the principle of conservation of energy does not depend merely on its accuracy as a statement of fact, nor even on the remarkable conclusions which may be deduced from it, but on the fertility of the methods founded on this principle....
>
> To appreciate the full scientific value of Helmholtz's little essay on this subject, we should have to ask those to whom we owe the greatest discoveries in thermodynamics and other branches of modern physics, how many times they have read it over and over, and how often during their researches they felt the weighty statements of Helmholtz acting on their minds like an irresistible driving-power.

[The main value, then, of this otherwise slight work lies in its program.]

While HELMHOLTZ's [careful and clear] reviews[7] show that he kept up with the developments of thermodynamics we are about to follow in detail, he did not contribute to them. Here and there he called attention to his priority for some of the statements made.

7G. JOULE's Summary of his Early Experiments

[Declarations such as MAYER's and HELMHOLTZ's could not convert scientists or laymen to belief in the uniform and universal Interconvertibility of Heat and Work. The truth of such an idea, counter as it is to daily

[6] J. C. MAXWELL, "Hermann Ludwig Ferdinand Helmholtz", *Nature* **15** (1877), 389–391 = *Papers* **2**, 592–598.

[7] *Fortschritte der Physik* **6** and **7** (1850/1), 1855; **9** (1853), 1856; **12** (1856), 1859. Some of these are cited here and there in this history.

experience and as it was to much of the theory of heat then received, might have been established by apt, widely ranging experiments and by clear interpretation of them. Both of these were attempted by one of those brilliant sports of science which it is Britain's glory to produce at unlikely times in unlikely places: JAMES PRESCOTT JOULE. While his name will be heard frequently in the rest of this tragicomedy, we will not follow his work in detail, for ours is not a history of experiment. Nevertheless we should sin by omission, should we fail to remark that at this point in the development of thermodynamic theory, experiment might have been decisive.

[We shall see in §7Iα why JOULE's experiments did not take command. In §10E we shall see that for thermodynamics further experiment was not really necessary anyway: From CARNOT's own axioms, modified only so as to make it possible for one ideal gas to have constant specific heats, an adroit mathematician could have proved that the theory of reversible engines *required* not only MAYER's Assertion but also the uniform and universal Interconvertibility of Heat and Work in cyclic processes. So much could have been established on the basis of experimental facts known in the year JOULE was born and of theoretical concepts made familiar shortly thereafter.

[That would not have made JOULE's experiments superfluous. Far from it! JOULE's stage was grander than ours. The thermodynamics of "reversible" processes in fluids cannot be regarded as more than a small part, and the simplest, of a general science of heat, a science which concerns mainly irreversible processes and which embraces the effects of electric and magnetic fields. It is that greater science, not classical thermodynamics, that gave man a grasp, even if often at first uncertain and sometimes partly incorrect, of a new concept that has reshaped his whole view of the phenomena of nature: *energy* and its conversions. That greater science JOULE's experiments, conceived with brilliant insight and presented to the reader in the easy clarity of colloquial English, helped to found, but mainly after the terminal date of our tragicomedy.]

JOULE began to disclose his researches on heat in 1840. Only in 1847 did he cast a general sum. He presented it as a lecture in the reading room of a church in Manchester and allowed it to be published only in a local newspaper[1].

> You will at once perceive that the living force of which we have been speaking is one of the most important qualities with which matter can be endowed, and, as such, that it would be absurd to suppose that it can be destroyed, or even lessened, without producing the equivalent of attraction through a given distance of which we have been speaking. You will therefore be surprised to hear that

[1] JOULE [1847]. JOULE [1845, *2*] had already announced his results concerning the heating of water by friction.

until very recently the universal opinion has been that living force could be absolutely and irrevocably destroyed at any one's option. Thus, when a weight falls to the ground, it has been generally supposed that its living force is absolutely annihilated, and that the labour which may have been expended in raising it to the elevation from which it fell has been entirely thrown away and wasted, without the production of any permanent effect whatever. We might reason, *à priori*, that such absolute destruction of living force cannot possibly take place, because it is manifestly absurd to suppose that the powers with which God has endowed matter can be destroyed any more than that they can be created by man's agency; but we are not left with this argument alone, decisive as it must be to every unprejudiced mind. The common experience of every one teaches him that living force is not *destroyed* by the friction or collision of bodies. We have reason to believe that the manifestations of living force on our globe are, at the present time, as extensive as those which have existed at any time since its creation, or, at any rate, since the deluge—that the winds blow as strongly, and the torrents flow with equal impetuosity now, as at the remote period of 4000 or even 6000 years ago; and yet we are certain that, through that vast interval of time, the motions of the air and of the water have been incessantly obstructed and hindered by friction. We may conclude, then, with certainty, that these motions of air and water, constituting living force, are not *annihilated* by friction. We lose sight of them, indeed, for a time; but we find them again reproduced. Were it not so, it is perfectly obvious that long ere this all nature would have come to a dead standstill. What, then, may we inquire, is the cause of this apparent anomaly? How comes it to pass that, though in almost all natural phenomena we witness the arrest of motion and the apparent destruction of living force, we find that no waste or loss of living force has actually occurred? Experiment has enabled us to answer these questions in a satisfactory manner; for it has shown that, wherever living force is *apparently* destroyed, an equivalent is produced which in process of time may be reconverted into living force. This equivalent is *heat*. Experiment has shown that wherever living force is apparently destroyed or absorbed, heat is produced. The most frequent way in which living force is thus converted into heat is by means of friction. Wood rubbed against wood or against any hard body, metal rubbed against metal or against any other body—in short, all bodies, solid or even liquid, rubbed against each other are invariably heated, sometimes even so far as to become red-hot. In all these instances the quantity of heat produced is invariably in proportion to the exertion employed in rubbing the bodies together—that is, to the living force absorbed. By fifteen or twenty smart and quick strokes of a hammer on the end of an iron rod of about a quarter of an inch in diameter placed upon an anvil an expert blacksmith will render that end of the iron visibly red-hot. Here heat is produced by the absorption of the living force of the descending hammer in the soft iron; which is proved to be the case from the fact that the iron cannot be heated if it be rendered hard and elastic, so as to transfer the living force of the hammer to the anvil.

The general rule, then, is, that wherever living force is *apparently* destroyed, whether by percussion, friction, or any similar means, an exact equivalent of heat is restored. The converse of this proposition is also true, namely, that heat cannot be lessened or absorbed without the production of living force, or its equivalent attraction through space. Thus, for instance, in the steam-engine it will be found that the power gained is at the expense of the heat of the fire,—that is, that the heat occasioned by the combustion of the coal would have been greater had a part of it not been absorbed in producing and maintaining the living force of the machinery. It is right, however, to observe that this has not as yet been demonstrated by experiment. But there is no room to doubt that experiment would prove the correctness of what I have said; for I have myself proved that a conversion of heat into living force takes place in the expansion of air, which is analogous to the expansion of steam in the cylinder of the steam-engine. But the most convincing proof of the conversion of heat into living force has been derived from my experiments with the electro-magnetic engine, a machine composed of magnets and bars of iron set in motion by an electrical battery. I have proved by actual experiment that, in exact proportion to the force with which this machine works, heat is abstracted from the electrical battery. You see, therefore, that living force may be converted into heat, and that heat may be converted into living force, or its equivalent attraction through space. All three, therefore—namely, heat, living force, and attraction through space (to which I might also add *light*, were it consistent with the scope of the present lecture)—are mutually convertible into one another. In these conversions nothing is ever lost. The same quantity of heat will always be converted into the same quantity of living force. We can therefore express the equivalency in definite language applicable at all times and under all circumstances. Thus the attraction of 817 lb. through the space of one foot is equivalent to, and convertible into, the living force possessed by a body of the same weight of 817 lb. when moving with the velocity of eight feet per second, and this living force is again convertible into the quantity of heat which can increase the temperature of one pound of water by one degree Fahrenheit.

[JOULE's statement is clear: *Heat* is the equivalent of living force apparently destroyed. *All* transformations in nature are included.

[All but one of JOULE's experiments refer to what today would be regarded as irreversible processes and thus would have at best indirect bearing on the early developments of thermodynamics.] The one exception[2] concerns the isothermal compression of air. For the following analysis of it I am indebted to Mr. C.-S. MAN.

In two series of experiments, dried air at barometric pressure p_a and at temperatures between 53.7°F and 58.8°F, 54.0°F and 62.6°F respectively, was compressed into a receiver of volume V_b through a pump working "at a moderate degree of speed". JOULE measured directly the "quantity of air compressed" V_a and the heat evolved C; thence he calculated numerical values of J by using the formula

$$J = \frac{p_a V_a \log \frac{V_b}{V_a}}{C} . \tag{7G.1}$$

[To obtain it, let ρ_a be the density of atmospheric air at pressure p_a and temperature θ, M_k the mass of air admitted into the receiver in the kth stroke, ρ_k the density of compressed air in the receiver after the kth stroke, and M_0 the original mass of air in the receiver when $\rho_0 = \rho_a$; then the work done in the kth stroke

$$L_k = -M_k \int_{\rho_a}^{\rho_{k-1}} \frac{p}{\rho^2} d\rho - (M_0 + M_1 + \cdots + M_k) \int_{\rho_{k-1}}^{\rho_k} \frac{p}{\rho^2} d\rho . \tag{7G.2}$$

Thus for n strokes the total work done

$$\begin{aligned} L &= L_1 + \cdots + L_n , \\ &= -(M_0 + M_1 + \cdots + M_n) \int_{\rho_a}^{\rho_n} \frac{p}{\rho^2} d\rho , \qquad (7G.3) \\ &= \int_{V_a}^{V_b} p dV , \\ &= p_a V_a \log \frac{V_b}{V_a} ; \end{aligned}$$

the last step follows by appeal to the ideal gas law (2A.1). JOULE obtained $(3)_4$ simply by applying PETIT's formula $(5K.1)_2$ without analysis.

[Applying to each isothermal subprocess the basic assumption of HOLTZMANN, namely

$$L(\mathscr{P}_\theta) = JC(\mathscr{P}_\theta) , \tag{7D.4$_r$}$$

which we may regard as a special instance of JOULE's far broader and less clear ideas, we obtain (1). JOULE simply assumes the interconvertibility that (7D.4) asserts. His results afford no test of it.]

---

[2] JOULE [1845, *1*, First experiment].

7H. The Bittersweet Indian Summer of the Caloric Theory: KELVIN's First Paper

In 1849 KELVIN published an [enthusiastic, rambling, and sometimes vague] *Account of Carnot's theory*[1]. [From the beginning the nemesis of thermodynamics took hold.] While CARNOT had written[2], "Heat can evidently be a cause of motion only by virtue of the changes of volume or of form which it produces in bodies," and DUHAMEL had incorporated "changes... of form" in a special case, KELVIN narrowed the subject by omitting changes of form. He wrote, "There are two, and only two, distinct ways in which mechanical effect can be obtained from heat. One of these is by means of the alterations of volume which bodies may experience through the action of heat; the other is through the medium of electric agency." [While CARNOT's mathematics had been obviously insufficient to treat the effects of change of form, had he wished to, KELVIN was an expert and creative mathematician of high order and might well have attacked the general problem. Instead, he ignored DUHAMEL's attempt and chose to follow in CARNOT's track. Again it was a turning point: CARNOT's limitations were made to seem essential to the whole doctrine.]

This paper is famous as the first to introduce the word "thermodynamic":

> 12. A *perfect* thermo-dynamic engine... is a machine by means of which the greatest possible amount of mechanical effect can be obtained from a given thermal agency.... CARNOT... proves the following proposition:—
>
> 13. *A perfect thermo-dynamic engine is such that, whatever amount of mechanical effect it can derive from a certain thermal agency; if an equal amount be spent in working it backwards, an equal reverse thermal effect will be produced.*

[KELVIN here seems to attribute to CARNOT proof that a cycle achieving maximum efficiency is necessarily reversible. I do not find that statement in CARNOT's treatise. Sound proof of it would be out of the question there, since such a proof would have to rest upon a structure embracing bodies susceptible of irreversible as well as reversible processes, while all of CARNOT's specific analysis appeals to the theory of calorimetry, which refers only to bodies such that all processes obey the reversal theorems (2C.7) and (2C.21).

[Like CARNOT and CLAPEYRON,] KELVIN was fascinated by the properties of saturated steam. Although the complications inherent in the use of steam were of great practical importance, [upon the general theory they cast

---

[1] W. THOMSON [1849].

[2] CARNOT [1824, p. 14].

obscurity, and in this account, which concerns only fundamentals, I consistently omit them.]

KELVIN, [like CARNOT and unlike CLAPEYRON], treats only of an ideal gas[3]. Through three pages (§§26–27) of manipulations of differentials KELVIN derives in differential form an expression for the work done by a simple cycle $\mathscr{C}$; his result is formally equivalent to

$$L(\mathscr{C}) = \iint_{\mathscr{A}} \frac{\partial p}{\partial \theta} dV d\theta \ , \qquad (5\text{L}.1)_{2\text{r}}$$

specialized to an ideal gas. He then writes, in effect,

$$dV = \frac{d\text{H}}{d\text{H}/dV} \ , \qquad (7\text{H}.1)$$

dH being the increment of heat. [Hence we see that

$$L(\mathscr{C}) = \iint_{\bar{\mathscr{A}}} \frac{1}{\Lambda_V} \frac{\partial p}{\partial \theta} d\text{H} d\theta \ , \qquad (7\text{H}.2)$$

in which $\bar{\mathscr{A}}$ stands for the image of $\mathscr{A}$ in the plane of the variables θ and H, and the integrand (defined as a function of V and θ) is regarded as a function of θ and H.] KELVIN gives this relation only in the special form appropriate to an ideal gas undergoing a simple Carnot cycle. Namely, [if the value of H_C is set arbitrarily at 0 at the beginning of the cycle,] then (his Equation (4))

$$L(\mathscr{C}) = \int_0^{C+} d\text{H} \int_{\theta-}^{\theta+} d\theta \frac{1}{\Lambda_V} \frac{\partial p}{\partial \theta} \ . \qquad (7\text{H}.3)$$

[To the modern reader these steps seem no more than blind juggling with differentials. That, indeed, they may have been. Nevertheless, they can be

---

[3] It is curious to follow KELVIN's reasoning. Sometimes he calculates a quantity first for a general equation of state but then specializes the formula at once to the case of an ideal gas. In this early paper KELVIN is strapped to air and steam and the experiments of REGNAULT. In §28 we read, "The preceding investigations, being founded upon [(2A.1)], would require some slight modifications, to adapt them to cases in which the gaseous medium employed is such as to present sensible deviations from [it]. REGNAULT's very accurate investigations shew that the deviations are insensible, or very nearly so, for the ordinary gases at ordinary pressures; although they may be considerable for ... carbonic acid under high pressure.... In cases when it may be necessary, there is no difficulty in making the modifications, when the requisite data are supplied by experiment." These words suggest that KELVIN here was thinking, not of the ideas in general terms, but of more complicated empirical equations of state. Because the modern reader is deterred just as much by details of equations too special as he is by considerations too vague to grasp, in the text above I present the forms KELVIN's results assume when his reasoning is applied to a general equation of state.

justified according to the Caloric Theory. Because of the constitutive inequality

$$\Lambda_V > 0 \tag{2C.5$_{1r}$}$$

CARNOT's heat function H_C is invertible locally for V when θ is fixed: $V = V_C(\theta, H)$. Thus *the Caloric Theory allows us to use θ and* H, the latter being the quantity of heat in a body, *as local co-ordinates* and so to describe sufficiently small cycles in the plane of θ and H. Therefore

$$\iint_{\mathscr{A}} \cdots dV d\theta = \iint_{\bar{\mathscr{A}}} \cdots \frac{\partial V_C}{\partial H} dH d\theta \,. \tag{7H.4}$$

But

$$\Lambda_V = \frac{\partial H_C}{\partial V} \,, \tag{5M.2$_{1r}$}$$

so

$$\frac{\partial V_C}{\partial H} = 1 \Big/ \frac{\partial H_C}{\partial V} = 1/\Lambda_V \,, \tag{7H.5}$$

and (2) follows.]

In the opening pages of the paper KELVIN somehow convinces himself of most of CARNOT's main assertions[4]. [However, he does not mention CARNOT's General Axiom.] Instead, he bases all his reasoning upon (3), or, rather, upon the statement from which he derived that: In an infinitesimal Carnot cycle (KELVIN's Equations (3) and (6))

$$L(\mathscr{C}) \approx \mu \Delta\theta \Delta H, \qquad \mu \equiv \frac{\partial p}{\partial \theta} \Big/ \Lambda_V \,. \tag{7H.6}$$

The notation μ is KELVIN's; he calls μ "CARNOT's coefficient" and "CARNOT's multiplier". The argument so far, [since it does not rest on CARNOT's General Axiom,] does not show that μ is a function of θ alone.

At this point KELVIN reproduces (§29) CARNOT's argument about the opposing Carnot cycles (above, §§5F–5H). He conclude that

> any two engines, constructed on the principles laid down above, ... must derive the same amount of mechanical effect from the same thermal agency.

Otherwise there would be "a residual amount of mechanical effect without any thermal agency, or alteration of materials, which is an impossibility in nature." [Clearly "thermal agency" may depend upon θ; KELVIN seems

---

[4] Most of the remarks of W. THOMSON [1849, §§5–24] merely describe Carnot cycles, the Caloric Theory, and reversibility.

to assume it may not depend upon V.] He concludes that μ in $(6)_2$ is a function of θ alone, the same for all bodies. Substitution of $(6)_2$ into (3) yields the work done in a Carnot cycle:

$$L(\mathscr{C}) = \left(\int_{\theta^-}^{\theta^+} \mu d\theta\right) C^+(\mathscr{C}) \ . \qquad (5L.7)_{2r}$$

[Hence μ is related as follows to CARNOT's F:

$$\mu(\theta) = F'(\theta) \ ; \qquad (5J.7)_{2r}$$

we recognize (6) and $(5L.7)_2$ as being neither more nor less than the Special CARNOT-CLAPEYRON Theorem and the equation in CARNOT's Special Axiom, respectively:

$$F'\Lambda_V = \frac{\partial p}{\partial \theta} \quad \text{and} \quad L(\mathscr{C}) = [F(\theta^+) - F(\theta^-)]C^+(\mathscr{C}) \ . \quad (5L.5)_r, (5I.6)_r$$

Thus KELVIN's argument includes a nearly obvious converse to CARNOT's: The Special CARNOT-CLAPEYRON Theorem suffices for the truth of CARNOT's Special Axiom. We have incorporated this statement into a more general one which we have labelled Scholion IV in §5L, above.]

In claiming (§14) that a Carnot cycle achieves a "degree of perfectibility which cannot be surpassed" and "gives all the mechanical effect that can possibly be obtained from the thermal agency employed" KELVIN seems to think that that statement follows from CARNOT's argument about the two engines. [As we have seen above in §§5F–5H, CARNOT used that argument only to infer that all Carnot cycles with the same operating temperatures have the same efficiency. To show that a Carnot cycle achieved maximum efficiency, CARNOT himself relied upon remarking that a Carnot cycle avoided "useless re-establishment of equilibrium in the caloric" (above, §§5D–5E),] which KELVIN does not mention. [Of course the claim is true, and we can prove it[5] by use of the very formulae KELVIN has provided here, namely (3) and the obvious extension (2) we have provided. We substitute (5L.5) into (2) and so obtain

$$L(\mathscr{C}) = \iint_{\mathscr{A}} F' d\theta d\mathrm{H} \ . \qquad (7H.7)$$

We may define τ as being the value of F:

$$\tau \equiv F(\theta) \ . \qquad (7H.8)$$

Since F is an increasing function, θ is a function of τ. Denoting by $\tilde{\mathscr{A}}$ the image of $\mathscr{A}$ in the plane of τ and H, we let the greatest and least values of

[5] TRUESDELL [1973, *1*].

It should be unnecessary to remark that the later and now generally accepted theory of CLAUSIUS forbids such a choice of co-ordinates, and nothing resembling the argument presented above could be used to prove the corresponding theorem in that theory.

H in $\tilde{\mathscr{A}}$ be H_{max} and H_{min}, and we let the greatest and least values of θ in the cycle be θ_{max} and θ_{min}. Then $\tau_{max} = F(\theta_{max})$ and $\tau_{min} = F(\theta_{min})$. The values H_{max}, H_{min}, θ_{max}, θ_{min} determine a unique simple Carnot cycle, which is the rectangle circumscribed upon $\tilde{\mathscr{A}}$ (Figure 5, in which $\mathscr{C}$ is the curve bounding $\tilde{\mathscr{A}}$). The area of that rectangle is greater than the area of $\tilde{\mathscr{A}}$ unless $\tilde{\mathscr{A}}$ is itself that rectangle. The area of the rectangle is $(\tau_{max} - \tau_{min})(H_{max} - H_{min})$. Therefore (7) yields

$$L(\mathscr{C}) \leqq (\tau_{max} - \tau_{min})(H_{max} - H_{min}) , \tag{7H.9}$$

and equality is achieved if and only if $\mathscr{C}$ is the boundary of the rectangle. Moreover, since H is an increasing function of V when θ is fixed,

$$C^+(\mathscr{C}) \geqq H_{max} - H_{min} , \tag{7H.10}$$

and equality is achieved if and only if the part of $\mathscr{C}$ on which H increases is a single simple arc. Therefore

$$L(\mathscr{C}) \leqq (\tau_{max} - \tau_{min})C^+(\mathscr{C}) , \tag{7H.11}$$

and equality holds if and only if $\mathscr{C}$ is a Carnot cycle.

[Thus we have established CARNOT's Claim I as stated in §5E. The attentive reader will recall that we have already proved that claim in §5M. The proof there is easier than that just above, but it uses a kind of reasoning which while entirely trivial is somewhat uncommon in the calculus as physicists used it 150 years ago. The proof we have just given employs the very apparatus KELVIN provides in the paper we are discussing. The price paid is

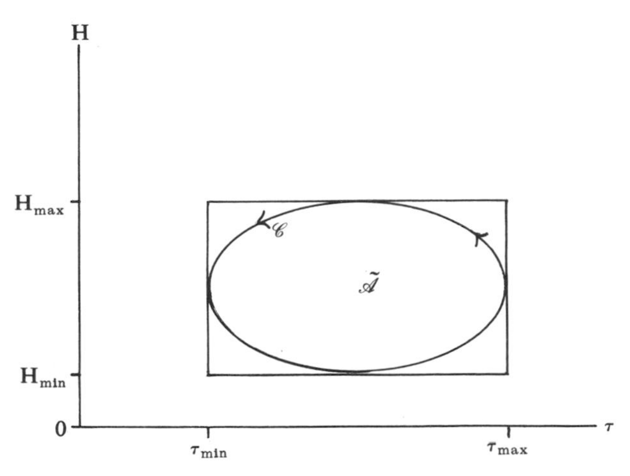

Figure 5. Cycle in the H–τ plane, with the corresponding Carnot cycle shown as the circumscribed rectangle. (Note that τ_{min} and τ_{max} correspond to θ_{min} and θ_{max}.)

essential use of the constitutive inequality $(2C.5)_1$, which the proof in §5M did not use. That is, the proof here is not so good as the one in §5M, but it shows less "hindsight" in that it is virtually read off from the formulae in one of the sources. It is strange that KELVIN did not notice this proof, since] τ as defined by (7) is exactly what he had the year before[6] called "the absolute temperature".

KELVIN concludes:

> 32. The complete theoretical investigation of the motive power of heat is thus reduced to the experimental determination of the coefficient μ....
>
> 33. The object of REGNAULT's great work... is the experimental determination of the various physical elements of the steam-engine; and when it is complete, it will furnish all the *data* necessary for the calculation of μ.

The main product of KELVIN's paper is two tables based insofar as possible upon data REGNAULT had gathered up to then in his "great work" upon saturated steam. The first gives values of μ for successive degrees of the air thermometer from 0° to 230° at intervals of 1°. The second does the same for $\int_0^\theta \mu(x)dx$, or, what is the same thing, for CARNOT's F. Thus KELVIN steps boldly in [where CARNOT dared not tread]: He uses CARNOT's theory to calculate the motive power of an engine with a *finite difference of operating temperatures*. [These tables of his are destined to work dire effects on the progress of thermodynamics. We shall encounter them again further on in this section and several times later in the course of our tragicomedy.

[As KELVIN was to write four years later[7]: "...by means of observations of any kind, whether on a single fluid or on different fluids," we may determine CARNOT's function "for any substance whatever...." If F really is a universal function, we need not face the complexities of experiment on different materials so as to determine it but for that end may use any conceivable, admissible theoretical model. If, on the other hand, we doubt that F be universal, we need tests, not with great detail and accuracy for the most useful substance in practice, but a table of measurements on many different substances in the same ranges of temperature. We must recall that steam is the only fluid for which abundant data was available then, so KELVIN had either to use steam as his only basis or to institute a series of experiments of his own. For air, for example, he could have measured K_p and then used CLAPEYRON's formula (5O.5), but we must recall that by 1849 no-one had projected a good method for measuring K_p accurately.

---

[6] THOMSON [1848]. For analysis of this work and of KELVIN's redefinition of "absolute temperature" in 1854 see §§11B and 11H, below.

[7] THOMSON [1853, *1*, §89].

[KELVIN's insistence on appeal to experiment will cost him priority in the second major discovery of the early thermodynamics, a discovery which lay ready to hand for him: the internal energy.

[Here there is a double irony. Only in this year, 1849, at last, does a mathematician capable of exploiting CARNOT's ideas take them up. While he does so in part, he limits his analysis to the sometimes misleading if important case of an ideal gas, then bends over backward to accommodate the latest data on steam, that most confusing of thermodynamic substances, and finally comes to doubt the Caloric Theory of heat. KELVIN has not yet seen the work of MAYER. At the very beginning (§8) he states that JOULE's experiments "seem to overturn the opinion commonly held that heat cannot be *generated*...." A page later he adds a footnote full of doubts: "When 'thermal agency' is thus spent in conducting heat through a solid, what becomes of the mechanical effect which it might produce?" [Far from finding unexpected from an expert on the conduction of heat this long overdue remark, we wonder why KELVIN was not the mathematician who could put FOURIER's theory and CARNOT's together. Blinded by the failures of his predecessors in theory,] KELVIN turns to experiment but cannot accept the results: "... the foundation of a solution of the difficulty has been actually found, in Mr JOULE's discovery of the generation of heat, by the internal friction of a fluid in motion. Encouraged by this example, we may hope that the very perplexing question in the theory of heat, by which we are at present arrested, will, before long, be cleared up." Moreover, "It might appear, that the difficulty would be entirely avoided, by abandoning CARNOT's fundamental axiom; a view which is strongly urged by Mr JOULE ...."[8] Nobody needs to be told after reading this that here lies KELVIN's near miss. He writes, "It is in reality to experiment that we must look—either for a verification of CARNOT's axiom, and an explanation of the difficulty we have been considering; or for an entirely new basis of the Theory of Heat." He had put another footnote a page earlier:

"So generally is CARNOT's principle tacitly admitted as an axiom, that its application in this case has never, so far as I am aware, been questioned by practical engineers." [A fatal mistake, this, for "practical engineers" are the last persons in the world from whom to expect searching questions!]

In the appendix, read nearly four months after the main text, just after the famous dictum "Nothing in the whole range of Natural Philosophy is more remarkable than the establishment of general laws by such a process of reasoning", KELVIN expresses "doubt... with reference to the truth of the axiom on which the entire theory is founded". KELVIN vacillates[9].

---

[8] *Cf.* also the doubts about JOULE's ideas KELVIN had expressed in a footnote to his earlier paper [1848].

[9] He who has read so far in this history will not be surprised to find lapses in KELVIN's faculty of criticism. In §45 he writes, "The fact of the gradual decrease of μ through a very extensive range of temperature... must be considered as a striking verification of the

[Despite his failure to get to the essence of the theory he is attempting to develop, KELVIN does succeed, almost in spite of himself, in opening a new avenue of thought. As we have seen in §5L, the relation $(6)_2$ is compatible with the Caloric Theory if and only if μ as determined by $(6)_2$ is the same function of temperature as is F', the derivative of the F in CARNOT's Special Axiom.

[However, we may regard $(6)_2$ in its own right. We may abandon the Caloric Theory but yet ask the following questions:

1. Is $(\partial p/\partial\theta)/\Lambda_V$ *a function of temperature alone* for *some* fluid body?
2. If so, is it a function of temperature alone *for all fluid bodies?*
3. If so, is the function μ so determined a *universal* function, the *same* for all bodies?

CARNOT's General Axiom makes the answers yes, yes, and yes; KELVIN, though he does not appeal to that axiom, expresses no doubt here. He could have attempted to determine μ by experiments on air, but then he would have had to conceive and perform those experiments. Instead], he chooses to check some of JOULE's views on the relation between heat and work, using as a basis the values of μ he had already determined by "REGNAULT's great work" on [that nemesis of thermodynamics], saturated steam [!]

To do so, KELVIN considers two known formulae regarding a body of ideal gas traversing an isothermal path $\mathscr{P}_\theta$: PETIT's evaluation of the work done and CARNOT's evaluation of the heat added, respectively

$$L(\mathscr{P}_\theta) = R\theta \log\frac{V_b}{V_a} \quad \text{and} \quad C(\mathscr{P}_\theta) = \frac{R}{\mu(\theta)}\log\frac{V_b}{V_a}\,. \qquad (5K.1)_2, (5K.5)_2$$

Dividing the former by the latter, he obtains[10] (his Equation (11) in §49)

$$\frac{L(\mathscr{P}_\theta)}{C(\mathscr{P}_\theta)} = \theta\mu(\theta)\,. \qquad (7H.12)$$

theory." CARNOT, indeed, had suggested that F' should decrease with θ (above, §5S), but neither he, CLAPEYRON, nor KELVIN had proved anything to that effect, so we wonder what "theory" KELVIN thinks he is verifying. Of CARNOT's main theorem (5K.5) on the heat gained in isothermal changes, he writes in §47, "This extremely remarkable theorem of CARNOT's was independently laid down as a probable experimental law by DULONG..., and it therefore affords a most powerful confirmation of the theory." As we shall see in §8B, this "powerful confirmation" is independent of "the materiality of heat".

[10] THOMSON [1852, *1*, §63] was to notice later the form (12) takes for a fluid obeying an arbitrary equation of state:

$$C(\mathscr{P}_\theta) = \frac{1}{\mu(\theta)}\int_{V_1}^{V_2}\frac{\partial p}{\partial\theta}(V,\theta)dV = \frac{1}{\mu(\theta)}\frac{d}{d\theta}L(\mathscr{P}_\theta)\,,$$

the volumes V_1 and V_2 being held constant. "This formula, established without any assumption admitting of doubt, expresses the relation between the heat developed by

§50 Hence we infer that

(1) The amount of work necessary to produce a unit of heat by the compression of a[n ideal] gas, is the same for all [ideal] gases at the same temperature.

(2) And that the quantity of heat evolved in all circumstances, when the temperature of the gas is given [*i.e.* constant], is proportional to the amount of work spent in the compression.

Substituting from Table 1 the values of μ at various temperatures, KELVIN calculates $\theta\mu(\theta)$ and hence the value of the left-hand side of (12). The results are far from being independent of θ. KELVIN compares them with three values obtained by JOULE:

> The largest of these numbers is most nearly conformable with Mr JOULE's views of the relation between such experimental "equivalents," and others which he obtained in his electro-magnetic researches; but the smallest agrees almost perfectly with the indications of CARNOT's theory....

[Here is one of our daemon's tricks.] As KELVIN is to tell us two years later[11], "It was suggested to me by Mr JOULE, in a letter dated December 9, 1848, that the true value of μ might be 'inversely as the temperatures from zero'...." That is, according to "Mr JOULE's opinion"

$$\mu = J/\theta \ . \qquad (5N.7)_r$$

The year thereafter, KELVIN[12] will tell us also how JOULE arrived at his conjecture:

> [It is] required to reconcile the expression derived from CARNOT's theory (which I had communicated to him) for the heat evolved in terms of the work spent in the compression of a gas, with the hypothesis that the latter of these is exactly the mechanical equivalent of the former, which he had adopted in consequence of its being, at least approximately, verified by his own experiments.

[Of course, that is the immediate conclusion we draw from (12) if we apply JOULE's claim that heat and work are universally and uniformly inter-

---

the compression of any substance whatever, and the mechanical work which is required to effect the compression, as far as it can be determined without hypothesis by purely theoretical considerations." To prove this formula we need only equate the right-hand sides of (5K.4) and (5L.2).

[11] THOMSON [1851, §42].

[12] THOMSON [1852, *1*, §65].

convertible in all circumstances. KELVIN seems not to have noticed that HELMHOLTZ had shown (5N.7), which we shall henceforth call the ***HELMHOLTZ–JOULE Determination***, to be a consequence of HOLTZMANN's Assertion applied to CARNOT's treatment of the Caloric Theory and had found it to be fairly consonant with data published by CLAPEYRON (above, §7F).] As KELVIN tells us politely, JOULE's own data spread too much to allow a reliable check on (5N.7). That is why KELVIN turned to REGNAULT's data and by use of it refuted JOULE's conjecture.

[Today the HELMHOLTZ-JOULE Determination is regarded as an established truth of classical thermodynamics. Why did KELVIN's appeal to experiment fail to confirm it? I am indebted to Mr. C.-S. MAN for analysis of KELVIN's calculation, which concerns special properties of vapors and thus does not of itself seem to form part of our tragicomedy. REGNAULT's *influence*, not REGNAULT's *data*, was the source of the error!] To carry through KELVIN's calculation, "the density of saturated vapour must be known" for "all temperatures between 0° and 230° cent. of the air-thermometer."

In his paper on an absolute thermometric scale[13], published the year before, KELVIN explained the difficulty here:

> M. Regnault announces his intention of instituting researches for this object; but till the results are made known, we have no way of completing the data necessary for the present problem, except by estimating the density of saturated vapour at any temperature (the corresponding pressure being known by Regnault's researches already published) according to the approximate laws of compressibility and expansion (the laws of Mariotte and Gay-Lussac, or Boyle and Dalton). Within the limits of natural temperature in ordinary climates, the density of saturated vapour is actually found by Regnault (*Études Hygrométriques* in the *Annales de Chimie*) to verify very closely these laws; and we have reason to believe from experiments which have been made by Gay-Lussac and others, that as high as the temperature 100° there can be no considerable deviation; but our estimate of the density of saturated vapour, founded on these laws, may be very erroneous at such high temperatures as 230°. Hence a completely satisfactory calculation of the proposed scale cannot be made till after the additional experimental data shall have been obtained; but with the data which we actually possess, we may make an approximate comparison of the new scale with that of the air-thermometer, which at least between 0° and 100° will be tolerably satisfactory.

[It is this erroneous belief that makes KELVIN's allegedly experimental

---

[13] THOMSON [1848].

determination of μ incorrect.] KELVIN is to admit as much, though only indirectly and not until 1854[14].

[This is irony indeed. KELVIN was reluctant to accept a hypothesis unsupported by experiment. He appealed to the best data he could get. It was insufficient. Then, as physicists will do, to supply wanting data he *appealed to authority* to make what could be little better than *another conjecture*, one that seemed to him secondary and innocuous: that saturated steam obeyed the laws of ideal gases, even at high temperatures. In itself such a guess merely reflects the belief, current in earlier periods and lingering on even in 1849, that nearly all gases are nearly ideal. Nevertheless, conjecture it was—if anything, still *less* supported by experiment than was the one KELVIN was trying to test—and it was fatal. Not only did it break the chain of his calculation allegedly based upon the results of experiment, not only did it vacate his claim to determine μ by appeal to nature itself, but also it tied his hands. In possession of all the elements out of which classical thermodynamics was just about to be formed, KELVIN was not the man destined to form it. His failure should serve as a classic warning to theorists of two kinds: those who bind themselves too tightly to the results of experiment, and those who make plausible conjectures about the results of unperformed experiments. In 1849 KELVIN was of both these kinds.]

7I. General Critique: Interconvertibility in 1849

Everything we have seen in this act that went in any essential way beyond CARNOT's work has concerned one or both of two of the claims set forth in §7A:

B. *Heat is only a kind of "force" or "energy"; hence heat and work are universally and uniformly interconvertible in all circumstances.*

C. *Heat and work are universally and uniformly interconvertible in isothermal processes.*

The former claim, which refers to total heat, is unsound. It forgets CARNOT's great lesson: In its power to do work, heat at high temperature is different from heat at low temperature. As CARNOT put it, "*Wherever a difference of temperature exists, there motive power can be produced* [*by*

[14] In §9D we shall read KELVIN's criticism of CLAUSIUS and spirited defense of his own calculation. *Cf.* THOMSON [1852, *1*, §§67–68] and also THOMSON [1851, §34]. Finally JOULE & THOMSON [1854, Theoretical Deductions, §2] accepted the deviation of steam from the laws of ideal gases and by use of their own new data concluded that $\mu\theta$ was in fact nearly constant when θ was measured by the air thermometer.

heat]." "the production of motive power is due... *to* [*the*] *transport* [*of heat*] *from a warm body to a cold body*...." *Cf.* above, §5D. GIBBS in the passage we have quoted more fully in Footnote 2 to §5D expresses CARNOT's general idea in specific terms that are acceptable in the later thermodynamics: "heat received at one temperature is by no means the equivalent of the same amount of heat received at another temperature.... But no such distinction exists in regard to work." Work, therefore, cannot be "equivalent" to anything that depends upon temperature. In particular, it cannot be "equivalent" to heat. Work is not dependent upon temperature, and it cannot be "equivalent" to something that is.

The latter claim, while indeed a special instance of the former, refers only to heat added at one temperature and does not suffer from the same obvious fault. It is tenable.

We now search the position in 1849 with regard to these two claims, in experiment first and then in theory.

α. *Experiment.* While REGNAULT was the leading experimentist of this period, he contributed nothing to any important issue. That leaves us with JOULE. For analysis of JOULE's experiments through 1850 and for locating previous criticisms of them I am greatly indebted to Mr. C.-S. MAN.

In §7G we have seen that JOULE at this time, at least if his own words can be taken at face value, espoused Claim B. Since Claim B is untenable, how can JOULE have thought he confirmed it? POINCARÉ[1] remarks that the early experiments of JOULE and others were conceived as "*verifying* an established principle"—established, that is, on the basis of preconceptions molecular and otherwise—while later students, abandoning the historical order of things, preferred to regard those same experiments as themselves "*establishing experimentally* the principle of equivalence". However, the experiments in question do not suffice to discharge the responsibility the tradition of physics has laid upon them. The unacceptable spread of JOULE's early results is summarized thus by MEYERSON[2]:

> The numbers of the English physicist vary within extraordinarily large limits; the average at which he arrives is 838 foot-pounds (for the quantity of heat capable of increasing the temperature of a pound of water by 1°F, which is about equivalent to 460 kilogrammeters to 1°C.); but the different experiments from which this average is drawn furnish results varying from 742 to 1,040 foot-pounds (or from 407 to 561 kilogrammeters)—that is, by more than a third of the lowest value—and he even notes an experiment which gives 587 lb. (322

[1] See §60 of H. POINCARÉ's *Thermodynamique* (1888/9), Paris, Georges Carré, 1892.

[2] See pp. 194–195 of E. MEYERSON's *Identity and Reality*, 3rd edition, 1926, translated by K. LOEWENBERG, London, George Allen and Unwin, 1930. The passage occurs also in the 2nd edition, 1912.

kilogrammeters) without seeing in it any source of particularly grave experimental errors. It is only in the postscript of this work that Joule tells of a series of experiments yielding as a result 770 foot-pounds (423 kilogrammeters), which approximates our present estimations. Moreover, if one considers that at this same moment Sadi Carnot and J. R. Mayer had already, each one for himself, calculated the equivalent of heat and had arrived at the figures of 370 and of 365 kilogrammeters (which is more than an eighth lower than Joule's value), it becomes really difficult to suppose that a conscientious scientist, relying solely on experimental data, could have been able to arrive at the conclusion that the equivalent must constitute, under all conditions, an invariable datum.

ROWLAND[3], writing earlier but still long after the fact, levels weightier charges:

> One very serious defect in Joule's experiments is the small range of temperature used, this being only about half a degree Fahrenheit, or about six divisions on his thermometer. It would seem almost impossible to calibrate a thermometer so accurately that six divisions should be accurate to one per cent, and it would certainly need a very skilful observer to read to that degree of accuracy. Further, the same thermometer "A" was used throughout the whole experiment with water, and so the error of calibration was hardly eliminated, the temperature of the water being nearly the same.

Indeed, JOULE obtained his final numbers in each experiment by simple averaging over many trials, carried out at various room temperatures, which anyone who has lived in Britain will expect to lie in a small, chilly interval, pretty close to freezing.

It is no wonder that JOULE's contemporaries, even his friend KELVIN, were reluctant to accept his early results. As we have seen in §7F, HELMHOLTZ rejected them at the very moment when accurate measurements over a broad range of temperatures could have exerted a decisive influence upon the course of thermodynamics. So late as February, 1850, RANKINE[4], who accepted *a priori* a *vis-viva* theory of heat such as to make heat only a form of mechanical energy (see below, §8G) and hence shared JOULE's views, rejected as being too large all the numerical values of J that JOULE had published up to then.

As KELVIN implies (*cf.* Footnote 14 to §7H above), JOULE *could not have induced his conjecture*, namely

$$\mu = J/\theta \,, \qquad (5N.7)_r$$

[3] ROWLAND [1880, p. 150].

[4] RANKINE [1850, §2].

from his experiments, as they did not allow the temperature to vary much. It was KELVIN, as KELVIN himself tells us, who communicated to JOULE the relation

$$\frac{L(\mathscr{P}_\theta)}{C(\mathscr{P}_\theta)} = \theta\mu(\theta) \ . \qquad (7H.12)_r$$

Thence JOULE inferred (5N.7) by applying his own *belief* in Claim B. *JOULE's conjecture was theoretical.* By his good luck, only Claim C was needed to get (5N.7), and Claim C, unlike Claim B, is tenable.

It is only hindsight, sharpened by the far greater definition of JOULE's later work, that makes the modern student expect JOULE to have been the hero of the new science, under whose banner all physicists ought have rallied.

β. Theory. Here only Claim C is worth further comment. We have seen that HOLTZMANN asserted it for ideal gases and that the earlier assertion of MAYER is equivalent to it, likewise for ideal gases, on the basis of the Doctrine of Latent and Specific Heats alone. KELVIN made a start at determining the position of HOLTZMANN's Assertion when CARNOT's *General* Axiom is assumed. Thus there is no chance of falling back onto the Caloric Theory. We now complete KELVIN's analysis, using only formulae already in the literature in 1849 and at KELVIN's easy disposal, had he chosen to regard them.

We shall establish connections among three central statements, *none of which presume the Caloric Theory of heat:*

1. The General CARNOT-CLAPEYRON Theorem:

$$\mu\Lambda_V = \frac{\partial p}{\partial \theta} \ . \qquad (5L.4)_r$$

2. HOLTZMANN's Assertion:

$$J\Lambda_V = p \ . \qquad (7D.1)_r$$

3. The HELMHOLTZ-JOULE Determination:

$$\mu = J/\theta \ . \qquad (5N.7)_r$$

We recall that μ in Assertions 1 and 3 derives from CARNOT's General Axiom, while Assertion 2 stands by itself, a separate and independent statement regarding Interconvertibility of Heat and Work in isothermal processes.

First we restate the General CARNOT-CLAPEYRON Theorem as applied to an ideal gas:

$$\mu\Lambda_V = p/\theta \ . \qquad (5L.6)_{1r}$$

Theorem 1: *Let the General CARNOT-CLAPEYRON Theorem be assumed. Then for ideal gases*

$$J\Lambda_V = p \quad \Leftrightarrow \quad \mu = J/\theta \ . \qquad (7I.1)$$

We have seen that CARNOT calculated $\mu\theta$ numerically for one particular value of θ by using the relation

$$\mu = R\frac{1 - \frac{\theta}{V}\frac{dV}{d\theta}}{\theta \mathrm{K}_p}, \tag{5O.5$_\mathrm{r}$}$$

which rests upon the General CARNOT-CLAPEYRON Theorem alone, an ideal gas being presumed. MAYER used his relation

$$J = \frac{R}{(1 - \gamma^{-1})\mathrm{K}_p} . \tag{7B.3$_\mathrm{r}$}$$

to calculate J. But Theorem 1 shows us that if we accept the General CARNOT-CLAPEYRON Theorem, MAYER could have obtained nothing else than $\mu\theta$ in this way. In this sense *the numerical calculations of CARNOT and MAYER (hence also those of CLAPEYRON and HOLTZMANN*[5]*) would have agreed exactly,* had all of them used the same experimental data.

However, the quantity J so calculated *did not necessarily have the meaning we attach to it today.* Its precise status may be determined as follows. Theorem 1 refers only to ideal gases, but CARNOT's General Axiom makes μ a universal function, the same for all fluids. The two together make HOLTZMANN's Assertion for ideal gases imply the HELMHOLTZ-JOULE Determination for μ in general, irrespective of particular fluids, so the General CARNOT-CLAPEYRON Theorem reduces to

$$J\Lambda_V = \theta\frac{\partial p}{\partial \theta}, \tag{7I.2}$$

a relation which is to become so familiar in the later thermodynamics that modern authors will regard Λ_V as a superfluous quantity and hence will not introduce it at all. Conversely, if (2) holds, we may compare it with (5L.4) and so recover (5N.7) and with it HOLTZMANN's Assertion (7D.1) for ideal gases. Thus we have established[6]

Theorem 2: *Let CARNOT's General Axiom be assumed. Then*

$$\begin{aligned} J\Lambda_V &= p \quad \text{for one ideal gas} \Leftrightarrow \\ \mu &= J/\theta \quad \text{for all fluids} \Leftrightarrow \\ J\Lambda_V &= \theta\frac{\partial p}{\partial \theta} \quad \text{for all fluids} . \end{aligned} \tag{7I.3}$$

The status of HOLTZMANN's Assertion with respect to CARNOT's theory is easy to establish. Comparing $(7\mathrm{I}.3)_2$ with (5N.6), using the statement in

---

[5] CLAUSIUS [1850, Introduction] was to remark that HOLTZMANN's calculation was equivalent to MAYER's.

[6] Theorems 1 and 2 are due essentially to CLAUSIUS [1850, §II, paragraph after his Equation (II.c)], but by no means clearly or completely.

the text just preceding the latter, and recalling also the statement in italics just before (5N.3), we obtain the following

Corollary: *Let CARNOT's Special Axiom be assumed. Then*

$$J\Lambda_V = p \quad \text{for one ideal gas} \quad \Leftrightarrow$$

$$G(\theta^+, \theta^-, C^+) = J \log \left(\frac{\theta^+}{\theta^-}\right) C^+ \quad \Leftrightarrow \tag{7I.4}$$

CARNOT's Special Axiom is dimensionally invariant.

Finally we determine the status of HOLTZMANN's Assertion with respect to CARNOT's general ideas. If it is assumed true for ideal gases, are they the only fluids for which it can hold? To answer this question, we need only put $(3)_1$ into $(3)_3$ and so obtain the partial differential equation

$$\theta \frac{\partial p}{\partial \theta} = p \, , \tag{7I.5}$$

the general solution of which is

$$p = f(V)\theta \; . \tag{7I.6}$$

Thus we have established

Theorem 3: *Let CARNOT's General Axiom be assumed. Then*

$$J\Lambda_V = p \text{ for one ideal gas} \quad \Leftrightarrow$$

$$J\Lambda_V = p \text{ for all fluids such that } p/\theta = f(V). \tag{7I.7}$$

This last statement may be expressed in terms of the classic experiments from which the concepts of "ideal gas" was distilled: the experiments of BOYLE as interpreted by TOWNELEY and POWER, and the results of AMONTONS, to which the opening sentences of this book refer. Namely, against the background of CARNOT's General Axiom, HOLTZMANN's Assertion is appropriate to fluids that satisfy the Law of AMONTONS, whether or not they satisfy also the Law of TOWNELEY and POWER.

Theorems 1, 2, and 3 establish the precise position of HOLTZMANN's Assertion (7D.1); equivalently, of MAYER's Assertion (7B.2).

Theorems 1 and 2 show that *MAYER's Assertion determines CARNOT's function* μ. Referring to Scholion IV in §5L, we know that without some further assumption, μ *does not determine the motive power of heat*. Such is the case also, then, for the assumptions of MAYER and HOLTZMANN. In particular, *neither implies that heat and work be uniformly interconvertible in cyclic processes*, whatever MAYER and HOLTZMANN may have meant by their accompanying statements. Both are compatible with *any theory that respects CARNOT's General Axiom*. From his equations we know that HOLTZMANN incorporated his assertion (7D.1) into the Caloric Theory and so obtained for the motive power expressions which CARNOT had already published and discussed. MAYER for lack of mathematics did not attempt to determine the motive power of heat. Had he done so, perhaps he, too, would have used the Caloric Theory. The work of CLAPEYRON and HOLTZMANN is sufficiently detailed to demonstrate abundantly that they accepted the Caloric Theory and thus could not have believed heat and work to be *universally and uniformly* interconvertible in cyclic processes. MAYER's work is so primitive as to leave us in doubt what he really did mean. His verbal claims are so broad as to be untenable.

Here we see a dichotomy, probably perpetual, between a history of scientific beliefs and a history of rational and experimental science. For the former, MAYER is the author of something called "the First Law of Thermodynamics", while HOLTZMANN is not. For the latter, MAYER and HOLTZMANN did no more than assert, each in his own way, that ideal gases interconvert heat and work uniformly and universally in isothermal processes. This claim of theirs neither implies the "First Law" nor excludes the Caloric Theory.

Theorems 1 and 2 express the meaning of the universal and uniform Interconvertibility of Heat and Work in an ideal gas undergoing an isothermal process. Theorem 3 shows that CARNOT's General Axiom will not allow such interconvertibility to all fluids. As KELVIN was to write[7], the quantity of work is not

> the simple mechanical equivalent of the heat, as it was unwarrantably assumed by MAYER to be,... [i]n violation of CARNOT's important principle, that thermal agency and mechanical effect, or mechanical agency and thermal effect, cannot be regarded in the simple relation of cause and effect, when any other effect, such as the alteration of the density of a body, is finally concerned.

We can put this in another way: The idea of a universal "simple mechanical equivalent of the heat" is *wrong*; it was *wrong* in 1845, too, as CARNOT's work made clear to those who, like KELVIN, could understand it; MAYER and JOULE's applications of it led nevertheless to tenable results. Falsehood may imply truth.

The particular simple mechanical equivalence in isothermal processes that MAYER, HOLTZMANN, and JOULE studied was referred by them, expressly or by implication, to *particular fluids*: the ideal gases. Theorem 3 delimits the class of fluids which CARNOT's General Axiom allows to enjoy this uniform interconvertibility. It is a special class. The simple and uniform equivalence in isothermal processes is *not even universal*.

The spectator innocent of thermodynamics, if such there be, may object: You have produced all this from CARNOT's General Axiom. What good is

---

[7] THOMSON [1852, *1*, §65].

that? Was not CARNOT's theory to be rejected? Here the answer is No! The mathematical developments which the spectator is to witness in Act IV and in much of Act V will be confined to theories *consistent with CARNOT's General Axiom*. That axiom in itself carries no prejudice for or against the Caloric Theory, no prejudice for or against the universal and uniform Interconvertibility of Heat and Work in cyclic processes. HOLTZMANN's Assertion, as HOLTZMANN's own use of it shows, is compatible with the Caloric Theory; it is compatible also, as we shall see in Act IV, with the universal and uniform Interconvertibility of Heat and Work in cyclic processes. The status of the constant J that figures in Theorems 1, 2, and 3 is that which those theorems give it. As yet, J is *not* the universal and uniform equivalent of a unit of heat in cyclic processes. So far, *no general equivalence* has been introduced, except in the vague and partly untenable assertions of MAYER and JOULE.

Who, then, is the discoverer of the "First Law of Thermodynamics"? I am not certain there was one, but I am certain that *before 1850 no "First Law" had been published by anyone*. If discoverer there were, certainly it was not MAYER, not HELMHOLTZ, not JOULE.

We are ready for the next act of our tragicomedy. Just after gifted thinkers have proposed a capital idea which they lacked the critical faculty to delimit and the mathematical skill to exploit, a great mathematician has buried his head in steam tables. KELVIN, failing to see that any acceptable theory of thermodynamics, in order to square with the simplest rough summary of experiments on gases, should allow the equation of state of an ideal gas to be compatible with constant specific heats, calls for still more experiment! A grievous error! As GIBBS[8], critical and level as always, was to remark long afterward, to create (in MAXWELL's words) "a science with secure foundations, clear definitions, and distinct boundaries" it was conceptual analysis that wanted: "The materials indeed existed for such a science..., such materials as had for years been the common property of physicists." To winnow, refine, and reorder these materials, the tragicomic muse of thermodynamics casts her aura and her curse upon a man who, like CARNOT, is a penetrating student of nature but a feeble mathematician: RUDOLF CLAUSIUS.

[8] "Rudolf Julius Emanuel Clausius", *Proceedings of the American Academy* (n.s.) **16** (1889), 458–465 = *Collected Works* **2**, 261–267 (1906).

8. Act IV. Internal Energy: the First Paper of Clausius. Entropy: the First Paper of Rankine

Prima che più entre,
sappi che se' nel secondo girone,
. . . e sarai mentre
che tu verrai ne l'orribil sabbione.
Però riguarda ben. . . .
Dante, *Inferno* XIII, 16–20.

8A. Clausius' Physical Concepts and Assumptions

After quoting Carnot's claim that to deny the existence of the heat function "would overthrow the whole theory of heat", Clausius wrote[1]:

> I am not aware, however, that it has been sufficiently proved by experiment that no loss of heat occurs when work is done; it may, perhaps, on the contrary, be asserted with more correctness that even if such a loss has not been proved directly, it has yet been shown by other facts to be not only admissible, but even highly probable. If it be assumed that heat, like a substance, cannot diminish in quantity, it must also be assumed that it cannot increase. It is, however, almost impossible to explain the heat produced by friction except as an increase in the quantity of heat. The careful investigations of Joule, in which heat is produced in several different ways by the application of mechanical work, have almost certainly proved not only the possibility of increasing the quantity of heat in any circumstances but also the law that the quantity of heat developed is proportional to the work expended in the operation. To this it must be added that other facts have lately become known which sup-

[1] Clausius [1850]. As Magie's translation is mainly adequate, I quote from it here and below, silently correcting it once in a while.

Clausius stated that he had not been able to see Carnot's book and was acquainted with its contents only through the accounts published by Clapeyron and Thomson.

port the view, that heat is not a substance, but consists in a motion of the least parts of bodies. If this view is correct, it is admissible to apply to heat the general mechanical principle that a motion may be transformed into work, and in such a manner that the loss of *vis viva* is proportional to the work accomplished.

These facts, with which Carnot also was well acquainted, and the importance of which he has expressly recognized, almost compel us to accept the equivalence between heat and work, on the modified hypothesis that the accomplishment of work requires not merely a change in the distribution of heat, but also an actual consumption of heat, and that, conversely, heat can be developed again by the expenditure of work.

CLAUSIUS cites with approval MAYER's evaluation of the mechanical equivalent of a unit of heat; he notices that HOLTZMANN's evaluation of it is the same, and he [justly] criticizes HOLTZMANN's further development for proceeding "exactly as CLAPEYRON did, so that in this part of his work he tacitly assumes that the quantity of heat is constant." CLAUSIUS describes THOMSON's doubts and refers to JOULE's experiments. After quoting THOMSON's refusal to abandon the heat function because of "innumerable other difficulties, insuperable without further experimental investigation, and an entire reconstruction of the theory of heat from its foundation," CLAUSIUS concludes,

I believe that we should not be daunted by these difficulties, but rather should familiarize ourselves as much as possible with the consequences of the idea that heat is a motion, since it is only in this way that we can obtain the means wherewith to confirm or to disprove it. Then, too, I do not think the difficulties are so serious as Thomson does, since even though we must make some changes in the usual form of presentation, yet I can find no contradiction with any proved facts. It is not at all necessary to discard Carnot's theory entirely, a step which we certainly would find it hard to take, since it has to some extent been conspicuously verified by experience. A careful examination shows that the new method does not stand in contradiction to the essential principle of Carnot, but only to the subsidiary statement *that no heat is lost*, since in the production of work it may very well be the case that at the same time a certain quantity of heat is consumed and another quantity transferred from a hotter to a colder body, and both quantities of heat stand in a definite relation to the work that is done.

CLAUSIUS lays down the following "first principle":

In all cases in which work is produced by the agency of heat, a quantity of heat is consumed which is proportional to the work done; and

> conversely, by the expenditure of an equal quantity of work an equal quantity of heat is produced.

CLAUSIUS does not mean us to accept this statement as broadly as it may sound. [By this time everyone knew that work could be done or consumed in an adiabatic process, at no expense of heat.] That is not all.

> If any body changes its volume, mechanical work will in general be either produced or expended. It is, however, in most cases impossible to determine this exactly, since besides the *external* work there is generally an unknown amount of *internal* work done. To avoid this difficulty, Carnot employed the ingenious method already referred to of allowing the body to undergo its various changes in succession, which are so arranged that it returns at last exactly to its original condition. In this case, if *internal* work is done in some of the changes, it is exactly compensated for in the others, and we may be sure that the *external* work, which remains over after the changes are completed, is all the work that has been done.

Thus we may express CLAUSIUS' principle as follows: *For every cycle* $\mathscr{C}$

$$L(\mathscr{C}) = JC(\mathscr{C}) \ . \qquad (7A.1)_{1r}$$

The constant J, which CLAUSIUS denotes by A^{-1}, is *the mechanical equivalent of a unit of heat* in cyclic processes. [CLAUSIUS has seen the defect in the unrestricted affirmations of MAYER, HELMHOLTZ, and JOULE. He sees that storage must be accounted for, and he sees that CARNOT's approach does allow for it *by considering cycles alone*, in which the overall effect of storage is null. Like CARNOT and CLAPEYRON], CLAUSIUS applies his basic assumption only to infinitesimal Carnot cycles.

Of course CLAUSIUS, like all his predecessors, accepts the theory of calorimetry [though he gives no evidence of seeing it in the simple and explicit completeness with which we have presented it in §2C, starting from the formal assumption

$$Q = \Lambda_V(V, \theta)\dot{V} + \mathrm{K}_V(V, \theta)\dot{\theta} \ . \qquad (2C.4)_r$$

Certainly all of his work fits easily into the framework as we have presented it[2].]

[2] CLAUSIUS' explanation of free and latent heat and of internal and external work is cloudy. On the other hand, anyone who reads with understanding CLAUSIUS' proof of his Equation (3) will see that there he does use (2C.4) and nothing else. His notations for Λ_V and K_V are

$$\left(\frac{dQ}{dv}\right) \text{ and } \left(\frac{dQ}{dt}\right) ,$$

CLAUSIUS employs also "an obvious subsidiary hypothesis" regarding ideal gases. After some motivating remarks which, though they include the phrase "there is no reason to think that...", he seems to regard as a proof, he arrives at "the law: *a permanent gas, when expanded at constant temperature, takes up as much heat as is consumed in doing external work during the expansion*. This law is probably true for any gas with the same degree of exactness as that attained by the laws of Mariotte and Gay-Lussac applied to it." The "law" (his Equation (9)) is

$$J\Lambda_V = p \ , \qquad (7D.1)_r$$

[namely, as the reference number indicates, HOLTZMANN's Assertion, which we have explained in §7E, and the logical status of which we have established in §7I].

These principles are contained in CLAUSIUS' introduction and §I. In §II he takes up CARNOT's ideas again. He gives [essentially CARNOT's] argument to indicate that a Carnot cycle delivers maximum motive power for given extremes of temperature. As transfer of heat by conduction need not produce mechanical effect, "the way in which the transfer of a certain quantity of heat between two bodies at the temperatures θ and τ can be made to do the maximum of work is so to carry out the process, as was done in the above cases, that two bodies of different temperatures never come in contact." [*Cf.* CARNOT's statement quoted above in §5D.]

"It is this *maximum* of work which must be compared with the heat transferred." [Again employing an argument which is essentially CARNOT's,] CLAUSIUS considers the state of affairs that results after completion of both a Carnot cycle and the reverse of a less efficient one, so adjusted as to annul in part the effect of the former. [As we have seen in §5G, CARNOT's construction was able to "re-establish things in their original state" because $C^+(\mathscr{C}) = C^-(\mathscr{C})$ according to the Caloric Theory. For CLAUSIUS, on the other hand, $C^+(\mathscr{C}) \neq C^-(\mathscr{C})$, and until a definite theory is constructed, we cannot know whether or not CARNOT's construction will re-establish everything in its original state. The requirements of "sound physics" have to be narrowed if they are to yield CARNOT's conclusion even in CLAUSIUS' theory.]

CLAUSIUS proceeds to do just this, but in terms of a somewhat different construction. He adjusts $\mathscr{C}_2$ so that $L_{B_2}(-\mathscr{C}_2) = -L_{B_1}(\mathscr{C}_1)$. After $\mathscr{C}_1$ and $-\mathscr{C}_2$ have been completed, no work has been done, yet if B_1 does more

---

respectively; his equation for an adiabatic process is

$$\left(\frac{dQ}{dv}\right)\delta v - \left(\frac{dQ}{dt}\right)dt = 0 \ ,$$

which in our notion would be $\Lambda_V dV + K_V d\theta = 0$; *etc.* This interpretation is confirmed by the footnote CLAUSIUS put on p. 29 of the reprint of this paper in his *Mechanische Wärmetheorie*, 1864.

work per unit heat than B_2 does in a Carnot cycle with the same operating temperatures, then after $\mathscr{C}_1$ and $-\mathscr{C}_2$ have been completed *some heat will have passed from the refrigerator to the furnace.*

> By repeating these two alternating processes, it would be possible, without any expenditure of force or other change, to get any amount of heat out of a *cold* body and into a *hot* one, and this contradicts the other behavior of heat, since heat everywhere strives to smoothe out such differences of temperature as occur and therefore to pass out of *hotter* bodies into *colder* ones.
>
> It seems, therefore, to be *theoretically* justified to retain the first and the really essential part of Carnot's assumptions ...
>
> On this assumption we may express the maximum of work that can be produced by the transfer of a unit of heat from the body A at the temperature θ into the body B at the temperature τ, as a function of θ and τ.

[Thus CLAUSIUS claims to motivate CARNOT's General Axiom. In CARNOT's applications of it $C^+(\mathscr{C}) = C^-(\mathscr{C})$, so the roles of C^+ and C^- are interchangeable. In Act II I have chosen C^+ as the variable in terms of which to state the axiom. CLAUSIUS chooses (in effect) C^-. With either choice it will turn out that $C^-(\mathscr{C})$ for a Carnot cycle $\mathscr{C}$ is determined by θ^+, θ^-, and $C^+(\mathscr{C})$; for the former choice, this statement is the burden of Theorem 9 in §10 of *Concepts and Logic*, and for the latter choice the same proof works with only trivial changes. Thus CLAUSIUS' statement is equivalent to Part I of CARNOT's General Axiom:

$$L(\mathscr{C}) = G(\theta^+, \theta^-, C^+(\mathscr{C})) \ , \qquad (5\mathrm{I}.1)_\mathrm{r}$$

$$G(x, y, z) > 0 \quad \text{if} \quad x > y > 0 \quad \text{and} \quad z > 0 \ . \qquad (5\mathrm{I}.2)_\mathrm{r}$$

[To me CLAUSIUS' argument suggests something a little different, namely, that *if two Carnot cycles with the same operating temperatures do the same amount of work, they emit the same amount of heat.* While thermodynamics can indeed be constructed on this basis, there is neither need for nor advantage in changing CARNOT's construction. If we are to narrow the requirements of "sound physics", we can do so directly from the construction CARNOT himself provided. With no appeal to any relation between heat and work, that construction leaves the quantity of heat in the furnace unchanged after positive work has been done. To obtain the desired prohibition, we may simply forget about the refrigerator and make "sound physics" require that *if work is to be done by heat, some heat must pass from a hot body to a cold one.* Then we draw CARNOT's conclusions just as well without use of the Caloric Theory.

[Such is the slipperiness of arguments based upon denial of "perpetual motion of the second kind".]

8B. Logical Content of CLAUSIUS' First Paper

Although CLAUSIUS restricts his mathematical theory to ideal gases and to vapors at their maximum density, "since these cases, in consequence of the extensive knowledge we have of them, are most easily submitted to calculation, and besides that are the most interesting," [most of his logical steps are valid for an arbitrary equation of state (2A.2), and we shall so present them[1]. As usual in this account, we shall omit all special consideration of vapors, and for two reasons. First, the properties of vapors by themselves serve only to illustrate the general formulae valid for an arbitrary equation of state. Second, much of the early work on steam takes account of the latent heat of vaporization and considers a mixture of steam and water. Those two features, one of which simplifies and the other complicates the theory, do not contribute to the conceptual structure of thermodynamics, so I simply leave this part of the story untold.]

CLAUSIUS' argument falls into five parts, largely independent.

α. *First line of reasoning.* CLAUSIUS first exploits the old, basic Doctrine of Latent and Specific Heats. Considering an infinitesimal Carnot cycle, he calculates the work done and the "heat consumed". His results, stated as his Equations (1) and (3), he substitutes into the statement of "our principle" of "the equivalence between heat and work" and so obtains his Equation (II.) [the general form[2] of which is

$$\frac{\partial p}{\partial \theta} = J\left(\frac{\partial \Lambda_V}{\partial \theta} - \frac{\partial \mathrm{K}_V}{\partial V}\right) . \Big] \tag{8B.1}$$

This is the general local statement of CLAUSIUS' *principle of uniform and universal Interconvertibility of Heat and Work* in cyclic processes undergone by fluid bodies.

CLAUSIUS states at once that his special case of (1) "may be brought into the form of a *complete* differential equation," by which a function $\mathrm{E}(V, \theta)/J$ may be defined. This function, which CLAUSIUS denotes by U, satisfies the relation (his Equation (II.a))

$$\dot{\mathrm{E}} = JQ + P , \qquad P \equiv -p\dot{V} . \tag{8B.2}$$

Thus "the total amount of heat received by the gas during a change of volume and temperature can be separated into two parts, one of which, E, which comprises the *free* heat that has entered and the heat *consumed* in doing *internal* work, if any such work has been done, has the properties which are commonly assigned to the total heat", namely, that it is the

[1] CLAUSIUS [1854] himself so presented them in his second paper.

[2] Appaerntly THOMSON [1851, Equation (2)] was the first to publish (1). For KELVIN's sanitary operations see also Footnotes 3 and 12 to §2C, above.

value of a function of V and θ; "while the other part, which comprises the heat *consumed* in doing *external* work, is dependent not only on the terminal conditions, but on the whole course of the changes between those conditions." [CLAUSIUS' interpretation of E as the free heat and the heat consumed "in doing *internal* work" really refers to the change of E in a process, not to the actual value of E. Today we call E the *internal energy* of the body whose constitutive functions are ϖ, Λ_V, and K_V.]

β. Second line of reasoning. Next CLAUSIUS takes up his "obvious subsidiary hypothesis", namely, [HOLTZMANN's] Assertion:

$$J\Lambda_V = p \ . \qquad (7D.1)_r$$

CLAUSIUS remarks, [as HOLTZMANN, who did not have CLAUSIUS' basic axiom of interconvertibility for cycles could not have done,] that this "subsidiary hypothesis" makes E and K_V functions of θ alone. "It is even probable that ... K_V ... is a constant."

γ. Third line of reasoning. The third course of deduction CLAUSIUS pursues refers to ideal gases [and rests upon the Doctrine of Latent and Specific Heats alone, making no use of his own basic axiom

$$L(\mathscr{C}) = JC(\mathscr{C}) \ .] \qquad (7A.1)_{1r}$$

By a [clumsy] proof, [in which $R\theta/V$ is carried through as an expression for p, only so as to cancel out after all,] he obtains from [HOLTZMANN's] Assertion (7D.1) the consequence

$$J(K_p - K_V) = R \ , \qquad (7B.2)_r$$

[which is MAYER's Assertion. We remember that HOLTZMANN had obtained the same result by unnecessary use of the Caloric Theory (§7E).] CLAUSIUS cites the experimental fact that R is "in so far different for the different gases that it is inversely proportional to their specific gravities." Thus "*the difference of the specific heats referred to the unit of volume is therefore the same for all gases.*" Dividing (7B.2) by K_V, CLAUSIUS concludes that $\gamma - 1$ is "*for the different gases inversely proportional to the specific heats of the same at constant volume, if these are referred to the unit of volume.*" He states that this law

> has been found by DULONG from experiment....
>
> If it is now assumed that the specific heat of gases at constant volume K_V is constant, which has been stated above to be very probable, the same follows for the specific heat at constant pressure, and consequently *the quotient of the two specific heats* $K_p/K_V = \gamma$ *must be constant.* This law, which Poisson assumed correct on the strength of the experiments of Gay-Lussac and Welter ... is therefore in

good agreement with our present theory, while it would not be possible in Carnot's theory as heretofore developed.

Continuing with special properties of ideal gases, CLAUSIUS shows that if γ = const., then the law of adiabatic change is (his Equations (16) and (15))

$$\frac{p}{p_0} = \left(\frac{V_0}{V}\right)^{\gamma} ,$$
$$\left(\frac{\theta}{\theta_0}\right)^{\gamma} = \left(\frac{p}{p_0}\right)^{\gamma-1} . \tag{8B.3}$$

"These equations agree precisely with those which have been developed by Poisson for the same case...." [Indeed, these formulae are equivalent to (3D.5).]

Next CLAUSIUS recalls that in an isothermal process

$$C(\mathscr{P}_\theta) = \int_{V_a}^{V_b} \Lambda_V(V, \theta)dV . \tag*{(5L.2)$_{1r}$}$$

For a body of ideal gas, if [HOLTZMANN's Assertion] (7D.1) holds, then (CLAUSIUS' Equations (18) and (19))

$$JC(\mathscr{P}_\theta) = \int_{V_a}^{V_b} p(V, \theta)dV = R\theta \int_{V_a}^{V_b} \frac{dV}{V} ,$$
$$= R\theta \log \frac{V_b}{V_a} = p_a V_a \log \frac{V_b}{V_a} . \tag{8B.4}$$

Thus CARNOT's theorem on progressions (5K.5) is recovered, but in a more specific form. From $(4)_4$ follows

> The well known law which Dulong proposed, ... *that all gases, if equal volumes of them are taken at the same temperature and under the same pressure, and if they are then compressed or expanded by an equal fraction of their volumes, either evolve or absorb an equal quantity of heat.*

However, $(4)_4$ is "much more general". It shows that "*if the original pressure is different in the different cases, the quantities of heat are proportional to it.*"

δ. *Fourth line of reasoning.* The fourth part of CLAUSIUS' argument, which is presented in a single short paragraph of his §II, starts from his reassertion of CARNOT's General Axiom, with C^+ replaced by C^-. [In a rather slipshod way] CLAUSIUS concludes that "the maximum of work which can be produced by the transfer of a unit of heat [in an infinitesimal Carnot cycle] ... may be expressed in the form $(1/C)d\theta$, where C is a function of θ only." [We recognize this result as being CARNOT's evaluation, namely

$$L(\mathscr{C}) \approx \mu(\theta)\Delta\theta C^+(\mathscr{C}) , \tag*{(5J.6)$_r$}$$

except that C^+ is replaced by C^- and CLAUSIUS' $1/C$ is what we denote by μ. The final outcome, obtained with CLAUSIUS' characteristic clumsiness, can be nothing but the General CARNOT-CLAPEYRON Theorem:

$$\mu\Lambda_V = \frac{\partial p}{\partial \theta} \text{ .]} \tag{5L.4$_r$}$$

This, indeed, CLAUSIUS obtains for the special case of an ideal gas; it is his Equation (IV.). All of CLAUSIUS' remarks make it clear that he, like CARNOT, regards μ as a *universal function of temperature, the same for all bodies*, just as J is a universal constant.

ϵ. *Fifth line of reasoning.* The last part of CLAUSIUS' argument reverts to use of his "subsidiary hypothesis" (7D.1) [HOLTZMANN's Assertion] for ideal gases. CLAUSIUS states that then (5L.4) is "not necessary" for "further determination" of Λ_V and K_V, but "we gain ... an opportunity to subject the results of the two principles to a comparative test." Indeed [*cf.* Theorems I and II in §7I], we have only to compare (5L.4) with (7D.1) to conclude that the two agree for an ideal gas if and only if [the HELMHOLTZ-JOULE Determination] holds:

$$\mu = J/\theta \text{ .} \tag{5N.7$_r$}$$

CLAUSIUS writes, "we see that they both express the same result, only the one in a more special ("bestimmter") way than the other," since the function μ was "only implied" ("nur angedeutete") by (5L.4).

8C. Critique: The Achievement of CLAUSIUS' First Paper

From the formulae CLAUSIUS obtains in this paper, once expressed in the generality his reasoning sufficed to obtain rather than the special forms suitable only for ideal gases or for vapors, *all of the thermodynamics of reversible processes in fluids* may be constructed. CLAUSIUS himself did not derive all the formulae of the subject, but those that he did not work out are easy consequences of those he did. Certainly he saw what to collect and retain from his predecessors' work:

1. The theory of calorimetry, freed of the Caloric Theory, with which it may have seemed connected though logically it never was.
2. CARNOT's General Axiom.
3. The universal and uniform Interconvertibility of Heat and Work in cyclic processes, of course contradicting the Caloric Theory.
4. HOLTZMANN's Assertion about ideal gases, leading to the HELMHOLTZ-

JOULE Determination of μ.

These ideas suffice to construct classical thermodynamics; *cf.* the Historical Scholion at the end of Chapter 11 of *Concepts and Logic.*

As had CARNOT before him, CLAUSIUS perceived the key role to be played by ideal gases. CLAUSIUS fully overcame the difficulties regarding them that had puzzled CARNOT. CLAUSIUS was the first to see that the LAPLACE-POISSON law of adiabatic change had to be reconciled with thermodynamics. He saw also that HOLTZMANN's Assertion ought to be compatible with thermodynamics and that MAYER's Assertion followed from it. Consequently he was the first to see that *thermodynamics should allow both specific heats of an ideal gas to be constant*[1]. He may have seen also that the Caloric Theory did not allow this possibility, although his statements in this regard are not accurate.

Not the least of CLAUSIUS' achievements is his remark, "It is even probable that ... K_V ... is a constant." No other student had suggested such a thing; it is further evidence of CLAUSIUS' physical insight and critical assessment of what was believed. K_V had never been measured directly[2]. From the day CARNOT's booklet appeared, it should have been obvious that for determining the relations between heat and work, *the nature of the specific heats of an ideal gas was of the essence*, yet REGNAULT with his ponderous,

---

[1] The early authors did not discuss the effect of a change from one empirical scale of temperature to another. They laid down one particular scale and stayed with it. For the general theory of change of scale see §11H, below. We note here a few aspects of present interest. First, the definition of K_V through (2C.4) makes it plain that the nature of K_V *depends upon the choice of temperature scale*. In particular, if K_V = const. with one choice of scale, $K_V \neq$ const. with most others. The statement that γ = const. *is invariant under change from one scale to another.* On the contrary, the definition of an ideal gas, MAYER's Assertion, and HOLTZMANN's Assertion are not. Furthermore, K_V = const. for one scale if and only if K_V is a function of temperature alone for all scales. The TOWNELEY-POWER-"BOYLE" law is invariant; the law of AMONTONS is not.

CLAUSIUS' conjecture that K_V = const. for all ideal gases nevertheless makes sense. Although he nowhere explains what he means by temperature, his calculations make it plain that he expects K_V to be defined in terms of the scale that makes an ideal gas have the equation of state (2A.1) (which is his Eq. (I.)), not the more general functional dependence $pV = f(\theta)$, which expresses the TOWNELEY-POWER-"BOYLE" law alone. We should say today that for CLAUSIUS θ is the "ideal-gas temperature". REGNAULT, whose experiments we shall cite in the next two footnotes, necessarily used an actual thermometer. His early work had established the air thermometer as superior to all others he tried, and thenceforth he used it alone. Thus his experimental conclusion is not quite the same as CLAUSIUS' conjecture. However, as we shall see in §11B, JOULE & KELVIN were to show that the deviations of temperatures according to the air thermometer from corresponding "absolute" ideal-gas temperatures were small in the ranges of temperatures then available in laboratories.

[2] So REGNAULT [1853] was to write. PARTINGTON in §1 of Chapter VII of the work cited above in Footnote 4 to §2C cites several measurements made between 1779 and 1840; he states that all gave values much too large. The oldest work he cites as having obtained an acceptable value is one published in 1880.

subsidized program of precise experiment did not publish anything about the specific heats until 1853! Even so[3], he measured only K_p. To CLAUSIUS, experiment seems to confirm[4] LAPLACE's calculation of the speed of sound; while that calculation does not require γ to be constant, CLAUSIUS may have thought it did. Alternatively, CLAUSIUS may have regarded the relation $c^2 \propto \theta$ as having been confirmed by experiment, and LAPLACE's theory delivers that relation only if $\gamma \equiv$ const. (*cf.* above, §§3F and 5T).

If γ is constant, CLAUSIUS' "subsidiary hypothesis" makes both specific heats of an ideal gas constant. Hence, no doubt, grew CLAUSIUS' remark that a constant K_V would be "probable". It turned out to be prophetic. REGNAULT was to find that for air, oxygen, carbon monoxide, and some other gases over a wide range of temperature and pressure K_p was indeed "sensibly" independent of both[5]. Since CLAUSIUS already regarded experiment as showing γ to be constant for nearly ideal gases, REGNAULT's conclusion, when it came, must have served to bear out his remark. So did the direct determinations of K_V, many years later.

Of course, CLAUSIUS' theory obviates all the objectionable features of CARNOT's yet retains its basic conceptual frame and such of CARNOT's results as seemed good, these being the theorem on progressions in isothermal processes (a consequence of (8B.4)) and the General CARNOT-CLAPEYRON Theorem (5L.4). From CLAUSIUS' standpoint CARNOT seems to have gotten some right results from partially wrong assumptions. This view has been disputed. Our analysis shows that if properly made specific, it is just.

Some of these achievements lay close to the surface, ready for the net of any fisherman who could see them. While CLAUSIUS' acknowledgments of his debt to his predecessors may seem sufficient, in fact they are scanty and in some cases misleading.

First, while dismissing HOLTZMANN's work, CLAUSIUS sets up as his "obvious subsidiary hypothesis" what is neither more nor less than HOLTZMANN's Assertion (7D.1). Although he could not have failed to notice it in the paper by HOLTZMANN that he criticized, CLAUSIUS lets his readers presume it is his own alone. Moreover, while his argument to show that it

---

[3] REGNAULT [1862, p. 4] himself was to write in 1853 that direct determination of K_V was "an important element for the physical theory of bodies". On 15 May, 1854, he deposited with the French Academy a paper on the measurement of K_V, but I cannot find any record of its having been published.

[4] REGNAULT [1862, p. 40], writing in 1853, was to disagree:

> I will show in the following memoir that the experimental procedure by which the ratio of specific heats has been found cannot inspire any confidence. Hence the explanation of Laplace is today only a hypothesis, very ingenious no doubt, but one which needs to be confirmed by experiment.

[5] REGNAULT [1853] [1862, pp. 109, 213, 301]. In the passage on p. 301 he goes so far as to conjecture that if a gas ceases to respect the ideal gas law, K_p will fail to be constant for it.

Of course REGNAULT found the behavior of carbon dioxide quite other than ideal.

implies MAYER's Assertion (7B.2) is better than HOLTZMANN's, CLAUSIUS lets his readers attribute to him not only the argument but even the fact it proves. CLAUSIUS' claims of confirmation by experiment, especially in footnote 2 and p. 44 of the reprint in his *Mechanische Wärmetheorie*, 1864, certainly lead the unwary to believe that (7B.2) was new when he published it and somehow special to his theory. Many pages later, just before his Equation (34), CLAUSIUS does state that both MAYER and HELMHOLTZ had used (7B.2) in their calculations. He fails to mention anywhere that (7B.2) is fully consistent with the Caloric Theory, as indeed HOLTZMANN's work abundantly shows. Here, no doubt, lies one reason why the tradition always elevates MAYER and ignores HOLTZMANN.

In his statements regarding the LAPLACE-POISSON law of adiabatic change CLAUSIUS is neither clear nor fully correct. He does not cite LAPLACE at all[6], and he does not represent the results of POISSON fairly, because he himself makes unnecessary assumptions not made by POISSON. The main difficulty here lies only partly in the obscurity of LAPLACE's writing and the shaky basis used by POISSON; mainly it is faulty logic, as we shall see in the succeeding section.

CLAUSIUS' treatment of what amounts to CARNOT's General Axiom is so brief that it has not attracted much notice. Its very brevity may indicate that CLAUSIUS claimed little for himself in it. While it is heavily influenced by the paper of CLAPEYRON, CLAUSIUS' argument is easier to follow because he uses the natural variables V, θ rather than V, p. This CARNOT had done before him, and indeed CLAUSIUS' presentation is close to CARNOT's but unlike it in being restricted to an infinitesimal difference of volume. CLAUSIUS had not seen CARNOT's treatise, and it is a sign of his physical sense that he turned away from the poor choice of variables he had found in the literature and unknowingly reverted to the usage of his great predecessor.

8D. Critique of CLAUSIUS' Reasoning

Although CLAUSIUS' derivations of his formulae for the work done and the heat consumed in an infinitesimal Carnot cycle are awkward, they lead to correct results. Indeed, with the aid of AMPÈRE's transformation of a

[6] We recall that while CARNOT had cited the appropriate book of the *Mécanique Céleste*, he had avoided any reference to the fact that LAPLACE had published therein a theory of heat which overlapped his own. In view of the scribal tradition of thermodynamicists, the total silence historical studies of the theory of heat until very recently have bestowed upon LAPLACE's work may well be explained as coming straight from the founders. The only exceptions I have seen are pp. 184–187 of W. WHEWELL's *History of the Inductive Sciences, from the earliest to the present time*, 3rd edition, Volume II, New York, D. Appleton, 1873, and Chapter 5 of FOX's *Caloric Theory*, cited above in Footnote 2 to §2A. The latter gives LAPLACE's work full and fair treatment.

line integral about an arbitrary simple closed circuit $\mathscr{C}$ into an integral over the region $\mathscr{A}$ that circuit incloses, we need only inspect the basic relations

$$L(\mathscr{P}) = \int_{\mathscr{P}} \varpi(V, \theta)dV \quad \text{and} \quad C(\mathscr{P}) = \int_{\mathscr{P}} [\Lambda_V(V, \theta)dV + \mathrm{K}_V(V, \theta)d\theta] \qquad (2C.20)_{3r}, (2C.6)_{2r}$$

to see at once that

$$L(\mathscr{C}) = \iint_{\mathscr{A}} \frac{\partial p}{\partial \theta} dV d\theta \ , \qquad (5L.1)_{2r}$$

$$C(\mathscr{C}) = \iint_{\mathscr{A}} \left(\frac{\partial \Lambda_V}{\partial \theta} - \frac{\partial \mathrm{K}_V}{\partial V}\right) dV d\theta \ , \qquad (8D.1)$$

with the convention that $\mathscr{C}$ is described clockwise. It is the differential approximations to these formulae that CLAUSIUS painfully carves out with his ds and δs, explained by pages of physical description. A glance at CLAUSIUS' basic assumption

$$L(\mathscr{C}) = JC(\mathscr{C}) \qquad (7A.1)_{1r}$$

now shows that its local equivalent is

$$\frac{\partial p}{\partial \theta} = J\left(\frac{\partial \Lambda_V}{\partial \theta} - \frac{\partial \mathrm{K}_V}{\partial V}\right) \ . \qquad (8B.1)_r$$

The simple and elegant statement (8B.1) is of supreme importance for thermodynamics.

Why does CLAUSIUS use infinitesimal Carnot cycles here? He is simply following in CLAPEYRON's tracks. If CLAPEYRON, who in turn was following CARNOT, had had to restrict himself to an infinitesimal difference of temperatures in order to get a definite answer, CLAUSIUS has no such need. Although the Carnot cycle was essential in CLAPEYRON's assumptions and for CLAPEYRON's reasoning, as it had been for CARNOT's, CLAUSIUS' assumptions refer to all cycles, and his calculation neither has to be restricted to Carnot cycles nor gains any advantage from being so. Nay rather, as the principle that heat and work are universally and uniformly interconvertible by cyclic processes is general, its local expression ought not be restricted to such bodies as admit Carnot cycles arbitrarily near to every point in the part of the quadrant over which their functions ϖ, Λ_V, and K_V are defined, so no proof of that expression that stops short after an argument based on use of Carnot cycles can be sufficient.

CLAUSIUS' derivation of his basic theorem on the internal energy,

$$\dot{\mathrm{E}} = JQ + P, \qquad P \equiv -p\dot{V} \ , \qquad (8B.2)_r$$

is equally obscure and even less convincing. We can obtain (8B.2) by merely writing CLAUSIUS' major result (8B.1) in the form[1]

$$\frac{\partial}{\partial \theta}(J\Lambda_V - p) = \frac{\partial}{\partial V}(J\mathrm{K}_V) \qquad (8D.2)$$

[1] Apparently THOMSON [1851, Equation (1)] was the first to see that CLAUSIUS' reasoning did in fact imply (8B.1) and to write it in the form (2).

and then applying CLAIRAUT's theorem on differential forms[2] to obtain a function $\mathrm{E}(V, \theta)$ such that

$$J\Lambda_V - p = \frac{\partial \mathrm{E}}{\partial V}, \qquad J\mathrm{K}_V = \frac{\partial \mathrm{E}}{\partial \theta}. \tag{8D.3}$$

This result[3], which is to become central in thermodynamics, leads at once to (8B.2).

So much for CLAUSIUS' first line of reasoning. The second line takes up the "obvious subsidiary hypothesis", namely, HOLTZMANN's Assertion:

$$J\Lambda_V = p. \tag{7D.1$_r$}$$

By looking at differential forms CLAUSIUS concludes that if that assertion as well as his basic principle holds, then E and K_V are functions of θ alone. A glance at (3), which CLAUSIUS does not write out, shows that *HOLTZMANN's Assertion* $J\Lambda_V = p$ *is not only sufficient but also necessary for* E *to be a function of* θ *alone*. This statement complements Theorem 3 in §7I.

As CLAUSIUS' third line of argument concerns only the Doctrine of Latent and Specific Heats when specialized by use of HOLTZMANN's Assertion, we might expect smooth sailing, but even here there are major flaws. These are not matters of mathematical proof but of physical concept and principle. Certainly it is a virtue in CLAUSIUS to suggest that both specific heats of an ideal gas may be constant and to explore the consequences of that possibility. When, on the other hand, he states that a constant ratio of specific heats "would not be possible in Carnot's theory", he is simply wrong. As we have seen in §5R, CARNOT's theory, if the specific heats of an ideal gas differ from each other, allows *either their ratio or their difference* to be constant, but not both. The important point is that "our present theory" does allow *both* specific heats to be constant. Indeed, had CLAUSIUS worked out the theory of calorimetry fully, he would have arrived at the relation

$$\mathrm{K}_p - \mathrm{K}_V = \frac{V\Lambda_V}{\theta}, \tag{2C.14$_3$}$$

[2] By this argument the function E is not proved to be single-valued unless the part of the V–θ quadrant over which it is defined is assumed to be simply connected. To prove that E is single-valued, we may appeal again to (7A.1); *cf.* the comment after Axiom VC in Chapter 15 of *Concepts and Logic*. Although commonplace in hydrodynamics, a point of this kind seems to be too subtle for works on classical thermodynamics and is never mentioned. Modern treatments avoid the difficulty by postulating from the start an energy principle in which a much more general E is single-valued by assumption. Results like (3) emerge then as special cases.

[3] Apparently THOMSON [1853, *1*, Equations (4) and (5)] was the first to write it out in print. He had described it in words in the passage cited in Footnote 1.

and by eliminating Λ_V between it and his Equation (II.) (our (8B.1), specialized to an ideal gas) would have obtained

$$\frac{R}{J} - (\mathrm{K}_p - \mathrm{K}_V) = \theta \frac{\partial}{\partial \theta}(\mathrm{K}_p - \mathrm{K}_V) - V \frac{\partial \mathrm{K}_V}{\partial V} . \tag{8D.4}$$

This condition is necessary and locally sufficient that the specific heats K_p and K_V of an ideal gas be compatible with "our fundamental principle". It shows that constant specific heats are admissible if and only if they obey MAYER's Assertion. In contrast, CARNOT's theory does not allow both specific heats to be constant. It is this distinction that CLAUSIUS may have striven to find, but surely he does not demonstrate it in his paper.

Again, it is CLAUSIUS' virtue to have seen that the LAPLACE-POISSON law of adiabatic change had to find its place within thermodynamics, but his treatment of that law suffers from a major conceptual blemish. His starting point to obtain it, namely, his Equation (13), incorporates HOLTZMANN's Assertion (7D.1), which is irrelevant as well as unnecessary. CLAUSIUS gives the reader the idea that he has added something here; on the contrary, he has blurred the facts. As we have seen in §3F, the LAPLACE-POISSON law follows from the theory of calorimetry alone, provided $\gamma =$ const. Any theory that allows to an ideal gas a constant *ratio* of specific heats delivers the LAPLACE-POISSON law for such a gas. CARNOT's theory is by no means excluded.

CLAUSIUS' proof of CARNOT's theorem on progressions associated with isothermal processes rests upon HOLTZMANN's Assertion alone, making no appeal to CARNOT's ideas. CARNOT's own proof, we recall from §5L, is general within the framework of his General Axiom. As CLAUSIUS in his fourth line of reasoning infers and adopts that axiom, our Theorem 1 in §7I makes CLAUSIUS' result a special case of CARNOT's.

The fourth line of argument is short and efficient. It is CARNOT's.

CLAUSIUS presents his fifth line of argument very compactly and with ill chosen, vague words. By saying that μ was "nur angedeutet" by the General CARNOT-CLAPEYRON Theorem (5L.4), and that HOLTZMANN's Assertion (7D.1) asserted the same thing "in bestimmterer Weise", he may mean only that (7D.1) is an obvious special case of the General CARNOT-CLAPEYRON Theorem

$$\mu \Lambda_V = \frac{\partial p}{\partial \theta} , \tag{5L.4)$_\mathrm{r}$ $$

yet none of CARNOT's reasoning is needed to get it. As, however, he knows that μ is a universal function, and as he adopts for it the HELMHOLTZ-JOULE Determination

$$\mu = J/\theta , \qquad (5\mathrm{N}.7)_\mathrm{r}$$

I think he must have seen that to obtain a universal function it suffices to determine that function for one special substance. For that reason I attribute to him in essence the statement asserted in Theorem 2 in §7I.

There is also a major gap that CLAUSIUS seems not to have noticed. He gives no reason why the constant J in his fundamental assumption

$$L(\mathscr{C}) = JC(\mathscr{C}) \qquad (7A.1)_{1r}$$

should be the same constant as the J in his subsidiary hypothesis,

$$J\Lambda_V = p \ . \qquad (7D.1)_r$$

They are logically independent statements; according to KELVIN's repeated objection, there was "no experimental evidence whatever" for the "pure hypothesis" (7D.1). The truth is, as KELVIN is to show in effect (see §9B, below), the two constants need *not* be the same if no assumption beyond CARNOT's General Axiom is brought in. HOPPE's theorem, given below in §9D, provides one sufficient assumption to this end. Namely, if thermodynamics based on CARNOT's General Axiom is to allow an ideal gas to have constant specific heats, as indeed CLAUSIUS stated it should, then for that gas both (7A.1) and (7D.1) *must* hold, with J for *both* given by MAYER's formula

$$J = \frac{R}{(1-\gamma^{-1})\mathrm{K}_p} \ . \qquad (7B.3)_r$$

Since both Js are universal constants, they must be the same for all fluids for which the latter J plays a role; the former, of course, pertains to all fluids without exception. CLAUSIUS may have thought that his "fundamental principle" applied to isothermal as well as to cyclic processes: "In all cases in which work is produced by the agency of heat, a quantity of heat is consumed which is proportional to the work done...." If so, he was in error, for, as his own theoretical work implies and as we have stated in Theorem 3 of §7I, the isothermal work done is *not* proportional to the work consumed except for fluids obeying the law of AMONTONS, or, equivalently, fluids such as to make E a function of θ alone.

Although CLAUSIUS' paper is not long, it would have been less confusing if still shorter. Because of its heavy use of results already known, its essential contents could more easily have been grasped if they had been presented in two parts so separated as to make the logical thread clear:

Part I. A succinct mathematical treatment of the theory of calorimetry, leading to the formula

$$C(\mathscr{C}) = \iint_{\mathscr{A}} \left(\frac{\partial \Lambda_V}{\partial \theta} - \frac{\partial \mathrm{K}_V}{\partial V}\right) dV d\theta \qquad (8D.1)_r$$

and to the results we have given in §2C, HOLTZMANN's and MAYER's Assertions and their relation, and the LAPLACE-POISSON theory of adiabatic change. These would have served as prolegomena to any thermodynamics; they would have provided an unequivocal, indisputable basis for all future discussion.

Part II. The affirmation of CARNOT's General Axiom and its consequence, the General CARNOT-CLAPEYRON Theorem, followed by rejection of the Caloric Theory and proposal of the principle of universal and uniform Interconvertibility of Heat and Work in cyclic processes:

$$L(\mathscr{C}) = JC(\mathscr{C}) \ . \qquad (7A.1)_{1r}$$

This part, very short and very telling, would have delivered first the constitutive restriction

$$\frac{\partial p}{\partial \theta} = J\left(\frac{\partial \Lambda_V}{\partial \theta} - \frac{\partial \mathrm{K}_V}{\partial V}\right) \qquad (8B.1)_r$$

and the existence of an internal energy such as to satisfy the relations

$$\dot{\mathrm{E}} = JQ + P, \qquad P \equiv -p\dot{V} \ , \qquad (8B.2)_r$$

$$J\Lambda_V - p = \frac{\partial \mathrm{E}}{\partial V} \ , \qquad J\mathrm{K}_V = \frac{\partial \mathrm{E}}{\partial \theta} \ . \qquad (8D.3)_r$$

Then by invoking HOLTZMANN's Assertion as an axiom it would have evaluated μ once and for all.

A clean, concise, and logical presentation of this kind, including CLAUSIUS' physical motivation just as he did present it, could have been given in two notes in the *Comptes Rendus*.

8E. CLAUSIUS' Comparisons with Experimental Data

About one third of CLAUSIUS' paper is devoted to details regarding experimental data, mainly on saturated steam. [While this part is tedious today, we easily see why CLAUSIUS must compose it. He had made the HELMHOLTZ-JOULE Determination, namely

$$\mu = J/\theta \ , \qquad (5N.7)_r$$

an essential part of his theory, yet KELVIN had already claimed that REGNAULT's data refuted it (§7H).]

First, as we have seen, CLAUSIUS takes CLAPEYRON's and KELVIN's calculations of μ at four temperatures, calculates the two sequences of ratios of μ^{-1} to the respective value for the lowest of the four temperatures, compares them with the corresponding ratios of the values θ, and concludes that this third sequence of three numbers

> diverges from the two others only as far as can be accounted for by the uncertainty of the data which underlie them....
>
> Such an agreement between results which are obtained from entirely different principles cannot be accidental; it rather serves as

a powerful confirmation of the two principles and the first subsidiary hypothesis annexed to them.

[Just why these ratios are the right thing to compare, CLAUSIUS does not tell us, nor does he seem to expect any reader to be convinced of this "powerful confirmation", for] he goes on to criticize KELVIN's calculation, the weak point of which KELVIN himself had pointed out (§7H, above): the density of saturated steam as a function of temperature. CLAUSIUS, convinced that the HELMHOLTZ-JOULE Determination is correct, uses it conversely to calculate empirical formulae for the behavior of saturated steam which would suffice to make the results of KELVIN's calculation conform with it. [It is a bold step.]

CLAUSIUS after pages of numerical work concludes that "*the work equivalent of the unit of heat is the lifting of something over 400 kilograms to the height of* 1^m." He compares this with the values 460, 438, and 425 that JOULE has found for J, the last being a rough mean indeed. He concludes:

> The agreement of these three numbers, in spite of the difficulty of the experiments, leaves really no further doubt of the correctness of the fundamental principle of the equivalence of heat and work, and their agreement with the number 421 confirms in a similar way the correctness of Carnot's principle, in the form which it takes when combined with the first principle.

[This is irony indeed. KELVIN has in effect rejected JOULE's conclusion as being based on insufficiently accurate experiments. CLAUSIUS adjusts the density of saturated steam by pure fudging so as to make it conform with the HELMHOLTZ-JOULE Determination and with otherwise the same data and so reaches a value of J not far from those JOULE has inferred from his measurements of three phenomena, two of which were altogether different from those classical thermodynamics considers.]

8F. Critique: CLAUSIUS' Bequest

There is no doubt that CLAUSIUS with this paper created classical thermodynamics. Compared with his work here, all preceding except CARNOT's is of small moment. CLAUSIUS exhibits here the quality of a great discoverer: to retain from his predecessors major and minor—in this case, from LAPLACE, POISSON, CARNOT, MAYER, HOLTZMANN, HELMHOLTZ, and KELVIN—what is sound while frankly discarding the rest, *to unite previously disparate theories*, and by *one simple if drastic change* to construct a complete theory that is new yet firmly based upon previous partial successes. That change, of course, is to replace the basic assumption $C^+(\mathscr{C}) = C^-(\mathscr{C})$ by

$$L(\mathscr{C}) = J[C^+(\mathscr{C}) - C^-(\mathscr{C})] \,. \qquad (7A.1)_{2r}$$

By no means disregarding the results of experiment, CLAUSIUS was the first theorist of thermodynamics who was not enslaved to them. Accepting every datum he felt justly established, he showed that if those which to him seemed dubious were to be rejected, experiments not yet performed would not *necessarily* contradict a theory so constructed as to reconcile what seemed fatal differences and remove what seemed fatal objections to all previous proposals.

Certainly KELVIN's position, holding doggedly to every datum of experiment until further experiment should correct it, was proper for the ordinary physicist, but KELVIN was not one of those. KELVIN knew where the weak link in his numerical calculation lay, the weak link that could (and in fact did) break the whole chain. Not only did he know it, he frankly pointed it out to all the world. Finally CLAUSIUS showed that a different hypothesis about saturated steam (for KELVIN's formula for its density was scarcely more than that) would reconcile with the new theory such experimental data as seemed reliable. It was superior insight and superior imagination, taking full advantage of what KELVIN had done to clear the way.

CLAUSIUS, like RANKINE, whose work we shall analyse presently, had another handle on the theory of heat. That was his kinetic theory of gases, to which, with still less basis in actual experiment, he adhered as an article of faith. Unlike RANKINE, CLAUSIUS kept his faith private until the time should be ripe for an evangelist. Both RANKINE's model and CLAUSIUS' model, contradictory with each other as they are in detail, led inevitably to a theory in principle "dynamical", as KELVIN was soon to call it. I think it was this purely ideal faith that gave both RANKINE and CLAUSIUS the confidence to go ahead, while KELVIN, not yet an atomist, wavered. As the event was to show, in molecular theory CLAUSIUS was not only the wiser man but also the better physicist.

We cannot say that CLAUSIUS completed the foundation of thermodynamics, even for substances compatible with the Doctrine of Latent and Specific Heats. CLAUSIUS made *essential use of the concept of ideal gas*, which carries with it the conceptual problem of what temperature really is. We know that he was already working on his kinetic theory, a molecular model which makes it easy to conceive and define an ideal-gas temperature. In §11H, below, we shall take up KELVIN's definition of a phenomenological absolute temperature.

Had CLAUSIUS been able to organize and clarify his work in some such way as our analysis in the preceding section has shown possible, he could have pulled thermodynamics out of the slough of obscurity in which CARNOT left it; he could have made of it a beacon of enlightenment. He did not do so.

CLAUSIUS confused two centuries of readers by writing (dC/dV) for the latent heat with respect to volume and $(dC/d\theta)$ for the specific heat at constant volume. In annotating the paper for republication in 1864 he ascribed the parentheses to EULER but remarked that the precaution of using them was really unnecessary, since even without them "usually no misunderstanding would be possible". This reassurance notwithstanding, he felt compelled to preface the reprint with a "mathematical introduction" of fifteen pages, all about inexact differentials, most of which, he states, he had published in a journal in 1858. In the footnote already mentioned CLAUSIUS wrote (letters conformed with ours)

$$dC = \left(\frac{dC}{d\theta}\right)d\theta + \left(\frac{dC}{dV}\right)dV \; ; \tag{8F.1}$$

this expression, which did not appear in the original paper, is neither more nor less than the basic statement of the Doctrine of Latent and Specific Heats:

$$Q = \Lambda_V(V, \theta)\dot{V} + \mathrm{K}_V(V, \theta)\dot{\theta} \; . \qquad (2C.4)_r$$

CLAUSIUS explains that $(dC/d\theta)$ and (dC/dV) are "to be regarded as completely determined functions of θ and V" and goes on to point out that "C itself" is a quantity of this kind if and only if

$$\frac{d}{dV}\left(\frac{dC}{d\theta}\right) = \frac{d}{d\theta}\left(\frac{dC}{dV}\right) . \tag{8F.2}$$

By "C itself" CLAUSIUS must mean the integral of the differential form (1) along some path, but he does not say so.

When CLAUSIUS integrates around an infinitesimal Carnot cycle, he does so by adding up terms involving four different differentials: dV, $\delta' V$, δV, and $d'V$. Here enters the horrid idea that thermodynamics involves a special calculus with all sorts of ds and δs and that something different can come out of a line integral if the line is approximated by a sequence of infinitesimal adiabats and isotherms, in defiance of the fundamental theorem of integral calculus. While CLAPEYRON, the engineer trained at the Ecole Polytechnique, had used his routine calculus efficiently, CLAUSIUS tries to adapt CLAPEYRON's steps to a case when he is not differentiating but integrating. Of course nobody ought to blame CLAUSIUS for not knowing the transformation of a line integral into a surface integral which AMPÈRE had published twenty-four years earlier in a research on electricity. Nevertheless, the tragicomic muse must laugh when modern specialists in thermodynamics state that CLAUSIUS had "very strict standards of mathematical rigor". Few mathematical physicists have shown so little sense of the right mathematics for the job.

CLAUSIUS' results are of supreme importance for thermodynamics. His way to them was like crawling through a thorn hedge. Mumbo-jumbo with mysterious differentials[1] became and is still the hallmark of thermodynamics.

---

[1] CARNOT himself had used two: d and δ.

It should be unnecessary to remark—though, alas, probably it is not—that those

The tradition[2], handed down in thousands of books, piously follows the traces left by CLAUSIUS' bloodstained palms and pasterns. For example, his miserable proof of the differential form of the relation

$$C(\mathscr{C}) = \iint_{\mathscr{A}} \left(\frac{\partial \Lambda_V}{\partial \theta} - \frac{\partial \mathrm{K}_V}{\partial V}\right) dV d\theta \qquad (8D.1)_r$$

has been reproduced again and again. Obscure logic, the painfully awkward calculus that dogs standard thermodynamics to this day, and vague expression have joined CLAUSIUS' splendid achievements to form his legacy. The tragicomic muse of thermodynamics chose to tell her votaries the truth in riddles they could only half decipher.

8G. RANKINE's First Paper

Before he could have seen the paper of CLAUSIUS we have just analysed, RANKINE had read to the Royal Society of Edinburgh his own first memoir on the subject[1]. In the introduction he adopts without reserve the *vis viva* theory of heat and bases his analysis on the "hypothesis of molecular vortices", namely,

That the elasticity due to heat arises from the centrifugal force of

---

atrocious differentials have no special appropriateness to the science of thermodynamics. (Of course, were this a social rather than a conceptual history, the differentials would be of the essence.)

Infinitesimals are always unnecessary but need be neither vague nor unrigorous so long as they are true differentials or well defined differential forms. For example, in the familiar relation $dy = f'(x)dx$ the symbol dx stands for an arbitrary increment of the independent variable x, and dy is the corresponding increment in the best linear approximation to the differentiable function f at x. The differential form $Adx + Bdy$ also makes sense in terms of integration along a curve if A and B are specified as functions of x and y. When, however, an author writes $dy = Adx$ and tells us no more than that about A, obscurity is certain to follow and error is likely in inexpert hands. Unfortunately this latter type of infinitesimal abounds in papers on thermodynamics and in the sanctimonious chants of today's experts in it.

Here I whisper sotto voce a tragicomic aside. In the year of CLAUSIUS' paper, 1850, KELVIN discovered the generalization of AMPÈRE's transformation which now is commonly called "STOKES' theorem" because KELVIN wrote it in a letter to STOKES and STOKES a few years later set it as an examination question. In his own work on thermodynamics, even that which was published in the very same year he discovered "STOKES' theorem", KELVIN never used this transformation. In §9B, below, we shall see how he obtained (8D.2) by dealing directly with line integrals. His proof is clean and efficient, but it lacks the clarity that (8D.1) provides.

[2] On the other hand, HELMHOLTZ [1855, pp. 568ff], in one of the first surveys of thermodynamics, stated politely that he hoped "to make it easier for the reader to follow" by deriving the main results in another way, which is very close to that of KELVIN (below, §9B).

---

[1] RANKINE [1850].

revolutions or oscillations among the particles of the atomic atmospheres; so that quantity of heat is the vis viva *of those revolutions or oscillations.*

[The terms reflect the unique combination of daring hypothesis with studious respect for tradition which RANKINE, a practising civil engineer, always displayed when he faced fundamental questions in natural philosophy.] The "elasticity" of an aeriform fluid is the eighteenth-century term for what we now call pressure. [To regard that pressure as the manifestation of both molecular spin and molecular vibration was to combine EULER's model[2] of a gas with the opinions current among chemists and physicists of RANKINE's own day.] In addition, RANKINE is ready to refashion his views in the light of recent experimental discoveries. [Here he exemplifies the ideal rather than the real or common engineer.] He is already altogether converted by JOULE's "valuable experiments to establish the convertibility of heat and mechanical power". However, it is to "the appearance of the experiments of M. REGNAULT on gases and vapours" that he attributes his having resumed studies he had "laid aside for nearly seven years, from the want of experimental data...."

The molecular vortices need not concern us here[3]. From his considerations regarding them RANKINE arrives at the basic relation[4] expressing the

---

[2] *Cf.* §1 of my paper cited in Footnote 1 to §2A. While he does not mention EULER, RANKINE [1851, *1*, ¶13] adopts EULER's critical assumption that the *linear* speed of rotation is the same for every atom: "the uniform velocity of motion of its parts". It is hard to see what physical idea could suggest this assumption.

[3] In a paper summarized in the introduction and read to the Royal Society of Edinburgh on the same day as the one we are analysing, RANKINE [1851, *1*] presented his investigation (¶1) "in detail in its original form," except for some "intermediate steps... modified in order to meet the objections of Professor WILLIAM THOMSON..., to whom the paper was submitted after it had been read...". Insofar as this paper concerns thermodynamics, it does not go beyond the one we are analysing.

[4] RANKINE [1850, Equations (3), (6), and (16)]. RANKINE, since he uses a mechanical model, is able to express all caloric quantities such as K_V and Λ_V in mechanical units. For easy comparison of his results with others' I have inserted the factor J where it must be if caloric units are used. For comparison with RANKINE's notations we must always put $J = 1$. On this understanding we compile the following comparison of notations, remarking that while our M stands for the mass of a fluid body, RANKINE's M denotes the mass of one atom.

Our	RANKINE's
θ	τ
θ_0	κ
R/M	$(CnM)^{-1}$
$\mathfrak{k}(\theta - \theta_0)$	Q
$\mathfrak{k}\dot{\theta}$	$\delta Q/dt$
$Q - \mathfrak{k}\dot{\theta}$	$-\delta Q'/dt$
p	P

Note that RANKINE's "absolute temperature" τ agrees exactly with the ideal-gas tem-

Doctrine of Latent and Specific Heats:

$$Q = \Lambda_V(V, \theta)\dot{V} + \mathrm{K}_V(V, \theta)\dot{\theta} \ . \qquad (2C.4)_r$$

The coefficients Λ_V and K_V are given as follows in terms of a constitutive function U of V and θ which RANKINE defines from his molecular model (his Equation (5)):

$$\begin{aligned} J\Lambda_V &= R(\theta - \theta_0)\left(\frac{1}{V} - \frac{\partial U}{\partial V}\right) , \\ J\mathrm{K}_V &= R\left[\frac{1}{N} - (\theta - \theta_0)\frac{\partial U}{\partial \theta}\right] = M\mathfrak{k}\left[1 - N(\theta - \theta_0)\frac{\partial U}{\partial \theta}\right] . \end{aligned} \qquad (8G.1)$$

Here θ_0 is a positive universal constant which RANKINE (just after his Equation (IX)) calls "the absolute zero of heat"[5]; $\mathfrak{k}$ is "the real specific heat" (his Equation (14)), a positive quantity depending on the nature of the substance, and independent of the temperature; also R and N are positive constitutive constants.

RANKINE lays down the principle of universal and uniform Interconvertibility of Heat and Work in cycles as a phenomenological axiom (his Equation (8)): In any cycle, "if, on the whole, any mechanical power has appeared, and been given out from the body, in the form of expansion, an equal amount must have been communicated to the body, and must have disappeared in the form of heat," and conversely. [This statement suggests that RANKINE's views may be as vague as those of MAYER, HELMHOLTZ, and JOULE, but the formal developments[6] he bases upon them are precise; as we shall see, they take proper account of storage of energy.]

---

perature θ, and for a body of ideal gas the R here turns out to be that which appears in the thermal equation of state $pV = R\theta$. RANKINE's "apparent" specific heats are reckoned per unit mass; in our notation they are K_V/M and K_p/M. Some more of RANKINE's notations will be needed by the reader who tries to follow his analysis. His b, presumably constant, which first appears in his Equation (I), is "the coefficient of atmospheric elasticity" of a particular substance; his μ is "the mass of the atmosphere of one atom"; and he introduces the following abbreviations:

$$\kappa \equiv Cn\mu b \ , \qquad N \equiv \frac{2\mu}{3kM} \ , \qquad \mathfrak{k} \equiv \frac{1}{CnMN} \ .$$

[5] RANKINE [1853, *3*, ¶53]: "the temperature corresponding to absolute privation of heat".

[6] The following passage from a later paper of RANKINE [1852, ¶9] makes clearer the ideas about heat he employs here:

> Let $\delta.Q$ represent, when positive, the indefinitely small quantity of heat which must be communicated to unity of weight of a substance, and when negative, that which must be abstracted from it, in order to produce the indefinitely small variation of temperature $\delta\tau$ simultaneously with the indefinitely small variation of volume δV. Let $\delta.Q$ be divided into two parts
>
> $$\delta Q + \delta Q' = \delta.Q \ ,$$
>
> of which δQ, being directly employed in varying the *velocity* of the particles, is the variation of the *actual* or *sensible* heat possessed by the body; while $\delta Q'$,

To calculate the heat supplied, RANKINE considers an infinitesimal cycle of alternating isochoric and isothermal changes. [Thus, unlike CLAUSIUS, RANKINE does not simply transpose CLAPEYRON's presentation of CARNOT's theory,] though he notes that (¶8) "the process followed ... is analogous to that employed by M. CARNÔT in his theory of the motive power of heat, although founded on contrary principles, and leading to different results." He goes on to criticize CARNOT's basic idea that "a body, having received a certain quantity of heat, is capable of giving out not only all the heat it has received, but also a quantity of mechanical power which did not before exist." His own view that heat is used up in giving out motion, he regards as justified by the *vis viva* theory of heat.

[Of course, as we know from AMPÈRE's transformation, the nature and size of the cycle makes no difference at all, and by correct mathematics nothing different from

$$\frac{\partial p}{\partial \theta} = J\left(\frac{\partial \Lambda_V}{\partial \theta} - \frac{\partial K_V}{\partial V}\right) \qquad (8B.1)_r$$

can come out.] By adding up strings of differentials spreading over a page RANKINE obtains exactly that result, in the special case when Λ_V and K_V have the forms (1). "The body being now restored in all respects to its primitive state, the sum of the two portions of power connected with change of volume, must, in virtue of the principle of *vis viva*, be equal to the sum of the four quantities of heat with their signs reversed." Hence [by use of (1) we show that] (RANKINE's Equation (9))

$$\frac{\partial p}{\partial \theta} = R\left(\frac{1}{V} - \frac{\partial U}{\partial V}\right) , \qquad (8G.2)$$

and consequently

$$U = \phi(\theta) + \int \left(\frac{1}{V} - \frac{1}{R}\frac{\partial p}{\partial \theta}\right) dV . \qquad (8G.3)$$

[Here we first encounter a favorite of RANKINE, the partial integral. The presence of the arbitrary function ϕ reminds us that $\int (\partial p/\partial \theta) dV$ is understood to mean an indefinite integral carried out when θ is fixed[7].] According

---

employed in varying their *orbits*, represents the amount of the mutual transformation of heat with expansive power and molecular action, or the variation of what is called the *latent* heat; that is to say, of a molecular condition constituting a source of power, out of which heat may be developed. ($\delta Q'$ in this paper corresponds to $-\delta Q'$ in my former papers.)

The table of notations in Footnote 4, above, explains RANKINE's meanings for his δQ and $\delta Q'$ in terms of our Q and conforms with the equation just given.

[7] In his later presentation of these statements RANKINE [1859, §246] takes ∞ as the lower limit of integration, "so as to correspond to the state of infinite rarefaction". CLAUSIUS [1862, just after Equation (11)] objected: "Why he chooses as initial volume

to RANKINE's molecular theory (¶7), "[t]he function U is one depending on molecular forces. The only case in which it can be calculated is that of a perfect gas.... Without giving the details of the integration" RANKINE states that for such a gas

$$U = \theta_0/\theta \ . \tag{8G.4}$$

"Now $\phi(\theta)$ being the same for all densities, is the value of U for the perfectly gaseous state, or θ_0/θ; for in that state, the integral = 0." RANKINE seems to regard this statement as determining ϕ once and for all:

$$\phi = \theta_0/\theta \ . \tag{8G.5}$$

Then (3) becomes

$$U = \frac{\theta_0}{\theta} + \int \left(\frac{1}{V} - \frac{1}{R}\frac{\partial p}{\partial \theta}\right) dV \ , \tag{8G.6}$$

so (RANKINE's Equations (11))

$$\begin{aligned} \frac{\partial U}{\partial V} &= \frac{1}{V} - \frac{1}{R}\frac{\partial p}{\partial \theta} \ , \\ \frac{\partial U}{\partial \theta} &= -\frac{\theta_0}{\theta^2} - \frac{1}{R}\int \frac{\partial^2 p}{\partial \theta^2} dV \ , \end{aligned} \tag{8G.7}$$

[and so (1) reduces to

$$\begin{aligned} J\Lambda_V &= (\theta - \theta_0)\frac{\partial p}{\partial \theta} \ , \\ JK_V &= R\left[\frac{1}{N} + (\theta - \theta_0)\left(\frac{\theta_0}{\theta^2} + \frac{1}{R}\int \frac{\partial^2 p}{\partial \theta^2} dV\right)\right] \ , \end{aligned} \tag{8G.8}$$

formulae that RANKINE will use in his next paper with no comment.]

Earlier in this paper RANKINE had written down a thermal equation of state of an imperfect gas (his Equation (VI)):

$$p = \frac{R\theta}{V}(1 - \Phi) + f \ , \tag{8G.9}$$

in which Φ is a function of θ_0/θ and V, while f is a function of V alone. RANKINE substitutes (9) into (7) and by integration expresses U in terms of

---

just the infinitely large volume, he does not say, although this choice is obviously not indifferent." Apparently none of the nineteenth-century thermodynamicists realized that the indefinite integral of a function of two variables with respect to one of them makes no sense unless the domain of the function is convex with respect to the variable integrated. For example, for a Van der Waals fluid RANKINE's integral with respect to V from ∞ does not exist at subcritical temperatures and volumes.

Φ. For an ideal gas Φ and f both vanish, (8) reduces to (RANKINE's Equation $(18)_3$)

$$J\Lambda_V = p\left(1 - \frac{\theta_0}{\theta}\right), \qquad JK_V = R\left[\frac{1}{N} + \left(1 - \frac{\theta_0}{\theta}\right)\frac{\theta_0}{\theta}\right], \tag{8G.10}$$

and (RANKINE's Equation $(19)_2$)

$$\gamma = 1 + N\frac{1 - \dfrac{\theta_0}{\theta}}{1 + N\dfrac{\theta_0}{\theta}\left(1 - \dfrac{\theta_0}{\theta}\right)}. \tag{8G.11}$$

"The value of θ_0 is unknown; and, as yet, no experimental data exist from which it can be determined. I have found, however, that practically, results of sufficient accuracy are obtained by regarding θ_0 as so small in comparison with θ, that θ_0/θ ... may be neglected in calculation." [Then from (10) and (11) follow both MAYER's Assertion

$$J(K_p - K_V) = R \tag{7B.2$_r$}$$

and HOLTZMANN's Assertion

$$J\Lambda_V = p\,; \tag{7D.1$_r$}$$

both specific heats are constant, and $N = \gamma - 1$.

The remainder of RANKINE's first paper concerns experimental data and calculation of the efficiency of a steam engine, taking account of the latent heat of evaporation.

8H. Critique of RANKINE's First Paper

RANKINE's results, like LAPLACE's in the Caloric Theory, are concealed by many preceding pages of elaborate, imperfectly presented, and totally speculative molecular theory.

A complete molecular theory should deliver the Interconvertibility of Heat and Work as a proved theorem. Perhaps RANKINE's theory could have done so, but he himself, like most early proponents of molecular models, mixes in phenomenological arguments when he needs them, and interconvertibility is a case in point. He simply assumes it outright, under the name "the principle of conservation of *vis viva*", so as to obtain the relation

$$\frac{\partial p}{\partial \theta} = R\left(\frac{1}{V} - \frac{\partial U}{\partial V}\right). \tag{8G.2$_r$}$$

We shall now analyse RANKINE's theory at three levels of generality. We consider the effect of his constant θ_0, the phenomenological structure of his theory, and the restrictions that are implied by the details of his molecular model.

α. *The constant* θ_0. How important is the constant θ_0? It is hard to say. RANKINE certainly took that universal minimum temperature seriously, and his molecular theory does not allow it to vanish. As he could not determine it, in all applications he supposed θ to be much larger than it. In effect, for those circumstances $\theta_0/\theta \approx 0$.

In our following analysis we shall leave θ_0 arbitrary and in most instances leave the reader to see for himself the effect of taking it as 0.

β. *Phenomenology*. Without prejudice of the function U, we reconsider RANKINE's determinations of Λ_V and K_V from it:

$$J\Lambda_V = R(\theta - \theta_0)\left(\frac{1}{V} - \frac{\partial U}{\partial V}\right), \qquad J\mathrm{K}_V = R\left[\frac{1}{N} - (\theta - \theta_0)\frac{\partial U}{\partial \theta}\right]. \tag{8G.1$)_{1,2\mathrm{r}}$}$$

As RANKINE himself showed in his rough way, they make the condition

$$\frac{\partial p}{\partial \theta} = R\left(\frac{1}{V} - \frac{\partial U}{\partial V}\right) \tag{8G.2$)_{\mathrm{r}}$}$$

necessary and sufficient that $L(\mathscr{C}) = JC(\mathscr{C})$ in every cycle $\mathscr{C}$. Adopting this condition, we see at once that

$$J\Lambda_V = (\theta - \theta_0)\frac{\partial p}{\partial \theta}\ ; \tag{8G.8$)_{1\mathrm{r}}$}$$

this is the form assumed by the General CARNOT-CLAPEYRON Theorem in RANKINE's theory. It yields $\mu = J/(\theta - \theta_0)$; if $\theta_0 = 0$, this last result reduces to the HELMHOLTZ-JOULE Determination. Moreover, (8G.1) and (8G.2) together imply that

$$\frac{\partial}{\partial \theta}(J\Lambda_V - p) = \frac{\partial}{\partial V}(J\mathrm{K}_V)\ , \tag{8D.2$)_{\mathrm{r}}$}$$

whence follow the relations

$$J\Lambda_V - p = \frac{\partial \mathrm{E}}{\partial V}\ , \qquad J\mathrm{K}_V = \frac{\partial \mathrm{E}}{\partial \theta}\ . \tag{8D.3$)_{\mathrm{r}}$}$$

and

$$\dot{\mathrm{E}} = JQ + P, \qquad P \equiv -p\dot{V}\ . \tag{8B.2$)_{\mathrm{r}}$}$$

If we regard U and p as *related through* (8G.2) *but otherwise arbitrary functions of V and θ, RANKINE's phenomenological structure extends CLAUSIUS' to fluids obeying an arbitrary equation of state*, to within choice of the constant θ_0. Furthermore, as we have seen, for ideal gases *RANKINE's theory proves as a theorem HOLTZMANN's Assertion*

$$J\Lambda_V = p\ , \tag{7D.1$)_{\mathrm{r}}$}$$

again on the supposition that $\theta_0 = 0$. RANKINE's argument to derive (8G.1) and (8G.2) had recourse to the principle of Interconvertibility of Heat and Work. The argument itself was phenomenological, making no use of his definition of his function U in terms of his molecular model. The argument we have just given proves the converse: The universal and uniform Interconvertibility of Heat and Work as expressed by CLAUSIUS' "First Law" (8B.2) *is a consequence* of RANKINE's formal structure.

That is not all. We can interpret RANKINE's U in general. By use of (8G.1) we calculate its rate of change $\dot{U}$ in an arbitrary process and obtain

$$\begin{aligned} \dot{U} &= \frac{\partial U}{\partial V}\dot{V} + \frac{\partial U}{\partial \theta}\dot{\theta}\,, \\ &= \left(-\frac{J\Lambda_V}{R(\theta-\theta_0)} + \frac{1}{V}\right)\dot{V} + \left(-\frac{JK_V}{R(\theta-\theta_0)} + \frac{1}{N(\theta-\theta_0)}\right)\dot{\theta}\,, \quad (8H.1) \\ &= -\frac{JQ}{R(\theta-\theta_0)} + \{\log[V(\theta-\theta_0)^{1/N}]\}^{\cdot}. \end{aligned}$$

That is, if H is defined as follows:

$$J\mathrm{H}/R \equiv \log[V(\theta-\theta_0)^{1/N}] - U + \text{const.} \qquad (8H.2)$$

then H is a function of V and θ, and

$$(\theta-\theta_0)\dot{\mathrm{H}} = Q\ . \qquad (8H.3)$$

To within questions concerning the troublesome constant θ_0, today a function H of this kind is called an *entropy* of the fluid body, and the function from which U is subtracted on the right-hand side of (2) is the entropy of an ideal gas whose ratio of specific heats is the constant $N + 1$. *RANKINE's function RU/J is the excess of the entropy of some ideal gas over the entropy of the fluid body in question.* For general considerations RANKINE's U will serve the same purposes as the entropy, with simple adjustments. Thus we may say that *RANKINE's phenomenological apparatus implies the basic constitutive restrictions of classical thermodynamics*, to within choice of θ_0. If $\theta_0 = 0$, the two sets of constitutive restrictions agree precisely. As he who has read *Concepts and Logic* will know, they imply the whole formal structure of the classical theory.

RANKINE's passage from (8G.2) to (8G.3) makes the function ϕ a constitutive function, restricted only by $(8G.1)_2$. On the contrary, RANKINE's sentence quoted above, just after (8G.4), seems to mean that he regards ϕ as a universal function, the same for all substances, so to determine it for "the perfectly gaseous state" determines it once and for all:

$$\phi = \theta_0/\theta\ . \qquad (8G.5)_r$$

That is the only way, it seems, to make $(8G.7)_2$ and $(8G.8)_2$ follow from what went before.

Abandoning this unjustified step of RANKINE's, let us now consider ϕ

in (8G.3) as a constitutive function, restricted only by RANKINE's general apparatus, which is summarized by his determinations of Λ_V and K_V, namely

$$J\Lambda_V \equiv R(\theta - \theta_0)\left(\frac{1}{V} - \frac{\partial U}{\partial V}\right), \qquad JK_V \equiv R\left[\frac{1}{N} - (\theta - \theta_0)\frac{\partial U}{\partial \theta}\right], \tag{8G.1$_{1,2r}$}$$

and by his partial determination of U without restriction to ideal gases:

$$U = \phi(\theta) + \int \left(\frac{1}{V} - \frac{1}{R}\frac{\partial p}{\partial \theta}\right) dV . \tag{8G.3$_r$}$$

Doing so, we evaluate Λ_V and K_V in terms of the thermal equation of state and the single constitutive function ϕ, a function of θ alone. The result is

$$\begin{aligned} J\Lambda_V &= (\theta - \theta_0)\frac{\partial p}{\partial \theta}, \\ \frac{JK_V}{R/N} &= 1 - N(\theta - \theta_0)\phi'(\theta) + \frac{\theta - \theta_0}{R/N}\int \frac{\partial^2 p}{\partial \theta^2}\,dV . \end{aligned} \tag{8H.4}$$

The former statement is RANKINE's $(8G.8)_1$; the latter frees RANKINE's relation

$$JK_V = R\left[\frac{1}{N} + (\theta - \theta_0)\left(\frac{\theta_0}{\theta^2} + \frac{1}{R}\int \frac{\partial^2 p}{\partial \theta^2}\,dV\right)\right] \tag{8G.8$_{2r}$}$$

from the special choice of ϕ RANKINE obtains from his molecular theory $\phi = \theta_0/\theta$, a choice he seems to regard as universal. If $\theta_0 = 0$, $(4)_1$ reduces to

$$J\Lambda_V = \theta\frac{\partial p}{\partial \theta}, \tag{7I.2$_r$}$$

destined to become celebrated as the final form of the CARNOT-CLAPEYRON Theorem. As we have seen in §7I, it is a consequence of HOLTZMANN's Assertion and CARNOT's General Axiom.

If we look only upon the positive aspects of the papers of CLAUSIUS and RANKINE published in 1850, we may summarize their achievements as follows:

1. *CLAUSIUS constructed the thermodynamics of ideal gases; for those gases he discovered the internal energy. He took HOLTZMANN's Assertion as one of his assumptions regarding ideal gases.*
2. *RANKINE obtained the basic constitutive restrictions of the thermodynamics of fluids; he expressed them in terms of a function that differs only inessentially from the entropy. He proved HOLTZMANN's Assertion as a theorem about ideal gases.*

CLAUSIUS was to come upon the entropy in his own way many years later, and years after that he was to coin the name.

The above summary, strictly correct as it is, does not represent RANKINE's work fairly. He himself seems not to have seen which parts of it were free of the special restrictions his molecular model implied, and only after some years was he to draw the conclusion $(4)_2$ or derive something really close to (3).

γ. Restrictions imposed by the "hypothesis of molecular vortices". RANKINE used his molecular model to determine U for ideal gases:

$$U = \theta_0/\theta \ , \qquad (8G.4)_r$$

With this choice of U we obtain from $(8G.1)_2$ the following expression for K_V:

$$J K_V = R\left[\frac{1}{N} + \frac{\theta_0(\theta - \theta_0)}{\theta^2}\right] . \qquad (8H.5)$$

With RANKINE's usual approximation $\theta_0/\theta \approx 0$ we conclude that $K_V =$ const. In this sense *RANKINE's model makes the specific heats of ideal gases constant.* In itself this limitation, which MAXWELL's later kinetic theory shares, is not objectionable. RANKINE, however, carries its effects over to all fluids through his general formula

$$J K_V = R\left[\frac{1}{N} + (\theta - \theta_0)\left(\frac{\theta_0}{\theta^2} + \frac{1}{R}\int \frac{\partial^2 p}{\partial \theta^2}\, dV\right)\right] . \qquad (8G.8)_{2r}$$

If in this formula we substitute $pV = R\theta$, we again obtain (5): *RANKINE's general theory does not permit an ideal gas to have specific heats that are not constant.* While a molecular model with constant specific heats may be adequate for some fluids, so general a restriction is dubious. Certainly it does not follow from CLAUSIUS' theory, which allows K_V for an ideal gas to be an arbitrary positive function of θ. CLAUSIUS[1] was to claim later that RANKINE's $(8G.8)_2$ was valid only for ideal gases. Perhaps he based this statement on nothing but a firm belief that his own results constituted the touchstone against which others' ought be tested. While I am not sure what CLAUSIUS meant, one thing is certain: The expression $(4)_2$, which follows from RANKINE's formal structure with no use of his molecular model, is general in the sense of CLAUSIUS' theory, and it reduces to RANKINE's result $(8G.8)_2$ *if and only if*

$$\phi = \frac{\theta_0}{\theta} \ . \qquad (8G.5)_r$$

Time has judged in favor of CLAUSIUS' theory and against RANKINE's restriction (8G.5). I am not sure that the objection to (8G.5) bears upon RANKINE's molecular vortices, for as far as I can see his passage from (8G.3) and (8G.4) to (8G.5) is just an error in his mathematical development

---

[1] CLAUSIUS [1862, §6].

of his phenomenological equations. I must leave to others the task of deciding whether or not RANKINE's molecular model forces his function ϕ to be universal rather than constitutive.

We must not believe that all consequences RANKINE is to draw from his special choice of ϕ will depend upon that choice. From $(4)_1$ we see that it has no effect on Λ_V and hence does not play any part in determining the heat added in an isothermal process. In §9A we shall see that RANKINE will use his formula $(8G.8)_2$ in another context where choice of ϕ makes no difference in the outcome. The results he will obtain will be just the same as if he had used his general structure alone, with no reference to his molecular vortices.

These facts must have been far from easy to see in 1850. In his criticism, already cited, CLAUSIUS is to dismiss all of RANKINE's work outright except insofar as it was specialized to ideal gases. Certainly CLAUSIUS was not one to seek out the best aspects of a competitor's work; his judgment here is superficial at best; but after all it is an author's duty to make his work clear to the reader, not the reader's duty to straighten out an author's mess. Perhaps RANKINE himself never subjected his ideas to criticism of the logical kind; certainly there is nothing in his printed works to suggest that he did. The early thermodynamicists were fertile in physical intuition and grandiose claims; in return, as regards the standards of mathematical hygiene established by HUYGENS, NEWTON, and the BERNOULLI-EULER school and maintained by CAUCHY and KIRCHHOFF at the very time when thermodynamics was being created, they seem to have been like the habitually unbathed, who notoriously cannot smell their own effluvia.

All of RANKINE's constants are of molecular origin, and RANKINE frequently refers to the model. There is nothing like a molecular theory to complicate simple phenomenological statements. To diminish the recognition history will afford to phenomenological discoveries, there is nothing like entwining them in a molecular theory that is later to be rejected[2]. CLAUSIUS, too, had a molecular theory, but wisely he withheld it until his phenomenal theory, which was independent of it, could become somewhat familiar and widely accepted. As he well knew, to rest gross concepts upon a molecular hypothesis risks loss of a truth because it turns out to have been derived from something wrong. The best function of a molecular theory is to refine the details of an accepted gross picture.

Had RANKINE separated his phenomenology from his vortices, his work would have been short, clear, and final. His basic phenomenological formulae (8G.1) and (8G.2) suffice to deliver the entire formal structure of classical thermodynamics, freed of CLAUSIUS' restriction to ideal gases and "subsidiary hypotheses" regarding them. Had RANKINE been content to analyse

[2] CLAUSIUS [1863, §5] justly defends his own presentation as superior to RANKINE's: "I laid very particular weight upon basing ... my development ... not upon special aspects of the molecular nature of matter but only upon general fundamental principles ...."

and develop those formulae, he would have earned the rank of first, best, and entire discoverer of classical thermodynamics, leaving to CLAUSIUS the honor of codiscovery of the thermodynamics of ideal gases. History was to be otherwise. RANKINE has gained respect but not factual recognition; his discoveries have never until now been disentangled from his illusions. To read one of RANKINE's papers today requires much patience and much training. CLAUSIUS' first paper, despite its poor organization, vague exposition, and insecure mathematics, is deservedly regarded as a classic second only to CARNOT's treatise.

9. Distracting Interlude: Explosion of Print

Qui vid' i' gente più ch'altrove troppa,
e d'una parte e d'altra, con grand' urli,
voltando pesi per forza di poppa.
Percotëansi 'ncontro; e poscia pur lì
si rivolgea ciascun, voltando a retro,
gridando: "Perché tieni?" e "Perché burli?"

DANTE, *Inferno* VII, 25–30.

9A. RANKINE's Second Paper

RANKINE's response[1] to CLAUSIUS' first paper was a [characteristically forthright and generous] "Fifth section" adjoined to his own first paper:

(40.) CARNÔT was the first to assert the law, *that the ratio of the maximum mechanical effect, to the whole heat expended in an expansive machine, is a function solely of the two temperatures at which the heat is respectively received and emitted, and is independent of the nature of the working substance.* But his investigations not being based on the principle of the dynamical convertibility of heat, involve the fallacy that power can be produced out of nothing.

(41.) The merit of combining CARNÔT's *Law*, as it is termed, with that of the convertibility of heat and power, belongs to Mr CLAUSIUS and Professor WILLIAM THOMSON; and in the shape into which they have brought it, it may be stated thus:—

The maximum proportion of heat converted into expansive power by any machine, is a function solely of the temperatures at which heat is received and emitted by the working substance; which function, for each pair of temperatures, is the same for all substances in nature.

This law is laid down by Mr CLAUSIUS, as it originally had been by CARNÔT, as an independent axiom; and I had at first doubts as to the soundness of the reasoning by which he maintained it. Having stated

[1] RANKINE [1851, *3*]. I regard his two preceding publications [1850] and [1851, *1*] as being what RANKINE himself states them to be: complementary parts of one and the same paper.

those doubts to Professor THOMSON, I am indebted to him for having induced me to investigate the subject thoroughly; for although I have not yet seen his paper, nor become acquainted with the method by which he proves CARNÔT's law, I have received from him a statement of some of his more important results.

(42.) I have now come to the conclusions,—First: *That* CARNÔT's *Law is not an independent principle in the theory of heat; but is deducible, as a consequence, from the equations of the mutual conversion of heat and expansive power, as given in the First Section of this paper.*

Secondly: *That the function of the temperatures of reception and emission, which expresses the maximum ratio of the heat converted into power to the total heat received by the working body, is the ratio of the difference of those temperatures, to the absolute temperature of reception diminished by the constant, which I have called* $\kappa = Cn\mu b$, and which must, as I have shewn in the Introduction, be the same for all substances, in order that molecular equilibrium may be possible.

To establish these conclusions, RANKINE considers a Carnot cycle. Again using a partial integral, he defines as follows a function F:

$$F(V, \theta) \equiv \int \frac{\partial p}{\partial \theta} dV \,. \tag{9A.1}$$

[Later[2] he is to call a purely formal generalization of this function the *heat-potential.*] Using F and the relation

$$J\Lambda_V = (\theta - \theta_0) \frac{\partial p}{\partial \theta} \tag*{(8G.8)$_{1r}$}$$

from his earlier paper, RANKINE evaluates as follows the heat absorbed and heat emitted in a Carnot cycle (his Equations (a) and (c), conformed with the notation of Figure 3 in §5C, above):

$$\begin{aligned} JC^+ &= (\theta^+ - \theta_0)[F(V_b, \theta^+) - F(V_a, \theta^+)] \,, \\ JC^- &= (\theta^- - \theta_0)[F(V_c, \theta^-) - F(V_d, \theta^-)] \,. \end{aligned} \tag{9A.2}$$

[Having in effect shown his molecular model to be consistent with the Doctrine of Latent and Specific Heats, namely

$$Q = \Lambda_V(V, \theta)\dot{V} + \mathrm{K}_V(V, \theta)\dot{\theta},] \tag*{(2C.4)$_r$}$$

he substitutes therein the relations

$$\begin{aligned} J\Lambda_V &= (\theta - \theta_0) \frac{\partial p}{\partial \theta} \,, \\ J\mathrm{K}_V &= R\left[\frac{1}{N} + (\theta - \theta_0)\left(\frac{\theta_0}{\theta^2} + \frac{1}{R}\int \frac{\partial^2 p}{\partial \theta^2} dV\right)\right] \,, \end{aligned} \tag*{(8G.8)$_r$}$$

---

[2] RANKINE [1853, *3*, ¶50].

which his preceding paper had implied, and so he obtains a differential equation for F as a function of θ along an adiabat. Writing $dV/d\theta$ for the derivative of the function f such as to make $V = f(\theta)$ be an equation for the adiabat, we may state RANKINE's result (just preceding his Equation (b)) as follows:

$$\begin{aligned} 0 &= R\left[\frac{1}{N} + (\theta - \theta_0)\frac{\theta_0}{\theta^2}\right] + (\theta - \theta_0)\left(\frac{\partial}{\partial\theta} + \frac{dV}{d\theta}\frac{\partial}{\partial V}\right)F\,, \\ &= R\left[\frac{1}{N} + (\theta - \theta_0)\frac{\theta_0}{\theta^2}\right] + (\theta - \theta_0)\frac{dF}{d\theta}\,. \end{aligned} \tag{9A.3}$$

Integration yields the changes in F along the adiabats of a Carnot cycle (RANKINE's Equations ((b) and (d)):

$$\begin{aligned} F(V_c, \theta^-) - F(V_b, \theta^+) &= \psi(\theta^+, \theta^-)\,, \\ F(V_d, \theta^-) - F(V_a, \theta^+) &= \psi(\theta^+, \theta^-)\,, \end{aligned} \tag{9A.4}$$

[the function ψ being easy to evaluate explicitly but not needed]. Hence (his Equation (64))

$$F(V_b, \theta^+) - F(V_a, \theta^+) = F(V_c, \theta^-) - F(V_d, \theta^-)\,. \tag{9A.5}$$

From (5) and (2) it follows that (his Equation (65)) in a Carnot cycle $\mathscr{C}$

$$\frac{C^-(\mathscr{C})}{C^+(\mathscr{C})} = \frac{\theta^- - \theta_0}{\theta^+ - \theta_0}\,. \tag{9A.6}$$

Because heat and work are uniformly and universally interconvertible in any cycle,

$$L(\mathscr{C}) = J[C^+(\mathscr{C}) - C^-(\mathscr{C})]\,. \tag{7A.1$_{2r}$}$$

[If $C^+(\mathscr{C}) > 0$, $(7A.1)_2$ implies that

$$\frac{L(\mathscr{C})}{JC^+(\mathscr{C})} = 1 - \frac{C^-(\mathscr{C})}{C^+(\mathscr{C})}\,. \tag{9A.7}$$

The dimensionless ratio on the left-hand side is the *efficiency* of the cycle, regarded as an engine.] By (6), then (his Equation (66)),

$$\frac{L(\mathscr{C})}{JC^+(\mathscr{C})} = \frac{\theta^+ - \theta^-}{\theta^+ - \theta_0}\,. \tag{9A.8}$$

RANKINE again (¶44) takes θ_0/θ as being 0, to "the nearest approximation we can at present make". [Then (6) becomes

$$\frac{C^-(\mathscr{C})}{C^+(\mathscr{C})} = \frac{\theta^-}{\theta^+}\,, \tag{9A.9}$$

and (8) becomes

$$\frac{L(\mathscr{C})}{JC^+(\mathscr{C})} = 1 - \frac{\theta^-}{\theta^+}\,.] \tag{9A.10}$$

RANKINE expects that (8) so approximated will be found to give nearly the same numerical results as the formula, "very different in appearance", arrived at by Professor THOMSON. He has not seen KELVIN's manuscript, but a numerical example KELVIN disclosed to him does indeed agree closely. Footnote: "From information which I have received from Professor THOMSON subsequently to the completion of this paper, it appears that his formula becomes identical" with (10), "on making the function called by him

$$\mu = J/\theta \,, \tag{5N.7)$_r$}$$

J being JOULE's equivalent. Mr JOULE also, some time since, arrived at this approximate formula in the particular case of a perfect gas." [In §§7H and 7I, above, we have discussed the origin of the HELMHOLTZ-JOULE Determination (5N.7).]

RANKINE interprets (10) for an actual steam engine and explains why a lesser efficiency will follow in practice. He then gives some examples, "ideal" and actual.

[This paper is famous for (9) and (10): *the classical emission-absorption ratio and the classical evaluation of efficiency, respectively*, for Carnot cycles. The proof may seem too special because RANKINE in using (3) has appealed to the relations

$$\begin{aligned} J\Lambda_V &= (\theta - \theta_0)\frac{\partial p}{\partial \theta}\,, \\ JK_V &= R\left[\frac{1}{N} + (\theta - \theta_0)\left(\frac{\theta_0}{\theta^2} + \frac{1}{R}\int \frac{\partial^2 p}{\partial \theta^2}\,dV\right)\right]\,, \end{aligned} \qquad (8G.8)_r$$

his much criticized inference from his theory of molecular vortices. In fact, however, he could just as well have used the statements

$$\begin{aligned} J\Lambda_V &= (\theta - \theta_0)\frac{\partial p}{\partial \theta}\,, \\ \frac{JK_V}{R/N} &= 1 - N(\theta - \theta_0)\phi'(\theta) + \frac{\theta - \theta_0}{R/N}\int \frac{\partial^2 p}{\partial \theta^2}\,dV\,, \end{aligned} \qquad (8H.4)_r$$

which we have seen to follow from his general relations

$$\begin{aligned} J\Lambda_V &= R(\theta - \theta_0)\left(\frac{1}{V} - \frac{\partial U}{\partial V}\right)\,, \\ JK_V &= R\left[\frac{1}{N} - (\theta - \theta_0)\frac{\partial U}{\partial \theta}\right]\,, \end{aligned} \qquad (8G.1)_{1,2r}$$

and

$$U = \phi(\theta) + \int \left(\frac{1}{V} - \frac{1}{R}\frac{\partial p}{\partial \theta}\right)dV\,. \qquad (8G.3)_r$$

Had he done so, he would have reached (4) in just the same way, but with a more general function ψ, which would have cancelled out in just the same way.

[As we shall see in §9E, two years later RANKINE is to perceive this greater generality but in terms of a more complicated concept.

[RANKINE's heat-potential F is easy to interpret. Comparing (1) with (8G.3) and then using the function H that we introduced in our discussion of RANKINE's first paper:

$$JH/R \equiv \log[V(\theta - \theta_0)^{1/N}] - U + \text{const.} \,, \qquad (8H.2)_r$$

we see that

$$\begin{aligned} F &= R(\log V - U) + \lambda(\theta) \,, \\ &= JH + \nu(\theta) \,, \end{aligned} \qquad (9A.11)$$

the functions λ and ν being constitutive. We have shown in §8H that H satisfies the relation

$$(\theta - \theta_0)\dot{H} = Q \qquad (8H.3)_r$$

and have concluded that to within choice of θ_0 the function H is what CLAUSIUS later, much later, was to rediscover and call the entropy. Therefore (11) shows that to within a constitutive function of temperature, *the heat-potential is the entropy*. Moreover, RANKINE knows how to use it! From (11) we see that for an isothermal process

$$\Delta F = J\Delta H \,. \qquad (9A.12)$$

Thus RANKINE's relations (2) reflect the fact that in an isothermal process at temperature θ^* the net gain of heat

$$C = (\theta^* - \theta_0)\Delta H \,. \qquad (9A.13)$$

RANKINE's relations (5) make the increments of H on the isothermal parts of a Carnot cycle have equal magnitude and opposite sign. If you like, *the working body gives to the refrigerator exactly the amount of entropy it has received from the furnace*—equivalently, *exactly the same difference of heat-potential*:

$$\begin{aligned} JC^+(\mathscr{C}) &= (\theta^+ - \theta_0)\Delta F = (\theta^+ - \theta_0)J\Delta H \,, \\ JC^-(\mathscr{C}) &= (\theta^- - \theta_0)\Delta F = (\theta^- - \theta_0)J\Delta H \,, \end{aligned} \qquad (9A.14)$$

and the key relation (6) follows. Consequently *RANKINE's arguments in deriving his efficiency formula* (8) *are general.*

[RANKINE always defines his quantities such as U and F in such a way as to make it obvious that they are the values of functions of V and θ. While U is defined in terms of his molecular model, it is effectively determined through (8G.3) by partial integration; partial integration defines F. RANKINE seems not to know about conditions of integrability. The price he pays is one no thermodynamicist of the period would have cared to recognize: His functions may exist only locally. Moreover, use of his U and F involve us with undetermined constitutive functions of θ. In the case of F that function cancels out in the particular passage we have just considered, but they will not do so in general. The entropy does not labor under this defect. It is a pity that RANKINE did not stay with the function U he had introduced in his first paper, because it differs from the entropy by a specific function, not an undetermined one.

[By the standards of his day we can only admire this passage in RANKINE's work and acclaim his achievement: *Once and for all, he calculates the efficiency of an arbitrary fluid body in a Carnot cycle.* From (10) we see at once that the function K in the equation of CARNOT's General Axiom in the reduced form

$$L(\mathscr{C}) = K(\theta^+, \theta^-)C^+(\mathscr{C}) \tag{5J.5$_{1r}$}$$

is determined:

$$K(\theta^+, \theta^-) = J\left(1 - \frac{\theta^-}{\theta^+}\right). \tag{9A.15}$$
]

9B. A Late Re-entrance, Stumbling: KELVIN's Second Paper

Acknowledging "the recent discoveries" made by MAYER and JOULE and citing the papers of CLAUSIUS and RANKINE published in the preceding year, KELVIN[1] in 1851 published another [verbose and rambling] essay on the theory of heat. At long last disabused of the Caloric Theory, he sets himself a threefold object (§6):

> (1) To show what modifications of the conclusions arrived at by CARNOT, and by others who have followed his peculiar mode of reasoning regarding the motive power of heat, must be made when the hypothesis of the dynamical theory, contrary as it is to CARNOT's fundamental hypothesis, is adopted.
>
> (2) To point out the significance in the dynamical theory, of the numerical results deduced from REGNAULT's observations on steam ... and to show that by taking these numbers ... in connexion with JOULE's mechanical equivalent of a thermal unit, a complete theory of the motive power of heat, within the temperature limits of the experimental data, is obtained.
>
> (3) To point out some remarkable relations connecting the physical properties of all substances, established by reasoning analogous to that

---

[1] THOMSON [1851].

of CARNOT, but founded in part on the contrary principle of the dynamical theory.

After explaining (§8) that the "*source of heat* will always be supposed to be a hot body at a given constant temperature," *etc.*, KELVIN writes (§9)

> The whole theory of the motive power of heat is founded on the two following propositions, due respectively to JOULE, and to CARNOT and CLAUSIUS.
>
> PROP. I. (JOULE).—When equal quantities of mechanical effect are produced by any means whatever from purely thermal sources, or lost in purely thermal effects, equal quantities of heat are put out of existence or are generated.
>
> PROP. II. (CARNOT and CLAUSIUS).—If an engine be such that, when it is worked backwards, the physical and mechanical agencies in every part of its motions are all reversed, it produces as much mechanical effect as can be produced by any thermodynamic engine, with the same temperatures of source and refrigerator, from a given quantity of heat.

KELVIN gives a "proof" (§§10–11) of Proposition I based upon "thermal motions" within the body and "molecular change or alteration of temperature" as well as reversibility. [While KELVIN's Proposition II would seem to assert that any reversible process achieves maximum efficiency, that is clearly not what it means to him. It imputes to the ideal engine much more than the modern sense of "reversible"; the characteristically vague phrase "physical and mechanical agencies in every part of its motions" engulfs also CARNOT's idea that bodies of unequal temperature should never be put in contact. Within the context of the theory of calorimetry, KELVIN's Proposition II seems to be nothing more than the equality in CARNOT's General Axiom, stated like a riddle. This guess of mine is supported by the use to which KELVIN puts Proposition II when he comes to §21, as we shall see. It is supported also by the argument KELVIN uses when he claims (§13) to "demonstrate" Proposition II, namely CARNOT's construction of the two engines (§5F, above). When both $\mathscr{C}_1$ and $-\mathscr{C}_2$ have been completed, the furnace contains just as much heat as it did before. In connection with the work of CLAUSIUS (above, §8A) we have seen that because the Caloric Theory is now rejected, the construction no longer serves to "re-establish things in their original state", for there is no reason to expect that the quantity of heat in the refrigerator should remain unchanged. In order to draw again CARNOT's conclusion (§§5D–5G) that for $\mathscr{C}_1$ and $-\mathscr{C}_2$ together to do positive work would be contrary to "sound physics", KELVIN, like CLAUSIUS before him, must narrow the meaning of that all too vague standard.] He sets forth as an axiom a new statement of what does not happen (§12):

> *It is impossible, by means of inanimate material agency, to derive mechanical effect from any portion of matter by cooling it below the temperature of the coldest of the surrounding objects.*

This stronger prohibition allows CARNOT's conclusion to be drawn again. The "surrounding objects" are now mentioned [but not provided with any properties specific enough to make them able to figure in a mathematical theory. The entry of "inanimate material agency" adds mystery to an already vague principle; BRIDGMAN[2] called it "a restriction so surprising as to be almost an admission of defeat"; but it may have been a wedge interpellated only to save miracles. Be that as it may, we see here another candidate for the title of "Second Law of Thermodynamics".]

> It must then be admitted that CARNOT's original demonstration utterly fails, but we cannot infer that the proposition concluded is false. The truth of the conclusion appeared to me, indeed, so probable, that I took it in connexion with JOULE's principle...as the foundation of an investigation of the motive power of heat in air-engines or steam-engines through finite ranges of temperature, and obtained about a year ago results, of which the substance is given in the second part of the paper at present communicated to the Royal Society....[T]he merit of first establishing the proposition upon correct principles is entirely due to CLAUSIUS....The following is the axiom on which CLAUSIUS' demonstration is founded:—
>
> *It is impossible for a self-acting machine, unaided by any external agency, to convey heat from one body to another at a higher temperature.*
>
> It is easily shown, that, although this and the axiom I have used are different in form, either is a consequence of the other. The reasoning in each demonstration is strictly analogous to that which CARNOT originally gave.

After remarking (§19) that CLAUSIUS had used "simply CARNOT's unmodified investigation of the relation between the mechanical effect produced and the thermal circumstances from which it originates, in the case of an expansive engine working within an infinitely small range of temperatures", KELVIN turns [at last] (§§20–21) to the mathematical theory, he effortlessly writes down as a line integral "[t]he total external effect, after any finite

[2] BRIDGMAN [1941, p. 5 of the 1961 ed.].

amount of expansion, accompanied by any continuous change of temperature":

$$L - JC = \int [pdV - J(\Lambda_V dV + \mathrm{K}_V d\theta)] \ , \tag{9B.1}$$

J being "the mechanical equivalent of a unit of heat". KELVIN interprets JOULE's proposition as meaning that the left-hand side of (1) is null in every cyclic process. Hence the integrand in (1) "must be the differential of a function of two independent variables", and KELVIN writes out both

$$J\Lambda_V - p = \frac{\partial \mathrm{E}}{\partial V} \ , \qquad J\mathrm{K}_V = \frac{\partial \mathrm{E}}{\partial \theta} \ , \tag{8D.3$_r$}$$

and

$$\frac{\partial p}{\partial \theta} = J\left(\frac{\partial \Lambda_V}{\partial \theta} - \frac{\partial \mathrm{K}_V}{\partial V}\right) . \tag{8B.1$_r$}$$

[This passage is the first and very nearly the last piece of clean mathematics in all of classical thermodynamics. The "function" to which KELVIN refers is E/J, which CLAUSIUS had produced in the special case of an ideal gas and had proved to satisfy (8B.2). KELVIN, like RANKINE before him, has *extended CLAUSIUS' results to a general equation of state*, but he makes little of that and does not mention it again.]

KELVIN goes on to say that "[t]he corresponding application of the second fundamental proposition is completely expressed" by (his Equation (3))

$$\mu\Lambda_V = \frac{\partial p}{\partial \theta} \ , \tag{5L.4$_r$}$$

that is, the General CARNOT-CLAPEYRON Theorem. [We have seen above that CARNOT's argument to prove it does not make any use of the Caloric Theory. Rather than repeat that proof in CLAPEYRON's form, however,] KELVIN (§21) remarks that for an infinitesimal Carnot cycle $\mathscr{C}$

$$\frac{L(\mathscr{C})}{C^+(\mathscr{C})} = \frac{\frac{\partial p}{\partial \theta}\Delta\theta\Delta V}{\Lambda_V \Delta V} = \frac{\frac{\partial p}{\partial \theta}}{\Lambda_V}\Delta\theta \ . \tag{9B.2}$$

The infinitesimal cycle

> may be considered as constituting a thermo-dynamic engine which fulfils CARNOT's condition of complete reversibility. Hence, by Prop. II., it must produce the same amount of work for the same quantity of heat absorbed in the first operation, as any other substance similarly operated upon through the same range of temperatures.

Hence the right-hand side of (2)

> must be the same for all substances, with the same values of θ and

$\Delta\theta$; or, since $\Delta\theta$ is not involved except as a factor, we must have [KELVIN's Equation (4)]

$$\frac{\dfrac{\partial p}{\partial \theta}}{\Lambda_V} = \mu \tag{9B.3}$$

where μ depends only on θ....

[As Mr. BHARATHA has kindly pointed out to me, this argument is superior to CLAPEYRON's in that it appeals directly to CARNOT's General Axiom, without use of (5J.5) and without imputing any smoothness to G. It provides μ but does not relate it to G through (5J.7). It uses the universality of G to infer in effect that on the isotherm $\theta = \theta_0$ the ratio $(\partial p/\partial \theta)/\Lambda_V$ has the same value at all points where $\Lambda_V(V, \theta_0) \neq 0$, and that that value is the same for every body whose constitutive domain intersects the isotherm $\theta = \theta_0$ at one point where $\Lambda_V \neq 0$.] At this time KELVIN was fluently dividing by Λ_V, [so he must have been tacitly adopting the traditional constitutive restriction

$$\Lambda_V > 0 \text{ .]} \tag{2C.5)$_{1r}$$

Thus, for him, (3) holds without exception. In annotating this passage in 1881 for the reprint in his collected works he remarked that eliminating $\partial p/\partial \theta$ between (8B.1) and (3) yields

$$\frac{J\left(\dfrac{\partial \Lambda_V}{\partial \theta} - \dfrac{\partial \mathrm{K}_V}{\partial V}\right)}{\Lambda_V} = \mu \text{ ,} \tag{9B.4}$$

"a very convenient and important formula". [That is, the ratio on the left-hand side of (4) has the properties we have just asserted for the ratio on the left-hand side of (3).] KELVIN's proof of 1881 involves use of CLAUSIUS' basic assumption $L(\mathscr{C}) = JC(\mathscr{C})$; [according to the Third Principal Lemma in Chapter 9 of *Concepts and Logic*, that assumption is unnecessary, but a more searching analysis of CARNOT's ideas is needed to prove that].

[Does the General CARNOT-CLAPEYRON Theorem express "the second fundamental proposition" completely? KELVIN does not quite say so. Theorem 11 or Corollary 11.2_{ext} of *Concepts and Logic* shows that in general the answer is no. Mr. BHARATHA has pointed out to me that the answer becomes yes[3]

---

[3] *Proof.* Set $h \equiv \exp \int (\mu/J)d\theta$, $g = Jh$. Then appeal to (8B.1) and (5L.4) to show that Equations (9.1) and (9.2) of *Concepts and Logic* are satisfied. Then use Corollary 11.2 in Chapter 10 to conclude that for a Carnot cycle

$$L(\mathscr{C}) = \frac{g(\theta^+) - g(\theta^-)}{h(\theta^+)} C^+(\mathscr{C}) = J\frac{h(\theta^+) - h(\theta^-)}{h(\theta^+)} C^+(\mathscr{C}) > 0 \text{ .}$$

Because μ is a universal function for KELVIN, so are h and g as defined here. Thus CARNOT's General Axiom holds.

on KELVIN's further assumption that his "first fundamental proposition" holds.]

KELVIN continues (§22)

> The very remarkable theorem that $(\partial p/\partial\theta)/\Lambda_V$ must be the same for all substances at the same temperature, was first given (although not in precisely the same terms) by CARNOT, and demonstrated by him, according to the principles he adopted. We have now seen that its truth may be satisfactorily established without adopting the false part of his principles. Hence all CARNOT's conclusions, and all conclusions derived by others from his theory, which depend merely on [the General CARNOT-CLAPEYRON Theorem (5L.4)] require no modification when the dynamical theory is adopted.

This statement applies in particular to the conclusions in §§43–53 of KELVIN's first paper (§7H, above).

> Also, we see that CARNOT's expression for the mechanical effect derivable from a given quantity of heat by means of a perfect engine in which the range of temperatures is infinitely small, expresses truly the greatest effect which can possibly be obtained in the circumstances; although it is in reality only an infinitely small fraction of the whole mechanical equivalent of the heat supplied; the remainder being irrecoverably lost to man, and therefore "wasted," although not *annihilated*.

On the contrary (§23), "when the range down to the temperature of the 'refrigerator' is finite," the expression for the work done "will differ most materially from that of CARNOT . . . ."

Next (§25) comes the revised evaluation of the efficiency of an ideal engine.

> We may suppose the engine to consist of an infinite number of perfect engines, each working within an infinitely small range of temperature, and arranged in a series of which the source of the first is the given source, the refrigerator of the last the given refrigerator, and the refrigerator of each intermediate engine is the source of that which follows it in the series.

[Figure 6 makes the argument clear. The given Carnot cycle $\mathscr{C}$ is subdivided by isotherms into a sequence of small Carnot cycles, of which the typical one $\mathscr{C}_\theta$ has the isothermal paths $\mathscr{P}_{\theta+\Delta\theta}$ and $-\mathscr{P}_\theta$. Then CARNOT's calculation (5J.6) applies. We write it here as follows:

$$L(\mathscr{C}_\theta) \approx \mu(\theta)C^+(\mathscr{C}_\theta)\Delta\theta \ . \tag{9B.5}$$

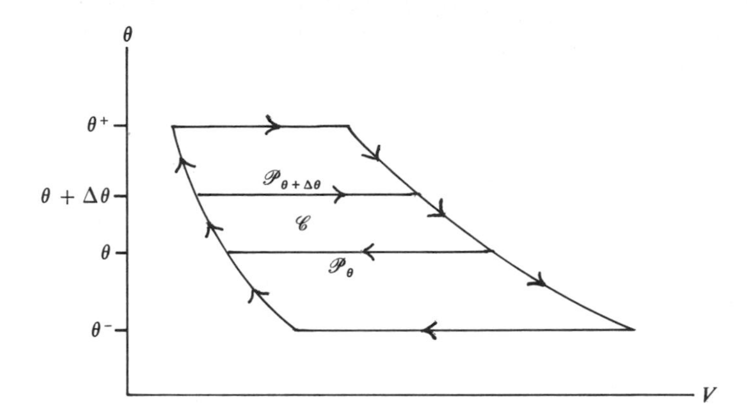

Figure 6. Construction to prove KELVIN's general evaluation of the motive power of a Carnot cycle

Appeal to the uniform and universal Interconvertibility of Heat and Work in Carnot cycles,

$$L(\mathscr{C}) = JC(\mathscr{C}) = J[C^{+}(\mathscr{C}) - C^{-}(\mathscr{C})] \; , \qquad \text{(7A.1)}_{\mathrm{r}}$$

yields

$$L(\mathscr{C}_{\theta}) = J[C^{+}(\mathscr{P}_{\theta+\Delta\theta}) - C^{+}(\mathscr{P}_{\theta})] \approx J\frac{dC^{+}(\mathscr{C}_{\theta})}{d\theta}\Delta\theta \; . \qquad \text{(9B.6)}$$

Comparing (5) with (6), we see that]

$$J\frac{dC^{+}}{d\theta} = \mu C^{+} \; . \qquad \text{(9B.7)}$$

Integration of this ordinary differential equation shows that

$$J \log C^{+}(\mathscr{C}_{\theta}) = \int^{\theta} \mu(x) dx \; . \qquad \text{(9B.8)}$$

Using (7A.1) again, we obtain KELVIN's result:

$$\frac{L(\mathscr{C})}{JC^{+}(\mathscr{C})} = 1 - \exp\left(-\frac{1}{J}\int_{\theta^{-}}^{\theta^{+}} \mu d\theta\right) \; . \qquad \text{(9B.9)}$$

[We should compare this formula with the counterpart, namely

$$L(\mathscr{C}) = \left(\int_{\theta^{-}}^{\theta^{+}} \mu d\theta\right) C^{+}(\mathscr{C}) \; , \qquad \text{(5L.7)}_{2\mathrm{r}}$$

which KELVIN had derived earlier from the Caloric Theory.] Later[4] he is to state that (9) "involves no hypothesis" [by which he means CLAUSIUS' "subsidiary hypothesis". It shows that in CLAUSIUS' theory as in CARNOT's

[4] THOMSON [1852, *3*].

theory, *CARNOT's function μ determines the motive power of heat*, but of course differently in the two. *Cf.* Scholion IV in §5L.]

From (6) KELVIN infers (§27) that "... as $\int_{\theta^-}^{\theta^+} \mu d\theta$ is increased without limit", $L(\mathscr{C}) \to JC^+(\mathscr{C})$.

> Thus we see that, although the full equivalent of mechanical effect cannot be obtained even by means of a perfect engine, yet when the actual source of heat is at a high enough temperature above the surrounding objects, we may get more and more nearly the whole of the admitted heat converted into mechanical effect, by simply increasing the effective range of temperature in the engine.

[The conclusion silently presupposes that as $\theta^+ - \theta^- \to \infty$, also $\int_{\theta^-}^{\theta^+} \mu d\theta \to \infty$. We know today that this is true, because J/μ must be a linear function of θ, but KELVIN writes as if his conclusions held for all choices of μ, which obviously they do not. It is one more example of his hasty mathematics.]

KELVIN asserts (§30),

> As yet no experiments have been made upon air which afford the required data for calculating the value of μ through any extensive range of temperature; but for temperatures between 50° and 60° Fahr., JOULE's experiments... afford the most direct data... which have yet been obtained....

KELVIN proceeds (§§31–41) to explain and tabulate quantities associated with the problem. One of them, as his formula (9) suggests, is $\int_{\theta^-}^{\theta^+} \mu d\theta$. He reminds us (§34) "that accurate experimental determinations of the densities of saturated steam at different temperatures may indicate considerable errors in the densities which have been assumed according to the 'gaseous laws,' and may consequently render considerable alterations in my results necessary...." As no pertinent experiments have been done in the interim, he sees no reason for supposing that the values of μ in his tables of 1848 and 1849 are "not the most probable that can be obtained in the present state of science".

[A ray of light enters] at §42 with the reference to JOULE's letter[5] of 1848

---

[5] CLAUSIUS [1856, *1*] is to reply with a characteristic claim of priority:

> Against this I must beg to urge,—*First*, that, as far as I am aware, it is usual, in determining questions of priority in scientific matters, only to admit such statements as have been *published*. And I believe that this custom ought to be

which we have quoted above, toward the end of §7H, as suggesting what we have called the HELMHOLTZ-JOULE Determination:

$$\mu = J/\theta \; . \tag{5N.7$_r$}$$

Then also[6] (Equation (13))

$$\frac{L(\mathscr{C})}{JC^+(\mathscr{C})} = 1 - \frac{\theta^-}{\theta^+} \; . \tag{9A.10$_r$}$$

Here we see RANKINE's major result (9A.10) recovered without reference to the molecular vortices. [Of course CLAUSIUS could have found this result, had he stopped to work it out from his assumptions; as we have shown in

---

conscientiously adhered to, especially in theoretical investigations; for it usually requires continued and laborious research in order to give to a thought, after it has been first entertained, and perhaps casually communicated to a friend, that degree of certainty which is necessary before venturing upon its publication. *Secondly*, that since Thomson does not say that Mr. Joule had proved the theorem, but only that he had offered it as an opinion, I do not see why this opinion should have the priority over that which Holtzmann had arrived at three years before.

This ugly passage presents only a part of the facts. First, it was CARNOT himself who first suggested (5N.7) (*cf.* §5S, above). Second, it was not HOLTZMANN but HELMHOLTZ, not in 1845 but in 1847, who first published (5N.7) as a consequence of relations HOLTZMANN had obtained. Third, (5N.7) may be read off from $(8G.8)_1$, which is clearly implied by results in RANKINE's paper published in 1850, and RANKINE was not tainted by ever having espoused the Caloric Theory. In reply THOMSON [1856] merely quoted in full the passage from his own work that CLAUSIUS had quoted in part.

[6] (KELVIN's footnote). "If we take $\mu = k/\theta$ where k may be any constant, we find

$$\frac{L(\mathscr{C})}{JC^+(\mathscr{C})} = \left(\frac{\theta^+ - \theta^-}{\theta^+}\right)^{k/J} ; \tag{A}$$

which is the formula I gave when this paper was communicated. I have since remarked, that Mr JOULE's hypothesis implies essentially" that $k = J$. "Mr RANKINE, in a letter dated March 27, 1851, informs me that he has deduced ... an approximate formula ... which ... I find agrees exactly" with the corresponding special case of (A). [*Cf.* §9A, especially (9A.10). Correct integration would have yielded

$$\frac{L(\mathscr{C})}{JC^+(\mathscr{C})} = 1 - \left(\frac{\theta^-}{\theta^+}\right)^{k/J}$$

instead of (A).

[This formula shows that CLAUSIUS did indeed make an assumption when he tacitly took k as being the same as J (above, §8D). In terms of the functions g and h used in *Concepts and Logic*, from Equation (9.2) and corollary 10.3 of that work we see that

$$g'/h = k/\theta, \qquad g' = Jh + \text{const.} \; ,$$

so

$$h = (\text{const.})\theta^{k/J} \; .$$

As we have seen in §8D, to conclude that $k = J$ it is *sufficient* to assume that one ideal gas has constant specific heats.]

§9A, RANKINE's argument is really not so special as it must have seemed when it was first published.] KELVIN still regards (5N.7) and its consequence (9A.10) as dubious, because (§45) the investigations of RANKINE and CLAUSIUS "involve fundamentally various hypotheses which may or may not be found by experiment to be approximately true; and which render it difficult to gather from their writings what part of their conclusions ... depend merely on the necessary principles of the dynamical theory."

Finally KELVIN decides to derive the consequences of the relations

$$\mu \Lambda_V = \frac{\partial p}{\partial \theta} , \tag{5L.4}_r$$

$$\frac{\partial}{\partial \theta}(J\Lambda_V - p) = \frac{\partial}{\partial V}(J\mathrm{K}_V) , \tag{8D.2}_r$$

and

$$J\Lambda_V - p = \frac{\partial \mathrm{E}}{\partial V}, \qquad J\mathrm{K}_V = \frac{\partial \mathrm{E}}{\partial \theta} . \tag{8D.3}_r$$

For example, first he notes the immediate consequence (Equation (14))

$$\frac{\partial \mathrm{K}_V}{\partial V} = \frac{\partial}{\partial \theta}\left(\frac{1}{\mu}\frac{\partial p}{\partial \theta} - \frac{p}{J}\right) . \tag{9B.10}$$

Then he derives the relation

$$\mathrm{K}_p - \mathrm{K}_V = -\Lambda_V \frac{\partial p}{\partial \theta}\Big/\frac{\partial p}{\partial V} \tag{2C.9}_{2r}$$

and by using (5L.4) obtains (Equation (16))

$$\mathrm{K}_p - \mathrm{K}_V = -\frac{1}{\mu}\left(\frac{\partial p}{\partial \theta}\right)^2\Big/\frac{\partial p}{\partial V} . \tag{9B.11}$$

which for ideal gases becomes (Equation (26))

$$\mathrm{K}_p - \mathrm{K}_V = R/(\mu\theta) . \tag{9B.12}$$

[Had CARNOT obtained the relation

$$\mathrm{K}_p - \mathrm{K}_V = \frac{R}{\theta F'} \tag{5Q.6}_r$$

clearly rather than in special cases presented verbally, KELVIN could have remarked that here, too, was a conclusion that carried over from the old, uncorrected theory.] Then (§52), "All the conclusions obtained by CLAUSIUS, with reference to air or gases, are obtained immediately from these equations" by adopting [the HELMHOLTZ-JOULE Determination]

$$\mu = J/\theta , \tag{5N.7}_r$$

which will make K_V independent of V, "and by assuming, as he does, that K_V ... is also independent of [the] temperature."

[KELVIN's criticism, which is not quite fair, shows how obscure CLAUSIUS' exposition was, even to one of the other three principal architects of the "dynamical" theory. CLAUSIUS' "subsidiary hypothesis" was HOLTZMANN's Assertion:

$$J\Lambda_V = p \; ; \qquad (7D.1)_r$$

equivalently, MAYER's Assertion:

$$J(K_p - K_V) = R \; . \qquad (7B.2)_r$$

Indeed, KELVIN's statement (11) makes this tantamount to assuming that μ is given by (5N.7). CLAUSIUS did not assume that either specific heat was constant for all ideal gases; he merely considered as an example the case when both were constant and thus tacitly *assumed that possibility to be consistent* with the theory he was constructing. Today we know that such an assumption is by no means innocent, for the Caloric Theory forbids it (above, §3F). CLAUSIUS and KELVIN may have known as much, but certainly they did not tell their readers so.

[This paper, like KELVIN's first one, disappoints a critical student. In its vacillation and obscurity the author of KELVIN's masterful and brilliant researches on electrostatics, hydrodynamics, and elasticity can scarcely be recognized. Sensing the weakness of CLAUSIUS' "subsidiary hypothesis" about ideal gases, KELVIN sees that the General CARNOT-CLAPEYRON Theorem

$$\mu\Lambda_V = \frac{\partial p}{\partial \theta} \qquad (5L.4)_r$$

and his own new evaluation (3) are free of it, but he lets them sit. The genius of thermodynamics always deflected its devotees from the essential logical structure and the key facts of experiment into a labyrinth of complicated and unenlightening details about steam.

[KELVIN has in his hands a brilliant chance to bring *known and accepted general theory* and *already accepted experiment* together, without a single glance at a steam table. Theory first. We apply the basic assumption of the "dynamical theory", namely

$$L(\mathscr{C}) = JC(\mathscr{C}) = J[C^+(\mathscr{C}) - C^-(\mathscr{C})] \; . \qquad (7A.1)_r$$

To calculate $C^+(\mathscr{C})$ and $C^-(\mathscr{C})$, we turn to CARNOT's evaluation of the heat added on an isothermal path $\mathscr{P}_\theta$ undergone by a body of ideal gas:

$$C(\mathscr{P}_\theta) = \frac{R}{\mu(\theta)} \log \frac{V_b}{V_a} \; , \qquad (5K.5)_{2r}$$

an evaluation KELVIN had confirmed in §46 of his first paper; as he has just remarked, §46 is one of the sections that remain valid. We thus show that

$$\frac{L(\mathscr{C})}{JC^+(\mathscr{C})} = 1 - \frac{\dfrac{1}{\mu(\theta^-)} \log \dfrac{V_c}{V_d}}{\dfrac{1}{\mu(\theta^+)} \log \dfrac{V_b}{V_a}} \; , \qquad (9B.13)$$

the notation being as in Figure 3 in §5C. The left-hand side is asserted to have the same value for all Carnot cycles whose operating temperatures are θ^+ and θ^-, and μ is asserted to be a universal function, the same for all bodies. As we have mentioned above in §7H, and as KELVIN himself was to see in 1853, we may evaluate it by considering any particular material we deem compatible with the theory or acceptable as a good representation of the facts of experiment. Now experiment. LAPLACE's theory of the speed of sound in ideal gases was generally accepted for air and some other fluids. As we have seen in §5T in our discussion of the work of CARNOT, the results of that theory do not fit experimental data unless $\gamma = \text{const.} > 1$.

CLAUSIUS (§8B) had already shown that such a gas was compatible with his results, but then he had used his "subsidiary hypothesis", which KELVIN refuses to adopt. No need. If we suppose that there is *one* ideal gas for which $\gamma = \text{const.} > 1$, by use of the LAPLACE-POISSON law of adiabatic change in the form

$$\frac{p}{p_0} = \left(\frac{V_0}{V}\right)^{\gamma} \tag{8B.3)$_{1r}$$

we see that in a Carnot cycle undergone by a body of that gas

$$\frac{V_a}{V_b} = \frac{V_d}{V_c}, \tag{9B.14}$$

a fact KELVIN was to notice a year later (*cf.* §9D, below); of course he could have read it off from RANKINE's (9A.5). Use of (14) reduces (13) to

$$\frac{L(\mathscr{C})}{JC^+(\mathscr{C})} = 1 - \frac{\mu(\theta^+)}{\mu(\theta^-)}. \tag{9B.15}$$

This result is compatible with KELVIN's own general formula (9) if and only if $\mu = J/(\theta + \text{const.})$. Since μ is universal, it is now evaluated once and for all, to within a constant. While this analysis does not quite establish the HELMHOLTZ-JOULE Determination (5N.7), it comes very close.

[It is strange that KELVIN did not think it worthwhile to write down the result of using the HELMHOLTZ-JOULE Determination to render the General CARNOT-CLAPEYRON Theorem specific. Of course it becomes

$$J\Lambda_V = \theta \frac{\partial p}{\partial \theta}; \tag{7I.2)$_r$$

we have seen that this statement was implied by RANKINE's paper of 1850 (above, §8G, especially $(8G.8)_1$), but there it was obscured by its apparent origin in the deportment of molecular vortices, and also RANKINE himself failed to notice it then. Later[7] KELVIN was to use (5N.7) to reduce (10) and (11) immediately to the forms we now find in textbooks:

$$\frac{\partial}{\partial V}(J\mathrm{K}_V) = \theta \frac{\partial^2 p}{\partial \theta^2}, \tag{9B.16}$$

$$J(\mathrm{K}_p - K_V) = \theta \frac{\left(\dfrac{\partial p}{\partial \theta}\right)^2}{-\dfrac{\partial p}{\partial V}}, \tag{9B.17}$$

[7] JOULE & THOMSON [1854, Theoretical Deductions, Equations (8) and (7) in §V].

but even then he did not record the centrally important relation (7I.2), which was reserved for CLAUSIUS, as we shall see below in §11E. Had KELVIN noticed (7I.2) then, he could have combined it with (16) to see at a glance that

$$\frac{\partial}{\partial\theta}\left(\frac{\Lambda_V}{\theta}\right) = \frac{\partial}{\partial V}\left(\frac{K_V}{\theta}\right) \tag{9B.18}$$

and so gone on to discover the entropy and its properties. *Cf.* RANKINE's near approach, described above in §§8H and 9A. In the next act (§10A) we shall encounter REECH's pro-entropy and witness his failure to reduce it.

[Two years later KELVIN is to have another chance and miss again. We shall discuss his work then in §9D.]

9C. A Voice Crying in the Wilderness: REECH's Return to First Principles

[The whimsy of the tragicomic muse next fell upon] REECH, a naval engineer who had already written a tract on the performance of marine steam engines in practice[1] and who was then in charge of test and analysis of such engines at the port of Lorient. [The opposite pole from the unmathematical and unpractised and ill-read MAYER,] REECH chose to express the results of his abundant experience and his study of previous work on the subject in terms of the formal calculus he had learnt at the Ecole Polytechnique. He did so in a [deadly] memoir 211 pages long, dense with equations, and published in a mathematical journal!

On November 17, 1851, REECH submitted to the Paris Academy "a very considerable work" entitled *Theory of the motive force of caloric*[2]. On December 1 of the same year he submitted an addendum, *Note on the theory of the dynamical effects of heat*[3], in which he referred to the recent theories of KELVIN and CLAUSIUS on the subject of his earlier communications. Between those two dates he published a short notice[4] of the memoir. [Perhaps one of the commissioners appointed to examine his work, among whom was CLAPEYRON, had directed his attention to the theories lately proposed by foreigners. That notwithstanding, we shall find in the enormous memoir

[1] *Mémoire sur les machines à vapeur et leur application à la navigation*, Paris, Bertrand, 1844. There is a copy in the British Museum. REECH states that the contents were submitted to the French Academy in 1838. I can find in this work no theory of heat engines and no reference to CARNOT or CLAPEYRON.

[2] Acknowledged on p. 540 of Volume 33 of the *Comptes Rendus*.

[3] Acknowledged on p. 602 of the same volume.

[4] REECH [1851].

he published two years later little influence of their work,] except that in its title, as in the title of his addendum, he used "heat" rather than "caloric". Deferring most of our analysis of the published memoir until §10A, we shall remark only upon what the preliminary notice reveals.

"Long ago I had grafted upon Mr. Clapeyron's memoir a very curious work regarding the production itself of caloric in the phenomenon of the combustion of a kilogram of carbon in a closed vessel with injection of air into this vessel, and I was prepared to publish my work when I learned that Mr. Regnault disputed" the equation $C^+(\mathscr{C}) = C^-(\mathscr{C})\ldots$; "encouraged by Mr. Regnault himself to sift the matter, I applied myself to reconsider minutely the reasoning of Messrs. Carnot and Clapeyron, and I was not slow to prove" that CARNOT's proposal should be replaced by

$$\begin{aligned} L(\mathscr{C}) &= \Gamma(\theta^+)C^+(\mathscr{C}) - \Gamma(\theta^-)C^-(\mathscr{C}) , \\ &= [\Gamma(\theta^+) - \Gamma(\theta^-)]C^+(\mathscr{C}) + \Gamma(\theta^-)[C^+(\mathscr{C}) - C^-(\mathscr{C})] . \end{aligned} \tag{9C.1}$$

Of course $\mathscr{C}$ is a Carnot cycle. The long memoir published two years later reveals how REECH had arrived at this conclusion[5].

REECH begins from first principles:

$$p = \varpi(V, \theta) , \qquad (2A.2)_{1r}$$

$$Q = \Lambda_V(V, \theta)\dot{V} + K_V(V, \theta)\dot{\theta} , \qquad (2C.4)_r$$

and the equation in CARNOT's General Axiom:

$$L(\mathscr{C}) = G(\theta^+, \theta^-, C^+(\mathscr{C})) . \qquad (5I.1)_r$$

[In this sense he is just when] he claims that he has deduced his result (1) "without making any hypothesis about the intimate nature of heat", and that (1) is "a superior relation which will dominate both Mr. Carnot's theory and the more recent theories of some other physicists." [REECH is the first and also the last of the early authors on thermodynamics to probe the logical structure of the subject. Alas, here too the tragicomic fury casts her spell by making him attempt to prove everything by running engines backward and forward against each other. Upon examination, many of his proofs may be seen to consist of nothing more than repeated applications of the reversal theorems:

$$C(-\mathscr{P}) = -C(\mathscr{P}) , \qquad L(-\mathscr{P}) = -L(\mathscr{P}) . \qquad (2C.7)_r , (2C.21)_r$$

and to be as loose as the verbiage of CLAUSIUS. Despite his fluency in differential calculus, REECH seems not to know how to handle the integrals. Furthermore, he reasons obscurely in terms of a net of adiabats and isotherms rather than the latent and specific heats directly. We might think he was

[5] REECH [1853, Chapter II, especially pp. 368–371].

attempting to appeal to principles more general than those of the Doctrine of Latent and Specific Heats,] but this is not so[6].

First, REECH shows that the equation in CARNOT's General Axiom, namely (5I.1), is equivalent to the equation in the following statement:

$$L(\mathscr{C}) = K(\theta^+, \theta^-)C^+(\mathscr{C}) \ , \quad K(x, y) > 0 \ \text{ if } \ x > y > 0, \quad K(x, x) = 0 \ ; \tag{5J.5$_r$}$$

so as to make the statement complete, we add the inequality into which the inequality in CARNOT's General Axiom, namely (5I.2), translates. For proof (p. 364) REECH considers n repetitions of the cycle $\mathscr{C}$ and asserts that the work done should be $nL(\mathscr{C})$. [Indeed, such is the case, and a proof of (5J.5) can be founded upon this fact[7], but the theorem follows at once by a simple and rigorous argument in terms of line integrals[8]. We consider a given Carnot cycle $\mathscr{C}$ (Figure 7), and we subdivide it into two others, say $\mathscr{C}_1$ and $\mathscr{C}_2$, corresponding to the same extremes of temperature. The construction is clear from Figure 7, in which ef is the adiabat descending at e from the isotherm $\theta = \theta^+$ to the isotherm $\theta = \theta^-$. We recall that $Q > 0$ all along

---

[6] In later parts of his memoir he uses latent heats and specific heats freely; in a later defense of his work REECH [1863, *2*] makes (2C.4) his explicit starting point; and finally in his textbook REECH [1868] develops the theory of calorimetry in detail before taking up thermodynamics at all.

[7] Dropping θ^+ and θ^- from the notation, we may write REECH's condition as $G(nC^+) = nG(C^+)$ for every positive number C^+ and every positive integer n. Taking $C^+ = 1/n$ shows that $G(1/n) = G(1)/n$. Repeated application of REECH's condition shows that if m and n are positive integers, then

$$G\left(\frac{m}{n}\right) = \frac{m}{n} G(1) \ .$$

Thus $G(x)$ coincides with $G(1)x$ whenever x is a positive rational number. Since the rationals are dense in the reals, the only continuous function G which has this property is $G(1)x$ itself.

Arguments of this kind go back at least to CAUCHY's lectures of 1823 at the Ecole Polytechnique and should have been familiar to REECH, though doubtless not to CARNOT or CLAPEYRON.

[8] In this day of dazzling enlightenment powered neither by learning nor by intelligence, an author who does what I have done here runs a double danger: to be reproached by the Historians of Science for having stopped copying and thus become present-minded, and to be reproached by thermodynamicists for merely unearthing old proofs without doing anything original. I claim that REECH's argument here was crude and awkward but substantially correct. I acknowledge the fact that without having read through REECH's sloppy proof, I should not have thought of the simple proof I present here, but I defy anyone to find that proof in REECH's text. On the other hand, mathematics being mathematics, I claim that REECH's result was implied all along, strictly and inescapably, by the premisses of CARNOT's theory without any further assumptions about heat and work and that my proof uses only mathematics widely available in the early nineteenth century. I see not only nothing inconsistent but also nothing either unhistorical or superannuated in maintaining all these views together. I treat REECH as a dead scientist whose works still live.

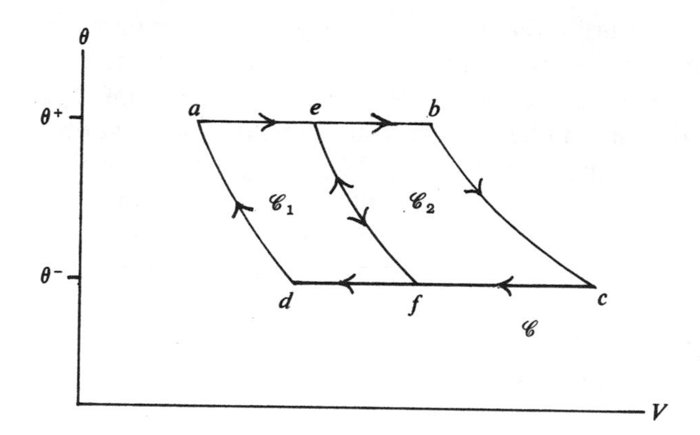

Figure 7

the line aeb, and hence the heats absorbed on the Carnot cycles $\mathscr{C}$, $\mathscr{C}_1$, and $\mathscr{C}_2$, respectively, are related thus:

$$C^+ = C_1^+ + C_2^+ \ . \tag{9C.2}$$

This relation follows from the definition

$$C^+ \equiv \frac{1}{2}\int_{t_1}^{t_2} (Q + |Q|)dt \geqq 0 \ , \qquad (2C.22)_{1r}$$

applied to each cycle in turn. Since the adiabat ef is traversed in opposite senses on $\mathscr{C}_1$ and $\mathscr{C}_2$, by using the relations

$$L(\mathscr{P}) = \int_{\mathscr{P}} \varpi(V, \theta)dV \ , \qquad L(-\mathscr{P}) = -L(\mathscr{P}) \ , \quad (2C.20)_{3r}, (2C.21)_r$$

we conclude that

$$L(\mathscr{C}) = L(\mathscr{C}_1) + L(\mathscr{C}_2) \ . \tag{9C.3}$$

Now we appeal to the equation in CARNOT's General Axiom, namely (5I.1); we substitute it into both sides of (3) and by use of (2) conclude that

$$G(\theta^+, \theta^-, w + u) = G(\theta^+, \theta^-, w) + G(\theta^+, \theta^-, u) \ . \tag{9C.4}$$

In virtue of CAUCHY's theorem on additive functions, all solutions of this functional equation that are continuous[9] in their third argument are linear in it. We have proved]

REECH's first theorem: *CARNOT's General Axiom is compatible with the theory of calorimetry only if it reduces to*

$$L(\mathscr{C}) = K(\theta^+, \theta^-)C^+(\mathscr{C}) \ , \quad K(x, y) > 0 \ \text{if} \ x > y > 0 \ , \quad K(x, x) = 0 \ . \qquad (5J.5)_r$$

[9] This theorem appears as Theorem 6 in Chapter 8 of *Concepts and Logic*; the proof there does not require G to be assumed continuous in its third argument and does take account of the fact that only positive arguments should be considered.

REECH's own proof that (5J.5) reduces to (1) [is complicated and incomplete]; also, as he notes, it fails in one case. [His theorem is easy to prove directly and without exception. To do so, we start from a given Carnot cycle $\mathscr{C}$ and form another corresponding to the same heat absorbed but a lower minimum temperature, say θ_0. We simply extend the adiabats bc and ad downward (Figure 8) until they intersect the isotherm $\theta = \theta_0$ at, say, g and h. Then $abcghda$ is a Carnot cycle, say $\mathscr{C}_4$, and so also is $dcghd$, say $\mathscr{C}_3$. The heats absorbed on these three cycles satisfy the relations

$$C^+(\mathscr{C}_4) = C^+(\mathscr{C}) \ , \qquad C^+(\mathscr{C}_3) = C^-(\mathscr{C}) \ . \tag{9C.5}$$

The latter relation follows from the reversal theorem

$$C(-\mathscr{P}) = -C(\mathscr{P}) \ . \tag*{(2C.7)$_r$}$$

and the fact that both $\mathscr{C}_4$ and $\mathscr{C}_3$ are Carnot cycles. Of course

$$L(\mathscr{C}_4) = L(\mathscr{C}) + L(\mathscr{C}_3) \ . \tag{9C.6}$$

In view of (5), by applying (5J.5) in turn to the three cycles $\mathscr{C}_4$, $\mathscr{C}$, and $\mathscr{C}_3$, we obtain

$$\begin{aligned} L(\mathscr{C}_4) &= K(\theta^+, \theta_0)C^+(\mathscr{C}) \ , \\ L(\mathscr{C}) &= K(\theta^+, \theta^-)C^+(\mathscr{C}) \ , \\ L(\mathscr{C}_3) &= K(\theta^-, \theta_0)C^-(\mathscr{C}) \ . \end{aligned} \tag{9C.7}$$

Substitution into (6) yields a functional equation for K:

$$K(\theta^+, \theta^-)C^+ = K(\theta^+, \theta_0)C^+ - K(\theta^-, \theta_0)C^- \ . \tag{9C.8}$$

Thus if we fix θ_0 at some value less than θ^-, we define Γ as follows:

$$\Gamma(\theta) \equiv K(\theta, \theta_0) \quad \text{provided} \quad \theta > \theta_0 \tag{9C.9}$$

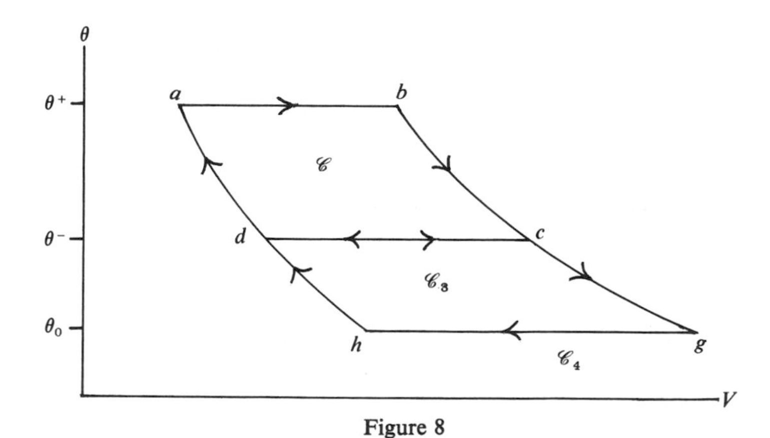

Figure 8

and write (8) as[10]

$$L(\mathscr{C}) = K(\theta^+, \theta^-)C^+(\mathscr{C}) = \Gamma(\theta^+)C^+(\mathscr{C}) - \Gamma(\theta^-)C^-(\mathscr{C}) , \qquad (9C.10)$$

thus proving]

REECH's second theorem: *CARNOT's General Axiom is compatible with the theory of calorimetry only if it reduces to* (1).

REECH infers from (1) the following ***corollary***: The Doctrine of Latent and Specific Heats being presumed, *CARNOT's Special Axiom*, namely

$$L(\mathscr{C}) = [F(\theta^+) - F(\theta^-)]C^+(\mathscr{C}) , \qquad (5I.6)_r$$

$$F(x) > F(y) \quad \text{if} \quad x > y > 0 , \qquad (5I.5)_{2r}$$

is a necessary and sufficient condition that CARNOT's General Axiom be compatible with the Caloric Theory. Indeed, necessity is immediate from (10). [I do not find REECH's proof of sufficiency convincing, but the statement is correct, for it is a consequence of Scholion I in §5J. The fact asserted is essential if we are to understand CARNOT's theory.]

REECH's note is cryptic. Not only does CARNOT's assumption that $C^+ = C^-$ correspond to one special case of REECH's result (1) but also the reader is expected to read off from $(1)_1$ the equivalence of the condition $\Gamma(\theta) = J$, a constant, to the assumption of CLAUSIUS: Heat is universally and uniformly interconvertible with work in cyclic processes. REECH does not mention these names, but he does mention the possibility that "by chance, experiment might make it come out" that $\Gamma(\theta) = \text{const.}$, though he regards this possibility as "truly very doubtful". This is the year 1851.

In Chapter 10 we shall consider the results REECH is to publish in the revised long memoir.

9D. KELVIN's Analysis of the Joule-Thomson Effect and Subsidiary Details

Some retrospective parts of KELVIN's next paper[1] have been mentioned above in §§7H and 7I. It is here (§65) that KELVIN writes of the statement

$$J\Lambda_V = p , \qquad (7D.1)_r$$

which we have called HOLTZMANN's Assertion,

---

[10] We must be careful not to think that C^+ and C^- can be varied independently. They cannot, since

$$C^+ = \int_{V_a}^{V_b} \Lambda_V(V, \theta^+)dV , \qquad C^- = \int_{V_d}^{V_c} \Lambda_V(V, \theta^-)dV ,$$

and the points a, b, c, and d are the vertices of a Carnot cycle. For example, if $\theta^- \to \theta^+$, it follows that $C^- \to C^+$, so both sides of (8) reduce to 0.

[1] THOMSON [1852, *1*].

> This, which will be called MAYER's hypothesis, from its having been first assumed by MAYER, is also assumed by CLAUSIUS without any reason from experiment; and an expression for μ the same as the preceding, is consequently adopted by him as the foundation of his mathematical deductions....

"The preceding", of course, is the HELMHOLTZ-JOULE Determination

$$\mu = J/\theta \ . \qquad (5N.7)_r$$

KELVIN obtains (§63) the integrated form of the General CARNOT-CLAPEYRON Theorem (5L.4) which we have recorded above in Footnote 10 to §7H.

Having refused for four years to accept JOULE's suggestion (5N.7), KELVIN faults CLAUSIUS for having assumed it "without any reason from experiment" (§65). He reminds the reader (§66) that "a complete test" of "MAYER's hypothesis" would be "the determination of the values of μ through a wide range of temperatures...with a single accurate determination of J.... Thus an experimental determination of the density of saturated steam for temperatures from 0° to 230° Cent. would complete the data...and would contribute more, perhaps, than any set of experimental researches that could at present be proposed, to advance the mechanical theory of heat." KELVIN tells us (§67)

> Mr. JOULE, when I pointed out these discrepancies to him in the year 1848, suggested that even between 0° and 100° the inaccuracy of the data regarding steam might be sufficient to account for them. I think it will be generally admitted that there can be no such inaccuracy in REGNAULT's part of the data....

As for CLAUSIUS' determination of an empirical formula for the density of steam sufficient to make μ conform with the HELMHOLTZ-JOULE Determination, KELVIN writes (§68)

> In this direction theory can go no further, for want of experimental data; although...it may be doubted whether such excessive deviations, in the case of steam, from the laws of a "perfect gas" are rendered probable by a hypothesis resting on no experimental evidence whatever.

Next (§72) KELVIN suggests the famous experiment that was to be interpreted as confirming the HELMHOLTZ-JOULE Determination $\mu = J/\theta$ while showing that HOLTZMANN's Assertion $J\Lambda_V = p$ may fail to be borne out by the behavior of real gases.

> Mr JOULE's second experiment on the same apparatus, in which he examined separately the external thermal effects round each of the

two vessels, and round a portion of the tube containing the small orifice (a stop-cock), has suggested to me a method which appears still simpler, and more suitable for obtaining an excessively delicate test of MAYER's hypothesis for any temperature. It consists merely in dispensing with the two vessels in JOULE's apparatus, and substituting for them two long spirals of tube (instead of doing this for only one of the vessels, as JOULE does in his third experiment with the same apparatus); and in forcing air continuously through the whole. The first spiral portion of the tube, up to a short distance from the orifice, ought to be kept as nearly as possible at the temperature of the atmosphere surrounding the portion containing the orifice, and serves merely to fix the temperature of the entering air. The following investigation shows what conclusions might be drawn by experimenting on the thermal phenomena of any fluid whatever treated in this manner.

73. Let p_1 be the uniform pressure of the fluid in the first spiral, up to a short distance from the orifice, and let p_2 be the pressure a short distance from the orifice on the other side, which will be uniform through the second spiral. Let θ be the constant external temperature, and let the air in both spirals be kept as closely as possible at the same temperature. If there be any elevation or depression of temperature of the fluid in passing through the orifice, it may only be after passing through a considerable length of the second spiral that it will again arrive sensibly at the temperature θ; and the spiral must be made at least so long, that the fluid issuing from the open end of it, when accurately tested, may be found not to differ appreciably from the primitive temperature θ.

74. Let $-C$ be the total quantity of heat emitted from the portion of the tube containing the orifice, and the second spiral, during the passage of a volume V_1 through the first spiral, or of an equivalent volume V_2 through the parts of the second where the temperature is sensibly θ. This will consist of two parts; one (positive) the heat produced by the fluid friction, and the other (negative) the heat emitted by that portion of the fluid which passes from one side to the other of the orifice, in virtue of its expansion.

The amount of work L, which "will be lost as external mechanical effect, and will go to generate thermal *vis viva*", is given by

$$L = \int_{V_1}^{V_2} p dV + p_1 V_1 - p_2 V_2 \,. \tag{9D.1}$$

The "quantity of heat thus produced" is L/J. Subtracting from this quantity "the amount previously found to be absorbed when the mechanical effect is all external" yields "the total quantity of heat emitted by that portion of tube which contains the orifice and the whole of the second spiral":

$$-C = \frac{1}{J}\left[\int_{V_1}^{V_2} p dV + p_1 V_1 - p_2 V_2\right] - \frac{1}{\mu}\int_{V_1}^{V_2} \frac{\partial p}{\partial \theta} dV \,. \tag{9D.2}$$

For an ideal gas (§75) $p_1 V_1 = p_2 V_2$, and (2) yields

$$-C = \left(\frac{1}{J} - \frac{1}{\mu\theta}\right) L \ . \tag{9D.3}$$

Thus (§76) "MAYER's hypothesis", which leads to the HELMHOLTZ-JOULE Determination

$$\mu = J/\theta \ , \qquad (5N.7)_r$$

implies that $C = 0$. KELVIN (§§77–80) regards this formula as lending itself to a good experimental test of "MAYER's hypothesis".

[Here we encounter both *the first attempt to treat by* thermodynamic theory *a process that is neither isothermal nor adiabatic nor cyclic and also the first attempt to take account of an "irreversible" effect* along with a reversible change. It is small wonder that on both counts the analysis is faulty. KELVIN seems to take over unchanged the ideas in CLAUSIUS' first paper yet allow the possibility that the fluid be something other than an ideal gas. From CLAUSIUS' results

$$\dot{\mathrm{E}} = JQ + P \ , \qquad P \equiv -p\dot{V} \ , \qquad (8B.2)_r$$

we see that if E is not a function of θ alone, the heat subtracted in an isothermal process is not just the mechanical equivalent of the work done on the body: KELVIN ought have added to the first integral on the right-hand side of (2) the quantity $\mathrm{E}(V_2, \theta) - \mathrm{E}(V_1, \theta)$. In §8D we have seen that CLAUSIUS' ideas if applied to a fluid having any equation of state would have led to the relation

$$J\Lambda_V - p = \frac{\partial \mathrm{E}}{\partial V} \ , \qquad (8D.3)_{1r}$$

which shows that HOLTZMANN's Assertion

$$J\Lambda_V = p \qquad (7D.1)_r$$

holds *if and only if* E is a function of θ alone. As KELVIN's formula (2) can hold only if HOLTZMANN's Assertion does, he cannot use one to test the truth of the other. Furthermore, to take account of the effects of fluid friction KELVIN should add further terms to his expression for the work done in pushing the gas through the spirals.] KELVIN seems to have grasped both these objections while this paper was passing through the press, for in a footnote to (2) he directs us to "a more comprehensive investigation, ... including a proof of this result", to which we turn now[2].

KELVIN first (§§84–85), after laying down the Doctrine of Latent and Specific Heats, *assumes* the truth of (8B.2) as an expression of the equivalence of heat and work. He *assumes* also that E, which he calls "the mechanical

---

[2] THOMSON [1853, *1*].

energy", is a function of V and θ. Thus he quickly derives the central formulae

$$\frac{\partial p}{\partial \theta} = J\left(\frac{\partial \Lambda_V}{\partial \theta} - \frac{\partial K_V}{\partial V}\right) \tag{8B.1)$_r$}$$

and

$$J\Lambda_V - p = \frac{\partial E}{\partial V}, \qquad JK_V = \frac{\partial E}{\partial \theta}. \tag{8D.3)$_r$}$$

[While for CLAUSIUS a major aim had been to *demonstrate* the existence of the internal-energy function, now the "First Law" is *stated* in terms of that function. Energy is become obvious. Subsequent work in thermodynamics has usually followed KELVIN in this regard. The idea of internal energy is easily adapted to thermodynamic systems far more general than any the pioneers considered.]

To explore the consequences of CLAUSIUS' thermodynamics *without* CLAUSIUS' "subsidiary hypothesis", KELVIN eliminates Λ_V between the General CARNOT-CLAPEYRON Theorem

$$\mu\Lambda_V = \frac{\partial p}{\partial \theta} \tag{5L.4)$_r$}$$

and $(8D.3)_1$. The result is

$$\frac{\partial E}{\partial V} = \frac{J}{\mu}\frac{\partial p}{\partial \theta} - p \tag{9D.4}$$

(KELVIN's Equation (4′)). J/μ is one and the same function of θ for all fluids. Once it be known, we can use (4) and $(8D.3)_2$ to determine E from p and K_V or p and K_V from E. "For example, let the fluid be atmospheric air, or any other subject to the 'gaseous' laws." Then for a body of ideal gas (his Equation (9))

$$E = R\left(\frac{J}{\mu} - \theta\right)\log\frac{V}{V_0} + J\int_{\theta_0}^{\theta} K_V(V_0, x)dx + E(V_0, \theta_0). \tag{9D.5}$$

[Since KELVIN is rejecting the HELMHOLTZ-JOULE Determination, E need not reduce to a function of θ alone.] Moreover, E and p together allow us to determine both K_p and K_V by using $(8D.3)_2$ and the calorimetric relation

$$K_p - K_V = -\Lambda_V \frac{\partial p}{\partial \theta}\bigg/\frac{\partial p}{\partial V}. \tag{2C.9)$_{2r}$}$$

KELVIN gives simple examples (§§90–92), leaving μ/J an arbitrary function of θ.

In the last three sections of the paper KELVIN corrects his former analysis of the Joule-Thomson effect. He considers the passage of a unit mass of air through the apparatus. He denotes by S "the mechanical value of the sound emitted from the 'rapids'", [that is, the amount of work that the body does so as to effect whatever dissipative processes necessarily accompany its transit]; if C is the heat added to the body, then (his Equation (16))

$$\mathrm{E}(V_2, \theta_2) - \mathrm{E}(V_1, \theta_1) = p_1 V_1 - p_2 V_2 - (-JC + S) \ . \qquad (9\text{D}.6)$$

> If the circumstances be arranged (as is always possible) so as to prevent the air from experiencing either gain or loss of heat by conduction through the pipe and stopcock, . . . $C = 0$; and if (as is perhaps also possible) only a mechanically inappreciable amount of sound be allowed to escape, we may take $S = 0$.

Then (6) reduces to (his Equation (17))

$$\mathrm{E}(V_2, \theta_2) = \mathrm{E}(V_1, \theta_1) - (p_2 V_2 - p_1 V_1) \ . \qquad (9\text{D}.7)$$

If the air is an ideal gas and $\theta_2 = \theta_1$, then $\mathrm{E}(V_2, \theta) = \mathrm{E}(V_1, \theta)$, "which is, in fact, the expression of MAYER's hypothesis, in terms of. . .mechanical energy. . . ." [Thus KELVIN has noticed, at last, this obvious and celebrated corollary of $(8\text{D}.3)_1$.] If, on the other hand, $\theta_2 \neq \theta_1$, we may calculate V_2 and V_1 from "the known laws of density of air," so from (7) we can calculate $\mathrm{E}(V_2, \theta_2) - \mathrm{E}(V_1, \theta_1)$.

If, instead (§96), "the air on leaving the narrow passage be. . .brought back exactly to the primitive temperature. . . ," and if S can be neglected, then use of (5) in (6) yields KELVIN's former result (2) as specialized to an ideal gas.

[In the explosion of tedious print we are here enduring, these few pages out of all KELVIN's effusions on thermodynamics seem doubly wonderful. Up to now *every thermodynamic analysis has rested upon the Doctrine of Latent and Specific Heats* and has shared its inherent limitations. Here, for the first time, we see *heating and working regarded as primitive concepts*. Indeed, in (6) we cannot expect to calculate C from the calorimetric relation

$$C = C(\mathscr{P}) = \int_{\mathscr{P}} [\Lambda_V(V, \theta) dV + \mathrm{K}_V(V, \theta) d\theta] \qquad (2\text{C}.6)_r$$

nor to regard S as an amount of work calculable from the definition

$$L \equiv \int_{t_1}^{t_2} p(t) \dot{V}(t) dt \ . \qquad (2\text{C}.19)_r$$

[KELVIN's result (6) is *the earliest example of a true balance of energy, heat, and work*, not bound to any particular class of constitutive relations. KELVIN's first derivation of (2) had rested upon use of a class of constitutive relations which in fact do not apply to irreversible processes. His analysis here shows that it need not have done so.

[In KELVIN's treatment we find also the first occurrence of the *enthalpy*: $X(V, \theta) \equiv E + pV$. It is the enthalpy that here is expended by the production of heat and loss of work through processes which need not be described in detail. The literature of thermodynamics has followed this way of looking at the matter. We shall see below in §10C that in this very same year, 1853, REECH will introduce the enthalpy formally in a more general thermodynamics and will show that the enthalpy when taken as a function of entropy and pressure is a thermodynamic potential.

[KELVIN still adheres to "the gaseous laws" expressed by the thermal equation of state $pV = R\theta$. The now accepted interpretation of the Joule-Thomson effect regards the experiment as demonstrating departure of a real gas from the behavior of an ideal one. To trace the development leading to this interpretation would carry us beyond the period of this tragicomedy, so I merely state the outcome. The enthalpy is calculated as a function of θ and p. Then $\dot{\theta}$ becomes proportional to $\dot{p}$ with a coefficient that is a known function of p and θ. The coefficient may be positive or negative or null. The ratio $\dot{\theta}/\dot{p}$ is interpreted as approximating $(\theta_1 - \theta_2)/(p_1 - p_2)$.]

Another note by KELVIN[3] presents for the first time [and astonishingly late in the day] the explicit calculation of the work done by an ideal gas in adiabatic expansion from the volume V_1 to the volume V_2, on the assumption that $\gamma = \text{const.}$:

$$L = \frac{R\theta_2}{\gamma - 1}\left[\left(\frac{V_2}{V_1}\right)^{\gamma-1} - 1\right] . \tag{9D.8}$$

KELVIN obtains this result straight off from the LAPLACE-POISSON Law $pV^\gamma = \text{const.}$, which shows him also that for a Carnot cycle labelled as in Figure 3 in §5C

$$\frac{V_a}{V_b} = \frac{V_d}{V_c} . \tag{9B.14$_r$}$$

Application of (9B.10) to (8) shows that the work done on the expanding adiabat is just annulled by the work lost on the contracting adiabat, so two uses of PETIT's formula

$$L(\mathscr{P}_\theta) = R\theta \log \frac{V_b}{V_a} \tag{5K.1$_{2r}$}$$

deliver the work done by the cycle:

$$L(\mathscr{C}) = R\theta^+ \log \frac{V_b}{V_a} - R\theta^- \log \frac{V_c}{V_d} . \tag{9D.9}$$

Another use of (9B.14) reduces (9) to the form

$$\begin{aligned} L(\mathscr{C}) &= R(\theta^+ - \theta^-) \log \frac{V_b}{V_a} , \\ &= \left(R\theta^+ \log \frac{V_b}{V_a}\right)\left(1 - \frac{\theta^-}{\theta^+}\right) , \\ &= L(\mathscr{P}_{\theta^+})\left(1 - \frac{\theta^-}{\theta^+}\right) , \end{aligned} \tag{9D.10}$$

[3] THOMSON [1852, *3*].

$L(\mathscr{P}_{\theta^+})$ being the work done in the expansion from V_a to V_b at the temperature θ^+. Thus much follows from nothing but the LAPLACE-POISSON law [and hence rests only upon the Doctrine of Latent and Specific Heats applied to an ideal gas having constant γ]. HOLTZMANN's Assertion in the form

$$L(\mathscr{P}_\theta) = JC(\mathscr{P}_\theta) \ , \qquad (7D.4)_r$$

converts (10) into

$$\frac{L(\mathscr{C})}{JC^+(\mathscr{C})} = 1 - \frac{\theta^-}{\theta^+} \ . \qquad (9A.10)_r$$

KELVIN remarks that his earlier derivation of this result from "MAYER's hypothesis" did not require γ to be constant, and that RANKINE also had arrived at it.

[KELVIN misses the point of what he has done. He has proved that (9A.10) follows from assuming that γ = const. and $J\Lambda_V = p$. But he already knew that $J\Lambda_V = p$ was equivalent to $K_p - K_V$ = const. (*cf.* §7E). Thus he assumes that both K_V and K_p are constant. What he has proved is that *an ideal gas with constant specific heats achieves the classical efficiency in Carnot cycles*. He could easily have gone on to prove] a theorem soon to be discovered by HOPPE[4]: *An ideal gas with constant specific heats interconverts heat and work uniformly and universally in cyclic processes*—this without applying any relation between heat and work, not even CARNOT's General Axiom! [Indeed, HOLTZMANN's Assertion $J\Lambda_V = p$ allows us to replace Λ_V by $R\theta/(JV)$ in the general formula for the heat absorbed on an isothermal path:

$$C(\mathscr{P}_\theta) = \int_{V_a}^{V_b} \Lambda_V(V, \theta) dV \ , \qquad V_b > V_a \ , \qquad (5L.2)_r$$

$$JC(\mathscr{P}_\theta) = R\theta \int_{V_a}^{V_b} \frac{dV}{V} = R\theta \log \frac{V_b}{V_a} \ . \qquad (9D.11)$$

Hence for a Carnot cycle

$$JC(\mathscr{C}) = R\theta^+ \log \frac{V_b}{V_a} - R\theta^- \log \frac{V_c}{V_d} \ . \qquad (9D.12)$$

If also the *ratio* of specific heats is constant, we may use KELVIN's result (9) and so conclude that $L(\mathscr{C}) = JC(\mathscr{C})$. Moreover, the J that appears here

[4] HOPPE [1856] established this fact somewhat awkwardly, using the unfortunate independent variables p and V. CLAUSIUS [1856, *3*], while pronouncing HOPPE's analysis elegant, explained why in his work of 1850 he had not assumed K_p and K_V constant: Even to assume that they were functions of θ alone was "counter to the then received views", and only later did REGNAULT's experiments lend credence to assuming them constant. CLAUSIUS misses the point here. It is neither necessary nor desirable to assume that K_p and K_V are constant for all ideal gases. If *one* ideal gas with constant specific heats is allowed as a *mathematical model*, that assumption contradicts the Caloric Theory and provides an *example* of a substance that *automatically* satisfies CLAUSIUS' basic assumption of 1850: $L(\mathscr{C}) = JC(\mathscr{C})$.

is the mechanical equivalent of a unit of heat in isothermal expansion of an ideal gas. The result established shows that *this same J is the mechanical equivalent of a unit of heat in Carnot cycles*; that is, the J that appears in CLAUSIUS' basic axiom (7A.1). HOPPE's theorem thus fills a gap in CLAUSIUS' analysis which we noted in §8D: *If one ideal gas has constant specific heats, the MAYER-HOLTZMANN equivalent and the CLAUSIUS equivalent are equal.* Both, of course, are the J obtained by use of MAYER's formula:

$$J = \frac{R}{(1 - \gamma^{-1})\mathrm{K}_p} \,. \tag{7B.3)$_r$}$$

[These observations provided one key to the program of *Concepts and Logic*. The other key was the idea that one example serves to determine a universal function. This applies to theory a statement KELVIN made in reference to experiment, a statement quoted above in §7H. To use this key, we *now*—and *only* now—invoke CARNOT's General Axiom. Then we at once determine the universal function μ by Theorem 2 in §7I, namely $\mu = J/\theta$. It is then an easy matter to demonstrate the "First Law" and the "Second Law" in their classical forms for reversible processes. To confirm this statement, the reader may consult *Concepts and Logic*.]

Appendix by C.-S. MAN: The JOULE–THOMSON Experiment

In the first of their communications to the Royal Society with the title "On the Thermal Effects of Fluids in Motion", KELVIN & JOULE[1] undertook to determine "the value of μ, CARNOT's function" as "[a] principal object of the researches." For that purpose they proposed to use the formula

$$\frac{J}{\mu} = \frac{J\mathrm{K}\delta + \int_{V_1}^{V_2} p\,dV + p_1V_1 - p_2V_2}{\int_{V_1}^{V_2} \partial p/\partial\theta \, dV} \,, \tag{1}$$

which is (9D.2), with C in (9D.2) written as $\mathrm{K}\delta$. Here "δ is the observed cooling effect; ... K the thermal capacity of a pound of the gas under constant pressure equal to that on the low-pressure side of the gas", and "$\mathrm{K}\delta$ is the heat that would have to be added to each pound of the exit stream of [gas] to bring it to the temperature of the bath". For convenience of exposition, in this appendix we follow KELVIN and adopt as empirical-temperature scale the centigrade temperature θ according to REGNAULT's standard air-thermometer.

A glance immediately reveals the fact that the experimental data delivered

[1] THOMSON & JOULE [1853, pp. 357–358].

by a porous-plug experiment with bath temperature θ_b, namely δ, p_1, and p_2, are insufficient to determine $\mu(\theta_b)$ from (1). To that end also J, K, V_1, V_2, and the functions $p(V, \theta_b)$, $\partial p/\partial\theta\ (V, \theta_b)$ "between the limits of pressure in the experiment" must be ascertained.

[By this time, a good value of J is available[2]. In April of this very year, 1853, REGNAULT[3] publishes the results of his long overdue experiments on specific heats of gases. Thus all requisite numbers and functions will be available, once the equation of state of the gas be known. However,] the equation of state as delivered by the "gaseous laws" is unacceptable here not only because "REGNAULT has shown that the [gaseous laws are] not rigorously true", but [more importantly] because JOULE & KELVIN's experiments "show that $\delta/\log(p_1/p_2)$ increases with p_1/p_2." Indeed, for KELVIN throughout the period of this tragicomical history the equation of state that follows from the gaseous laws is, for a pound of gas,

$$pV = p_0V_0(1 + \alpha_0\theta) \ ; \tag{2}$$

p_0V_0 is "the product of the pressure...into the volume...of the gas at 0° Cent."; $\alpha_0 = 0.003665$ is "the standard coefficient of expansion of atmospheric air" for the standard air-thermometer. Substitution of (2) into (1) yields

$$\frac{J}{\mu} = \frac{1}{\alpha_0} + \theta + \frac{JK\delta}{\alpha_0 p_0 V_0 \log(p_1/p_2)} \ . \tag{3}$$

Data which show $\delta/\log(p_1/p_2)$ increasing with p_1/p_2 will not provide a consistent estimate for μ. "Hence in reducing the experiments," KELVIN & JOULE observe, "a correction must be first applied to take into account the deviations, as far as they are known, of the fluid used, from the gaseous laws, and then the value of μ may be determined [through (1)]."

As JOULE & KELVIN remark in §III of their second communication[4] on the same subject, their preliminary experiments[5], for which the cooling effect δ is "very slight", are sufficient to show that $J/\mu \approx (1/\alpha_0) + \theta$ approximately, [say to within one or two percent,] at least for the range of temperature of their experiments (from about 0° to 77°C). [Hence the result which they expect to get is quantitative, say accurate to "within less than two or three tenths of a degree"; appeal to (1) shows that the scatter of their data of 1854 would exclude anything sharper. If on the one hand REGNAULT's "great work" of 1847 does show definitively that no real gas obeys the "gaseous laws", it seems not to provide data sufficient to deliver for any real gas an empirical equation of state good enough to calculate J/μ from (1) with an error less than two or three tenths of a degree.]

---

[2] JOULE [1850].
[3] REGNAULT [1853].
[4] JOULE & THOMSON [1854].
[5] JOULE & THOMSON [1852].

In proposing the use of (1) to evaluate μ (with air as the gas in the experiment), KELVIN & JOULE also remark:

> An expression for $[\int_{V_1}^{V_2} p dV]$ for any temperature may be derived from an empirical formula for the compressibility of air at that temperature, and between the limits of pressure in the experiment.

[Thus they are not yet aware of the difficulty.]

In their second communication[6] to the Royal Society JOULE & KELVIN point out the difficulty inherent to their previous proposal (p. 348):

> The direct use of this equation [i.e., (1)] for determining J/μ requires, besides our own results, information as to compressibility and expansion, which is as yet but very insufficiently afforded by direct experiments, and is consequently very unsatisfactory....

Moreover, although they report data of experiments on air, carbon dioxide and hydrogen, only those for air are "of sufficient accuracy" to warrant using them in the evaluation of μ. For air, JOULE & KELVIN's main experimental result is (their Equation (20)):

$$\delta = 0.26 \frac{p_1 - p_2}{\Pi} \quad \text{at } 17°\text{C} \; ; \tag{4}$$

$p_1 - p_2$ is the difference of the pressures (in lb/sq in.) of air before and after passing through the porous plug, and Π is a conversion factor, changing units of pressure from lb/sq in. to atm. It follows immediately that with (4) alone, only the value of J/μ at 17°C can be obtained directly through (1).

Nevertheless JOULE & KELVIN go on to outline two plans for estimating the value of μ at 16°C according to their previous proposal (§III). Instead of using (4), which is obtained through various averagings—in fact the bath temperature used to get (4) ranges from about 12°C to about 20°C for an individual experiment—they select and use the data from eight of their experiments at 16°C.

[Because the authors themselves seem not to take either of their plans seriously, and because I find numerous errors in their numerical calculation, I will not describe these plans in detail here, nor even quote the results. It suffices to say that] one plan makes use of RANKINE's equation of state for air, which is "a formula obtained with such insufficient experimental data as Mr. RANKINE had for investigating the empirical forms which his theory left undetermined." In the other plan, to calculate the integrals in (1) JOULE & KELVIN use, for one pound of air, an equation of state of the form

$$pV = H(1 + \bar{\alpha}\theta) \; ; \tag{5}$$

---

[6] JOULE & THOMSON [1854].

H is "the product of the pressure into the volume of a pound of air, at 0° Cent.", and $\bar{\alpha}$ denotes "a certain mean coefficient of expansion suitable to the circumstances of each individual experiment". Thus (1) becomes

$$\frac{J}{\mu} = \frac{1}{\bar{\alpha}} + \theta + \frac{JK\delta + p_1V_1 - p_2V_2}{\alpha_0 H \log(p_1/p_2)} \; ; \tag{6}$$

really the coefficient of the denominator of the last fraction should be $\bar{\alpha}$, but JOULE & KELVIN substitute α_0 for it, "since the numerator...is so small, that the approximate value may be used for the denominator." To estimate the values of $\bar{\alpha}$ and $p_1V_1 - p_2V_2$ for the several experiments, a further problem arises for want of data. To evade the problem, JOULE & KELVIN estimate $p_1V_1 - p_2V_2$ "by using REGNAULT's experimental results on compressibility of air as if they had been made, not at 4°75, but at 16° Cent."; as for $\bar{\alpha}$, which should represent some "mean coefficient of expansion" for the range of pressure p_1 to p_2 "at the particular temperature of the experiment", they use REGNAULT's observations regarding the effect of variations of density on "the mean [pressure coefficient of] expansion from 0° to 100°". [In following this plan, JOULE & KELVIN have to resort to various arbitrary assumptions, which make it virtually impossible to estimate the error thus introduced.]

In consequence of the difficulty inherent to their original program, in §V of their second communication JOULE & KELVIN adopt another approach, one result of which is the table "Comparison of Air-thermometer with Absolute Scale" at the end of §IV.

If the "absolute scale" (their Equation (6), see §11B, below)

$$\mathrm{T} = \frac{J}{\mu}$$

is used instead of the empirical-temperature scale θ, rearrangement of (1) yields (their Equation (15))

$$\delta = \frac{1}{JK}\left\{\int_{V_1}^{V_2}\left(\mathrm{T}\frac{dp}{d\mathrm{T}} - p\right)dV + (p_2V_2 - p_1V_1)\right\} . \tag{7}$$

JOULE & KELVIN's idea is to use their experimental data on δ and results of other experiments to determine an equation of state for a pound of air:

$$p = \varpi(V, \mathrm{T}) \; . \tag{8}$$

Then substitution of (8) into the formula (their Equation (9))

$$\Theta = 100\,\frac{p - p_i}{p_s - p_i}, \qquad V = \text{const.} \; , \tag{9}$$

which defines the centigrade temperature according to a constant-volume air-thermometer (p_s and p_i denote the pressure of the thermometer at the steam point and the ice point, respectively), will give Θ as a function of T, with the volume V of a pound as parameter. Obviously this includes the result expressing θ, the centigrade temperature according to the standard air-thermometer, as a function of T, the inverse of which gives the evaluation of CARNOT's function "for any temperature".

[Experimental data alone cannot determine *a posteriori* an equation of state. As a matter of fact, a class of equations of state with undetermined coefficients has to be proposed to start with. Experimental data are then used to determine these coefficients. Thus the proposed class of approximate constitutive functions is crucial.]

JOULE & KELVIN thus explain how they choose the class of equations of state they adopt (p. 357):

> [w]e adopt the form to which Mr. RANKINE was led by his theory of molecular vortices, and which he has used with so much success for the expression of the pressure of saturated steam and the mechanical properties of gases; with this difference, that the series we assume proceeds in descending powers of the absolute thermo-dynamic temperature, while Mr. RANKINE's involves similarly the temperature according to what he calls "the scale of the perfect gas-thermometer."

In fact, they adopt a simplified version of RANKINE's formula, assuming for a pound of air (their equation (17)) that

$$pV = A\mathrm{T} + B + (C + D/\mathrm{T} + G/\mathrm{T}^2)\Phi/V\ , \tag{10}$$

> where A, B, C, D and G are all constants to be determined by the comparison with experimental results, and Φ denotes a particular volume corresponding to a standard state of density, which it will be convenient to take as 12.387 cubic feet, the volume of a pound when under the atmospheric pressure Π... [at the ice point]. The series is stopped at the fifth term, because we have not at present experimental data for determining the coefficients for more.

From the start, "all the terms following [B] in the series" are presumed to be "very small fractions of pV".

> The experimental data which we have, and find available, are (1) the results of REGNAULT's observations on the coefficients of expansion at different constant densities, (2) the results of his observations on the compressibility, at a temperature of 4°.75 Cent., and (3) our own experimental results now communicated to the Royal Society. These are expressed within their limits of accuracy (at least for pressures of from

one to five or six atmospheres, such as our experiments have as yet been confined to), by the following equations [their Equations (18) and (19)]:

$$\alpha = 0.003665 + \frac{0.0000441}{3.81}\left(\frac{\Phi}{V} - 1\right), \tag{11}$$

or

$$\alpha = 0.00365343 + 0.000011575\,\Phi/V \tag{12}$$

$$p_1V_1 - p_2V_2 = 0.008163\,\frac{p_1 - p_2}{\Pi}\,p_1V_1, \text{ at temperature } 4^\circ\!.75 \text{ Cent.}, \tag{13}$$

and by (4) above. [I do not find (11), (12), or (13) in REGNAULT's "great work" of 1847. Probably JOULE & KELVIN devise for each of the above two formulae a format which serves their purpose in the calculation, and then determine the coefficients that best fit REGNAULT's data. Checking (12) against REGNAULT's data (with Φ/V ranging from 0.1444 to 4.8100)[7], I find it fits those data fairly well (error within $\pm 0.1\%$) except for the two with the smallest values of Φ/V, namely 0.1444 (error $\approx \pm 0.2\%$) and 0.2294 (error $\approx +0.13\%$). An error of 0.2% would certainly be intolerable if accuracy to "within less than two or three tenths of a degree" were sought for the absolute temperature of the ice point.]

From (4), (7), (10), (12), (13) and the definition (their Equation (13))

$$\alpha \equiv (p_s - p_i)/100p_i\ ,\ V = \text{const.}, \tag{14}$$

JOULE & KELVIN (i) putting $T_{4.75} = T_0 + 4.75$, $T_{17} = T_0 + 17$, where T_0, $T_{4.75}$, and T_{17} denote the absolute temperature corresponding to the ice point, the hotness represented by $\theta = 4.75°C$, and the hotness represented by $\theta = 17°C$, respectively, then (ii) "neglecting squares and products of the small quantities C, D, G,"* and then (iii) putting

$$\frac{\Phi}{V} = \frac{p(AT_0 + B)}{\Pi(AT + B)} \tag{15}$$

whenever it is necessary to convert Φ/V into p/Π, obtain (their Equations (26) and (30))

$$B = 0\ ,$$
$$1/T_0 = 0.00365343\ , \tag{16}$$

or

$$T_0 = 273^\circ\!.72\ ,$$

$(16)_2$ being the value of α when $V = +\infty$. Three linear equations (their Equations (27), (29), and (31)) involving the four unknowns A, C, D, G remain. A further linear equation (p. 359)

---

[7] REGNAULT [1847, p. 110].

*[By "the small quantities C, D, G" JOULE & KELVIN really mean terms with coefficients C, D, G, namely $C\Phi/V$, $D\Phi/TV$, $G\Phi/TV^2$. Moreover what they really mean to assume is that these terms are "small quantities" as compared with pV, for a range of temperature which includes 0° to 17° C and for a range of pressure say from one-half to a few atmospheres.]

is afforded by a determination of the density of air, which has been most accurately given by REGNAULT, ..., a result which is expressed by the equation

$$\Pi\Phi = 26224 .$$

...Hence we have the equation

$$A\mathrm{T}_0 + B + C + D/\mathrm{T}_0 + G/\mathrm{T}_0^2 = 26224 . \tag{17}$$

Calling 26224, H, ...we may simplify the treatment of the four equations by taking the approximate value H for $A\mathrm{T}_0$, in three of them, [their Equations (27), (29), (31)] without losing accuracy, and we may afterwards use (17) to determine the exact value of [$A\mathrm{T}_0$].

Using the value 1390 for J and 0.238 for K, JOULE & KELVIN thus finally obtain (their Equation (39))

$$pV = H\left\{\mathrm{T} - \left(0.0012811 - \frac{1.3918}{\mathrm{T}} + \frac{353.2}{\mathrm{T}^2}\right)\Phi/V\right\} ; \tag{18}$$

by substituting (18) into (9), they calculate the second column in the table of §IV.

[By checking the numerical calculations I find that the numerical coefficients in (18) are erroneous. So are the entries in the second column of their table, with still more computational errors added.]

JOULE & KELVIN claim that they also obtain the third column of their table from (18) but do not indicate how they do so. [Judging from what they have done earlier and also later in the same section in the calculation of specific heats, I think it likely that by using (15) and (16) they convert the right-hand side of (18) from a function of Φ/V and T into a function of p/Π and T. By substituting the subsequent equation into (their Equation (10))

$$\vartheta = 100\frac{V - V_i}{V_s - V_i} , \ p = \text{const.} , \tag{19}$$

where V_s and V_i denote the volume of a pound at the steam point and the ice point respectively, they would have expressed ϑ as a function of T and p/Π; here, as is apparent from the defining equation (19), ϑ denotes the centigrade temperature according to a constant-pressure air-thermometer. In view of the numerous errors in numerical calculation I have found so far, I do not think it worthwhile to check my conjecture by comparing explicit results that follow from it with the entries in the third column of the table.]

That portion of §V of their paper of 1854 which we have just described in the text and another portion of the section that deals with the calculation of specific heats are excised in the volumes of collected papers of JOULE and KELVIN. A note of KELVIN's of 1882 gives the reason: At the conclusion of

Part IV of the series of papers bearing the same title, they have derived "a better and simpler empirical formula...from more comprehensive experimental data."

Some years after it appeared, JOCHMANN[8] criticized the work of JOULE & KELVIN we have described above. He based his objections on the following grounds:

(i) "When they set up [(18)], they make no use at all of Mr. Thomson's own definition of absolute temperature, which never enters the calculation."

(ii) "In the calculation, the purpose of which should be neither more nor less than to determine the difference between the two scales, the absolute temperatures that correspond to the temperatures 4.75°C and 17°C are set equal to $T_0 + 4.75$ and $T_0 + 17$."

(iii) In the calculation, the value of T_0 turns out to be the reciprocal of α in (12) for $V = +\infty$. The formula (12) is used in the determination of the coefficients of (18). "However, [(12)] agrees with experiment with tolerable accuracy only for pressures greater than one atmosphere; at lower pressures it fails to agree so significantly that for example, even when $\Phi/V = 0.1444$, Regnault's observed value of $\alpha = 0.0036482$ is smaller than the limiting value assumed by Joule and Thomson."

[JOULE & KELVIN's calculation is first and foremost based on the *a priori* form (10) of the equation of state. The most important implicit assumption underlying this choice is that KELVIN's absolute temperature T coincides with the ideal-gas temperature. This fact will make JOCHMANN's first objection at best nominal. KELVIN's definition $T = J/\mu$ does enter the calculation through (7). Hence JOCHMANN's first objection is entirely groundless. However, JOCHMANN's second and third objection are basically sound.

[The value of JOULE & KELVIN's work described above does not lie in the results that it in fact obtains. As we have remarked above, as far as qualitative conclusion is concerned, the JOULE-THOMSON experiments of 1854 cannot add anything new to the result of their preliminary experiments of 1852. On the other hand, if we wish to compare the scale of the air-thermometer with the absolute scale, neither the experiments of JOULE & KELVIN nor those of REGNAULT are sufficiently accurate to provide results of some permanence.

[Correction of the gas-thermometer naturally divides into two parts: (i) to determine the absolute temperature of some fixed point, say T_0 for the ice point; (ii) for other hotnesses to determine the corrections which convert the gas scale to the thermodynamic scale. If RANKINE[9] anticipates JOULE & KELVIN in being the first to attempt to estimate the value of T_0, it is JOULE & KELVIN's paper of 1854 which formulates this very problem in its entirety and provides a framework of ideas which later workers will use and modify.]

---

[8] JOCHMANN [1860, pp. 98–100].

[9] RANKINE [1853, *3*]. Of course, the determination of the "absolute zero" has a long history, but it was RANKINE who first attempted to determine "the absolute zero of the perfect gas thermometer".

9E. RANKINE's Further Effusions

RANKINE, feeling that the assumptions in his first work had left it "doubtful whether the conclusions deduced from the hypothesis were applicable to any substances except those nearly in the state of perfect gas", next[1] examines his theory of molecular vortices in greater generality and concludes that its results are "applicable to all substances in all conditions...." His thermal equation of state is now (his Equation (1))

$$p = F(V, \mathfrak{k}(\theta - \theta_0)) + f(V) \,. \tag{9E.1}$$

In Footnote 6 to §8G, above, we have described his views on the nature of heating at this time. [We have seen above in §8H that RANKINE's phenomenological structure of 1850 was already both complete and general, so little more can be expected for our tragicomical history from refinement of the molecular model.] RANKINE introduces the internal energy explicitly (his Equation (27)):

$$\mathrm{E} \equiv M\mathfrak{k}(\theta - \theta_0) + R\theta_0\left(\log\theta + \frac{\theta_0}{\theta}\right) + \int\left[(\theta - \theta_0)\frac{\partial p}{\partial\theta} - p\right]dV \,. \tag{9E.2}$$

He concludes that $\Delta\mathrm{E}$ is "the total amount of power which must be exercised" upon a body to make it pass from one value of (V, θ) to another (¶10):

> This quantity consists partly of expansive or compressive power, and partly of heat, in proportions depending on the mode in which the intermediate changes of temperature and volume take place; but the total amount is independent of these changes.
>
> Hence, *if a body be made to pass through a variety of changes of temperature and volume, and at length be brought back to its primitive volume and temperature, the algebraical sum of the portions of power applied to and evolved from the body, whether in the form of expansion and compression, or in that of heat, is equal to zero.*
>
> This is one form of the law, proved experimentally by Mr. JOULE, of the equivalence of heat and mechanical power. In my original paper on the Mechanical Action of Heat, I used this law as an axiom, to assist in the investigation of the equation of latent heat. I have now deduced it from the hypothesis on which my researches are based—not in order to prove the law, but to verify the correctness of the mode of investigation which I have followed....
>
> The train of reasoning in this article is the converse of that followed by Professor WILLIAM THOMSON of Glasgow, in article 20 of his paper on the Dynamical Theory of Heat, where he proves from JOULE's law that the quantity corresponding to $d\mathrm{E}$ is an exact differential.

---

[1] RANKINE [1852].

[While we must respect these words as expressing an early statement of the meaning of internal energy, RANKINE's definition (2) is not a happy one. Indeed, from it we see that $\partial E/\partial \theta$ is the function JK_V as given by RANKINE's insufficient formula $(8G.8)_2$. In §8H, above, we have shown that RANKINE's phenomenological apparatus of 1850 was not subject to this limitation. RANKINE has not yet seen what is the matter here.]

On January 5, 1853, RANKINE read before the Philosophical Society of Glasgow a short and ambitious paper[2] called *On the general law of the transformation of energy*. The aim of this work is stated after his summary of it in a second paper[3], which he read twelve days later before the Royal Society of Edinburgh:

> (52.) We have now obtained a system of formulae, expressing all the relations between heat and expansive power, analogous to those deduced from a consideration of the properties of temperature, by Messrs CLAUSIUS and THOMSON, and from the Hypothesis of Molecular Vortices in the previous sections of this paper; but, in the present section, both the theorems and the investigations are distinguished from former researches by this circumstance;—that they are independent, not only of any hypothesis respecting the constitution of matter, but of the properties, and even of the existence, of such a function as Temperature; being, in fact, simply the necessary consequences of the following
>
> DEFINITION OF EXPANSIVE HEAT
>
> *Let the term* EXPANSIVE HEAT *be used to denote a kind of Physical Energy convertible with, and measurable by, equivalent quantities of Mechanical Power, and augmenting the Expansive Elasticity of matter, in which it is present.*

By this time the Interconvertibility of Heat and Work is beginning to seem self-evident because (¶47) "physical power cannot be annihilated, nor produced out of nothing," and RANKINE contends even that the law was "virtually, though not expressly, admitted by those who introduced the term Latent Heat into scientific language...." [At this point thermodynamics begins to create a fictitious history for itself.]

The short paper begins as follows:

> ACTUAL, or SENSIBLE ENERGY, is a measurable, transmissible, and transformable condition, whose presence causes a substance to tend to

---

[2] RANKINE [1853, *1*].
[3] RANKINE [1853, *3*].

change its state in one or more respects. By the occurrence of such changes, actual energy disappears, and is replaced by

POTENTIAL OR LATENT ENERGY; which is measured by the product of a change of state into the resistance against which that change is made.

(The *vis viva* of matter in motion, thermometric heat, radiant heat, light, chemical action, and electric currents, are forms of actual energy; amongst those of potential energy are the mechanical powers of gravitation, elasticity, chemical affinity, statical electricity, and magnetism.)

The law of the *Conservation of Energy* is already known, viz.:—that the sum of all the energies of the universe, actual and potential, is unchangeable.

The object of the present paper is to investigate the law according to which all *transformations of energy*, between the actual and potential forms, take place.

Let V be the magnitude of a measurable state of a substance;

U, the species of potential energy which is developed when the state V increases;

P, the common magnitude of the tendency of the state V to increase, and of the equal and opposite resistance against which it increases; so that—

$$dU = PdV; \quad \text{and} \quad P = \frac{dU}{dV} \quad \cdots \quad \text{(A.)}$$

Let Q be the quantity which the substance possesses, of a species of actual energy whose presence produces a tendency of the state V to increase.

It is required to find how much energy is transformed from the actual form Q to the potential form U, during the increment dV; that is to say, the magnitude of the portion of dU, the potential energy developed, which is due to the disappearance of an equivalent portion of actual energy of the species Q.

[In this passage the spectator sees the first appearance of the term "potential energy". RANKINE, heretofore the scrupulous engineer and natural philosopher, now utters pronouncements about "all the energies of the universe"; he has been poisoned by thermodynamics; he suddenly begins to wallow in orgies of vagueness wilder than any up to now encountered, a vagueness which MAXWELL[4], a generation later, is to ridicule; unfortunately this phantasmagoric obscurity is destined to become standard in textbooks.] RANKINE's avowed axiom is

GENERAL LAW OF THE TRANSFORMATION OF ENERGY:—

The effect of the whole Actual Energy present in a substance, in

---

[4] MAXWELL [1878].

causing Transformation of Energy, is the sum of the effects of all its parts.

[The following physical assumptions and mathematical manipulations are cloudy enough in themselves but made more so by RANKINE's failure to state what is a function of what, and what a derivative such as dU/dV is to mean if the "state" V is anything other than a single real variable.] The body of the paper evaluates the efficiency of what seems to be a Carnot cycle and asserts, "This principle is applicable to all possible engines, known and unknown."

To understand the part of RANKINE's ideas that bear upon our tragicomedy, we turn now to the long paper, for it concerns mainly those modest quantities to which the early thermodynamics really refers: a fluid body of volume V, subject to pressure p. In giving an account of this work we use the letter E_0/M to replace[5] RANKINE's letter Q, which stands (¶48) for the mechanical equivalent of "the absolute quantity of thermometric heat" or the "Sensible heat (which retains its condition)" in unity of mass. Then RANKINE's operative assumption is

$$JQ = \dot{E}_0 + \dot{S} + p\dot{V} \ , \tag{9E.3}$$

in which S is "Latent heat, or heat which disappears in overcoming molecular action" or "the potential of molecular action", and the working $p\dot{V}$ is the rate of change of "Latent heat equivalent to the visible mechanical effect". [Of course this is not the latent heat we have denoted all along by Λ_V;] RANKINE in ¶ 49 calls Λ_V "[t]he coefficient of latent heat of expansion at constant heat".

[In §8H we have shown that RANKINE's first work, if we simply leave aside the molecular vortices and interpret his basic formal statements phenomenologically, implies for a body obeying an arbitrary equation of state the full structure of classical thermodynamics—energy, entropy, and all. We might expect that after having seen CLAUSIUS' work RANKINE himself would have reached this same conclusion. Perhaps he did, but at the same time he chose to abandon the traditional ideas of heat and temperature. His explanation of what he does[6] is so regal as to make the reader wonder

---

[5] RANKINE's notations and those in terms of which I describe his work may be interconverted as follows:

Here	RANKINE
E_0/M	Q
E/M	Ψ
$(\mathrm{E} - E_0)/M$	S

[6] RANKINE [1853, *3*, ¶48]:

To determine the portion of the mechanical power PdV which is the effect of heat, let the total heat of the body, Q, be now supposed to vary by an indefinitely

if he be adjusting his assumptions so as to get an answer of desired form.] He concludes that p is the value of a function of E_0 and V and that "the whole mechanical power for the expansion ... due to the whole heat possessed by the body" is $E_0(\partial p/\partial E_0)\dot{V}$, so $[E_0(\partial p/\partial E_0) - p]\dot{V}$ is the power "expended in overcoming molecular attraction." RANKINE defines a "potential of molecular action" S such as to make this power equal to $\partial S/\partial V$:

$$S \equiv \int \left(E_0 \frac{\partial p}{\partial E_0} - p\right) dV + \phi(E_0) \ . \tag{9E.4}$$

Substitution into (3) yields (his Equation (72))

$$JQ = \left[1 + E_0 \int \frac{\partial^2 p}{\partial E_0^2}\, dV + \phi'(E_0)\right] \dot{E}_0 + E_0 \frac{\partial p}{\partial E_0}\, \dot{V} \ . \tag{9E.5}$$

RANKINE asserts (¶48) "This formula expresses completely the relations between heat, molecular action, and expansion ...." Hence, concludes RANKINE (his Equations (73) and (74)),

$$J\Lambda_V = E_0 \frac{\partial p}{\partial E_0} \ , \qquad \frac{J\mathrm{K}_V}{M\mathfrak{k}} = 1 + E_0 \int \frac{\partial^2 p}{\partial E_0^2}\, dV + \phi'(E_0) \ . \tag{9E.6}$$

RANKINE next [translates into his new language his work of 1851]. Replacing θ by E_0 in (9A.1), he gives the resulting F the name *heat-potential*. [Thus (9A.11) holds, except that the new F is (N/R) times the old one.] RANKINE obtains a slightly weaker result (his Equation (77)): In an adiabatic process

$$-\dot{F} = \frac{1 + \phi'(E_0)}{E_0}\, \dot{E}_0 \ , \tag{9E.7}$$

which he interprets as follows:

---

small quantity dQ. Then the mechanical power of expansion PdV will vary by the indefinitely small quantity

$$dQ \times \frac{dP}{dQ} dV \ .$$

This is the development of power for the expansion dV, caused by each indefinitely small portion dQ of the total heat possessed by the body; and consequently, the whole mechanical power for the expansion dV due to the whole heat possessed by the body Q, is expressed as follows:—

$$Q \frac{dP}{dQ} \cdot dV \ . \tag{67.}$$

and this is the equivalent of the heat transformed into mechanical power, or the latent heat of expansion of unity of weight, for the small increment of volume dV, at the volume V and total heat Q.

When the quantity of heat in a body is varied by variation of volume only, the variation of the heat-potential depends on the heat only, and is independent of the volume.

RANKINE now gives his analysis of the Carnot cycle in the general form [which we have shown already to be the essence of his treatment of 1851], leading to (9A.14). Of course (9A.14) follows, and from it (9A.10), both of them with θ replaced by E_0. Thus (RANKINE's Equation (79))

$$\frac{L(\mathscr{C})}{JC(\mathscr{C})} = 1 - \frac{E_0^-}{E_0^+} \ . \tag{9E.8}$$

[As everyone knows now, this approach to the Carnot cycle, appealing directly to properties of entropy, is the neatest of all.]

Finally, by comparing (5) with (3) RANKINE calculates $\dot{E}_0 + \dot{S}$ and concludes that (his Equation (80))

$$\mathrm{E} = E_0 + S = E_0 + \phi(E_0) + \left(E_0 \frac{\partial}{\partial E_0} - 1\right) \int p dV \ . \tag{9E.9}$$

[Thus, at last, RANKINE has freed himself from the limitations which, as we have seen, may have been implied by his theory of molecular vortices or, more likely, the result of his misinterpreting what his mathematics had delivered. His new formula (6) is just (8H.4) in strange dress, employing the mysterious new quantity E_0, "the absolute quantity of thermometric heat", and (8H.4), we know, he could have read off from his formal apparatus in 1850. Likewise, (9) frees the too special result (2) he had published in 1852, again with E_0 replacing $\mathfrak{k}(\theta - \theta_0)$.

[But what is E_0?] RANKINE continues:

(53.) Still abstaining from the assumption of any mechanical hypothesis, let us proceed a step beyond the investigation of the foregoing articles, and introduce the consideration of *temperature*; that is to say, of an arbitrary function increasing with heat, and having the following properties.

Definition of Equal Temperatures.

Two portions of matter are said to have equal temperatures, when neither tends to communicate heat to the other.

Corollary.

All bodies absolutely destitute of heat have equal temperatures.

The ratio of the real specific heats of two substances, is that of the quantities of heat which equal weights of them possess at the same temperature.

Theorem.

The ratio of the real specific heats of any pair of substances, is the same at all temperatures.

After giving what he regards as a proof of this theorem, RANKINE concludes that if θ is the temperature "according to the scale adopted", then

$$E_0 = \mathfrak{k}[\psi(\theta) - \psi(\theta_0)] \ , \tag{9E.10}$$

the function ψ being "the same for all substances".

"Now, in the notation of Professor THOMSON, we have [¶55]

$$\frac{\psi(\theta) - \psi(\theta_0)}{\psi'(\theta)} = \frac{J}{\mu} \text{,''} \tag{9E.11}$$

and RANKINE's value of the efficiency of a Carnot cycle falls into agreement with KELVIN's evaluation (9B.9).

RANKINE turns now to the question of how temperature and heat should be connected (¶56):

> The results of the investigations in the preceding part of this section are consistent alike with all conceivable hypotheses which ascribe the phenomena of heat to invisible motions amongst the particles of bodies.
>
> Those investigations, however, leave undetermined the relation between temperature and quantity of heat, except in so far as they show that it must follow the same law of variation in all substances.
>
> By adopting a definite hypothesis, we are conducted to a definite relation between temperature and quantity of heat; which, being introduced into the formulæ, leads to specific results respecting the phenomena of the mutual transformation of heat and visible mechanical power; and those results, being compared with experiment, furnish a test of the soundness of the hypothesis.
>
> Thus, the hypothesis of molecular vortices, which forms the basis of the investigations in the first five sections of this paper, and in a paper on the centrifugal theory of elasticity, leads to the conclusion, that, if temperature be measured by the expansion of a perfect gas, the total quantity of heat in a body is simply proportional to the elevation of its temperature above the temperature of absolute privation of heat; ....

That is,

$$\psi(\theta) = \theta, \qquad E_0 = \mathfrak{k}(\theta - \theta_0) \ . \tag{9E.12}$$

[RANKINE's results, for example (6) and (8), tempt us to take (9) as a definition and so interpret his ψ as being the absolute-temperature scale KELVIN is soon to introduce (*cf.* below, §§11B and 11H), but RANKINE is too obscure for us to be certain where E_0 comes from in the first place.]

The remainder of the paper conerns the Joule-Thomson effect. Comparison with the results of the experiment convince RANKINE that his θ_0 is about 2 centigrade degrees above his "absolute zero of temperature".

He proposes a form of the pressure function to represent the deviations found from the ideal gas law. In an appendix he mentions that REGNAULT had "found the specific heat of air to be sensibly constant" over a wide range of temperatures and pressures. *Cf.* §8C, above.

RANKINE's next paper[7] is a long miscellany. ¶1 describes the already well known "indicator diagram" of WATT, and most of ¶¶1–31 concern geometrical calculation of various quantities in the p–V quadrant. [The diagrams are both unnecessary and confusing, and RANKINE's arguments based on them often are unsound or presume unnecessary properties of substances, such as that various congruences of curves may be prolonged to infinity.] Again E_0 is "the total *sensible* or *actual heat* present in the body"; a curve on which E_0 constant is an "*isothermal curve* of E_0"; "*a Curve of No Transmission*" is what RANKINE is soon[8] to rename "an adiabatic curve"; the curve of points at which "the substance is absolutely destitute of heat" is "the *Curve of Absolute Cold* ..., at once an isothermal curve and a curve of no transmission". [To within the question of the difference between E_0 and θ, discussed above, Figure 6 in RANKINE's ¶9 shows a pattern of adiabats and isotherms now familiar as being appropriate to an ideal gas with constant ratio of specific heats.] Also "the *efficiency* of the engine is expressed by the ratio of the heat converted into notive power to the whole heat expended ..."; *cf.* also ¶21. [For a cycle this definition would seem to be

$$\text{efficiency} \equiv \frac{L(\mathscr{C})}{JC^{+}(\mathscr{C})} \;;\Big] \tag{9E.13}$$

but RANKINE goes on to calculate efficiencies in terms of "the actual heat[s]" at which "heat is being received" or "is carried off by conduction."

Since for RANKINE $\mathrm{E} = E_0 + S$ (*cf.* (9)), he now (¶8) interprets E as being "the sum of the actual energy of heat, and the potential energy of molecular action...." In ¶10 he introduces[9] as follows "a *thermodynamic function*" (his Equations $(11)_1$ and $(12)_1$):

$$\dot{\mathrm{H}}_{\mathrm{R}} = \frac{Q}{E_0}\,, \qquad \mathrm{H}_{\mathrm{R}} \equiv \int \frac{Q\,dt}{E_0}\;; \tag{9E.14}$$

[If we take E_0 as being proportional to θ, then $\mathrm{H_R}$ is what CLAUSIUS was later to call the "entropy", as the lynx-eyed GIBBS was to remark[10].]

---

[7] RANKINE [1854].

[8] RANKINE [1859, §239].

[9] RANKINE's notation here is F, not to be confused with the "heat potential", which has appeared above in (9A.1) and (7).

[10] GIBBS [1873, *1*, first footnote on p. 2 of the reprint] [1873, *2*, footnote on p. 52 of the reprint].

According to Professor M. J. KLEIN, it was from the former paper of GIBBS that MAXWELL and, following him, the later British authors came to understand what CLAUSIUS the Heraclitean had meant by entropy. See §2 and especially Footnote 28 of

From $(14)_1$ it is clear that $\mathrm{H_R}$ *is constant along the adiabats.* RANKINE uses his earlier result (5) to show that (his Equation $(12)_2$)

$$J\mathrm{H_R} = \int \frac{1 + E_0 \int \frac{\partial^2 p}{\partial E_0^2} dV + \phi'(E_0)}{E_0} dE_0 + \int \frac{\partial p}{\partial E_0} dV . \quad (9E.15)$$

[RANKINE's cavalier calculus is confusing. The second integral is a partial one; the undetermined constitutive function ϕ reflects the fact that the thermal equation of state does not determine K_V uniquely. We understand (15) better if we write it in terms of the integral of an exact differential:

$$\mathrm{H_R} = \int \left(\frac{\mathrm{K}_V/(\mathfrak{k}M)}{E_0} dE_0 + \frac{\Lambda_V}{E_0} dV \right) . \quad (9E.16)$$

That this differential is exact, follows at once from the relation

$$\frac{\partial}{\partial V}\left(\frac{\mathrm{K}_V/(\mathfrak{k}M)}{E_0}\right) = \frac{\partial}{\partial E_0}\left(\frac{\Lambda_V}{E_0}\right) , \quad (9E.17)$$

which we can read off from (6). Hence in a simply connected region of the V–E_0 plane there is a function $\mathrm{H_R}$ of V and E_0 such that

$$\frac{\mathrm{K}_V}{\mathfrak{k}M} = E_0 \frac{\partial \mathrm{H_R}}{\partial E_0} , \qquad \Lambda_V = E_0 \frac{\partial \mathrm{H_R}}{\partial V} . \quad (9E.18)$$

RANKINE never stops to ask if the integral in the definition (15) makes sense; although he occasionally states that some expression is "obviously an exact differential", he gives no evidence of knowing how to handle such a thing, and he never appeals to a condition of integrability; nevertheless, in exhibiting (15) he in his own way effectively shows that the function $\mathrm{H_R}$ does exist locally. Moreover, if we take the lower limit of integration in the second integral of (15) as $\psi(\theta)$, by substituting (15) into (18) we recover (6) with the same lower limit of integration. Thus we cannot deny that RANKINE did in fact calculate a function $\mathrm{H_R}$ that extends to an arbitrary equation of state his results regarding his function U in his first paper (§§8G–8H), and that $\mathrm{H_R}$ reduces when $E_0 = \mathfrak{k}\theta$ to what is today called the entropy.]

KLEIN's paper, "Gibbs on Clausius", ***Historical Studies in the Physical Sciences* 1** (1969), 127–129. That MAXWELL [1871, pp. 186–188] had missed this point, is further evidence of RANKINE's "inscrutable" presentation; it is evidence also that MAXWELL had not studied CLAUSIUS' papers carefully: Indeed, he stated that CLAUSIUS had introduced the term "entropy" in his paper of 1854, while in fact neither the name nor the quantity appears there. *Cf.* §§11C–11H, below.

In later editions of the book of MAXWELL [1891, especially the footnote on p. 189] I cannot find any indication that MAXWELL accepted GIBBS's attribution of the entropy to RANKINE. That may be because GIBBS, perhaps from a desire to avoid offense to the jealous old CLAUSIUS, omitted to specify the *date* of RANKINE's work, although any careful reader by checking GIBBS's citation of RANKINE would find that it refers to the year 1853, while his citations of CLAUSIUS both refer to 1865!

In ¶¶14–15 RANKINE defines and discusses "*Curves of Free Expansion*". [From his Equation (17A) we see that these are curves of constant *enthalpy*; *cf.* KELVIN's work of 1853, described in the preceding section; for REECH's introduction of a generalized enthalpy, also in 1853, see below, §10C.] The remainder of the paper concerns specific properties of engines and vapors, the numerical calculation of efficiencies, and the hypothesis of molecular vortices.

[RANKINE's brilliant and original beginnings, which we have analysed in §§8G–8H, have now matured into a torrent of words and symbols so obscure that even other thermodynamicists[11] considered them so. This is obscurity indeed! To close our study of the work of this most appealing of Scottish engineers], I first quote MAXWELL[12] "... though the construction and distribution of his vortices may seem to us as complicated and arbitrary as the Cartesian system, his final deductions are simple, necessary, and consistent with the facts." [With regret I must add that this judgment applies only to RANKINE's earliest work. His middle work, that which we have just described, fully deserves] MAXWELL's judgment of his "Second Law": "inscrutable". [RANKINE was closer to clarity when he depended upon his molecular vortices. Hypothetical as those vortices were, and intricate as were the calculations into which they led him, they kept him specific. When he attempted to present phenomenological concepts directly, he took refuge in fancy words to describe assumptions that seem to have served only to let him work backward toward the results that in his first work he had somehow extracted from his model.]

[RANKINE presented the first complete and general formal structure for classical thermodynamics, introduced the entropy of a body susceptible only of reversible processes, showed how to use the entropy to analyse a Carnot cycle, and obtained the final formula for the efficiency of such a cycle. Nevertheless the tradition of thermodynamics, while it mentions RANKINE's name with respect, attributes at most one of these discoveries to him. Perhaps readers of his papers have asked, Are those things really there, or have I

---

[11] CLAUSIUS [1865, §5], MAXWELL [1878]. The editor of RANKINE's collected papers seems to have made no attempt even to correct obvious misprints in equations, let alone to render the notations consistent within a single paper or to help the reader connect the notations in one paper with those in another.

[12] MAXWELL [1878]. In reviewing a particular paper of RANKINE [1853, *1*] when it appeared, HELMHOLTZ [1856, *2*] after objecting to one proof added "it is very hard to work through Mr. RANKINE's papers sufficiently to reach an opinion about them. Mr. RANKINE has found in his way many results which are recognized as correct from entirely different starting points by other investigators. But it is usually impossible to follow him step by step in his efforts to prove his theorems, so the reader gets the impression that he found his results more through a kind of correct mechanical instinct than through rigorous mathematical analysis. Either that, or he has left out so many intermediate steps in his logical train that the reader almost has to rediscover that train afresh."

located them because I knew what to look for? There are θ_0 and E_0 to puzzle over: "the temperature corresponding to absolute privation of heat" and "the absolute quantity of thermometric heat". Is it only retrospective generosity that makes us forget θ_0 and replace E_0 by $\mathfrak{k}\theta$?]

9F. KELVIN's Analysis of the "Anomalous" Behavior of Water

In 1854 JOULE & KELVIN published a major paper on "the Joule-Thomson effect" and related matters. [It is safe to assume that the part called "Theoretical Deductions" was written by KELVIN alone.] KELVIN begins[1] with [what might seem to be one more defense for his not having accepted the HELMHOLTZ-JOULE Determination]

$$\mu = J/\theta \ . \qquad (5N.7)_r$$

"Mayer's hypothesis" cannot be general:

SECTION I. *On the Relation between the Heat evolved and the Work spent in Compressing a Gas kept at constant temperature.*

> This relation is not a relation of simple mechanical equivalence, as was supposed by MAYER in his 'Bemerkungen ueber die Kräfte der Unbelebten Natur, in which he founded on it an attempt to evaluate numerically the mechanical equivalent of the thermal unit. The heat evolved may be less than, equal to, or greater than the equivalent of the work spent, according as the work produces other effects in the fluid than heat, produces only heat, or is assisted by molecular forces in generating heat, and according to the quantity of heat, greater than, equal to, or less than that held by the fluid in its primitive condition, which it must hold to keep itself at the same temperature when compressed. The *à priori* assumption of equivalence, for the case of air, without some special reason from theory or experiment, is not less unwarrantable than for the case of any fluid whatever subjected to compression. Yet it may be demonstrated[2] that water below its temperature of maximum density (39°·1 FAHR.), instead of evolving any heat at all when compressed, actually absorbs heat, and at higher temperatures evolves heat in greater or less, but probably always very small, proportion to the equivalent of the work spent; while air, as will be shown presently, evolves always, at least when kept at any temperature between 0° and 100° Cent., somewhat more heat than the work spent in compressing it could alone create.

---

[1] JOULE & THOMSON [1854, Theoretical Deductions, §I].

[2] KELVIN's footnote here is misleading: He lets the reader think he discussed the behavior of water in the paper which our text above analyses. That is not so.

[Certainly there could be no more decisive proof that HOLTZMANN's Assertion cannot hold for all fluids: For water in the range of its anomalous behavior, Λ_V and p are of opposite sign!] KELVIN leaves us to guess how this "may be demonstrated". [I conjecture that he looked at the calorimetric relation

$$\mathrm{K}_p - \mathrm{K}_V = -\Lambda_V \frac{\partial p}{\partial \theta} \Big/ \frac{\partial p}{\partial V} , \qquad (2\mathrm{C}.9)_{2\mathrm{r}}$$

which he had derived two years earlier in the generality required. Of course he assumed the constitutive inequality

$$\frac{\partial p}{\partial V} < 0 , \qquad (2\mathrm{A}.5)_{1\mathrm{r}}$$

but obviously he is here relinquishing the second heretofore standard constitutive inequality:

$$\frac{\partial p}{\partial \theta} > 0 . \qquad (2\mathrm{A}.5)_{2\mathrm{r}}$$

From $(2\mathrm{C}.9)_2$ it is clear that

$$\mathrm{K}_p = \mathrm{K}_V \quad \Leftrightarrow \quad \Lambda_V \frac{\partial p}{\partial \theta} = 0 . \qquad (9\mathrm{F}.1)$$

When a fluid undergoes an isobaric process,

$$\frac{\partial p}{\partial V} \dot{V} + \frac{\partial p}{\partial \theta} \dot{\theta} = 0 . \qquad (9\mathrm{F}.2)$$

Thus if V passes through a minimum, $\partial p/\partial \theta = 0$ there. For water being cooled isobarically near 4° C at atmospheric pressure the point V, θ therefore passes from a region where $\partial p/\partial \theta > 0$ into one where $\partial p/\partial \theta < 0$. At the dividing point, then, $\partial p/\partial \theta = 0$, and consequently $\mathrm{K}_p = \mathrm{K}_V$ there. Experiments had not been done to determine K_p and K_V in such circumstances: All experimental data and experience then available referred to other conditions, and in them

$$\mathrm{K}_p > \mathrm{K}_V . \qquad (2\mathrm{C}.10)_{2\mathrm{r}}$$

If we dare extend this conclusion to water at points where $\partial p/\partial \theta \neq 0$, we conclude that at those points

$$\Lambda_V \frac{\partial p}{\partial \theta} > 0 . \qquad (9\mathrm{F}.3)$$

Then in water being cooled isobarically near 4°C at atmospheric pressure the point V, θ passes from a region where $\Lambda_V > 0$ into one where $\Lambda_V < 0$.] That is what KELVIN wrote in the passage we have just quoted:

"... water below its temperature of maximum density (39°·1 FAHR.), instead of evolving any heat at all when compressed [isothermally], actually absorbs heat ...."

[The foregoing "demonstration" involves the same sin as KELVIN had committed in his disastrous evaluation of μ (above, §7H): It extends into a presently inaccessible range the results of experiment elsewhere. KELVIN may have reasoned instead in a purely theoretical way. Because KELVIN rejects HOLTZMANN's Assertion for ideal gases as having no foundation in experiment, he certainly cannot begin from the central formula

$$J\Lambda_V = \theta \frac{\partial p}{\partial \theta} . \qquad (7I.2)_r$$

It is more like him to have used as the basis of his argument his favorite starting point, the General CARNOT-CLAPEYRON Theorem:

$$\mu\Lambda_V = \frac{\partial p}{\partial \theta} . \qquad (5L.4)_r$$

He may have thought that CARNOT's argument to derive this theorem showed also that $\mu > 0$. As we have remarked in §5J, in fact that argument proves only the weaker inequality $\mu \geqq 0$. We cannot exclude the possibility that $\mu = 0$ on a set of temperatures with empty interior. KELVIN knows very well that μ is a universal function, the same for all bodies. Therefore, to prove that $\mu(\theta_0) > 0$ it is sufficient to find *one* body in whose constitutive domain there is a point on the isotherm $\theta = \theta_0$ at which $\Lambda_V \neq 0$ and $\partial p/\partial \theta \neq 0$. For example, if HOLTZMANN's Assertion $J\Lambda_V = p$ holds for *one* ideal gas, then $\mu(\theta) > 0$ for *all* θ corresponding to points in the constitutive domain of that gas. Alternatively, we may assume that *one* ideal gas has distinct specific heats. Once we know that $\mu > 0$, it follows from (5L.4) that Λ_V and $\partial p/\partial \theta$ change sign together, which is just what KELVIN asserted in regard to water.

[While KELVIN may have used an argument of this kind, the early developments in thermodynamics do not suffice to render it tight, since the whole analysis of Carnot cycles from CARNOT's time on has *assumed* that $\Lambda_V > 0$! Even the existence of μ has been demonstrated only under this assumption! Fallacy of this kind is only standard in works on thermodynamics. However, this particular vicious circle can be set aside through a more careful treatment[3].]

---

[3] Two points are at issue:

1. Proof that "MAYER's hypothesis" is special. In §7I we have given a rigorous proof that such is the case even in the class of "all fluids" considered in thermodynamics before 1854, that is, fluids for which $\Lambda_V > 0$. KELVIN's treatment of 1854 by its unnecessary appeal to the "anomalous" behavior of water misses the point at issue.

2. The thermodynamics of the "anomalous behavior" of water. First considering a particular body having constitutive domain $\mathscr{D}$, we set aside some preliminaries.

A. It is easy to *state* CARNOT's General Axiom in the same words but leave aside

Whatever his reasoning may have been, KELVIN is the first to see that $\Lambda_V < 0$ if $\partial p/\partial\theta < 0$. This passage, published in 1854, contains the first and last reference to the "anomalous" behavior of water we shall encounter in this tragicomical history.

[From this point onward thermodynamics must abandon altogether the heretofore standard adscititious inequalities

$$\frac{\partial p}{\partial \theta} > 0\ , \qquad \Lambda_V > 0\ , \qquad (2A.5)_{2r}, (2C.5)_{1r}$$

while the other heretofore standard inequalities, namely

$$K_p > K_V\ , \qquad \gamma > 1\ , \qquad (2C.10)_{2r}, (2C.17)_r$$

are weakened to

$$K_p \geqq K_V\ , \qquad \gamma \geqq 1\ . \qquad (9F.4)$$

Appealing to the calorimetric relation

$$K_p - K_V = -\Lambda_V \frac{\partial p}{\partial \theta}\bigg/\frac{\partial p}{\partial V} \qquad (2C.9)_{2r}$$

and to the remaining adscititious inequality $\partial p/\partial V < 0$, we conclude that

$$\Lambda_V \frac{\partial p}{\partial \theta} \geqq 0\ . \qquad (9F.5)$$

This same statement may be proved directly from the general expression for the work done by a simple cycle:

$$L(\mathscr{C}) = \iint_{\mathscr{A}} \frac{\partial p}{\partial \theta}\, dV d\theta\ , \qquad (5L.1)_{2r}$$

and CARNOT's assertion that $L(\mathscr{C}) > 0$ if $\mathscr{C}$ is a Carnot cycle. See Lemma 1 in Chapter 8 of *Concepts and Logic*. Then the inequalities (4) follow as proved theorems, not assumptions.

[On the other hand, I have not found any textbook or other exposition of thermodynamics published before 1977 that makes these matters clear

---

the old requirement that $\Lambda_V > 0$. Carnot cycles in regions of $\mathscr{D}$ where $\Lambda_V < 0$ or even where Λ_V is not of one sign are possible. Their forms, however, are not at all like the one sketched in Figure 3 in §5C. One is shown in the figure in Footnote 4 to §5M. Others are shown in Figure 9 in Chapter 7 of *Concepts and Logic*.

B. In a *region* of $\mathscr{D}$ where $\Lambda_V < 0$ we can still follow CARNOT's analysis of his cycle if we make some minor modifications. We conclude that the *General CARNOT-CLAPEYRON Theorem still holds*.

This much is all that is needed if we are to conclude that if $\mu > 0$, then Λ_V and $\partial p/\partial\theta$ have the same sign. It does not suffice to exclude the possibility that $\mu = 0$ at isolated points. To do so, we have given some possible arguments in the text. There is no indication that KELVIN perceives how subtle this point is. *Cf.* the discussion of absolute temperature in §11H, below, and the more explicit treatment of TRUESDELL [1979].

and explicit; *a fortiori*, none that derives them mathematically within the framework of the founders' ideas. That was one of the reasons why I undertook the work presented in *Concepts and Logic*. The reader of that book will see that the formal structure of classical thermodynamics, including the inequalities stated just above, does follow precisely if we limit attention to points in what Mr. BHARATHA & I define as the *normal set* of a body's constitutive domain.]

9G. General Critique: The Disastrous Effects of Experiment upon the Development of Thermodynamics, 1812–1853

Nobody needs to be told that theories of nature must grow from experience, that experiment can sharpen, refine, and extend experience and correct our conception of it, that a scientific theory of an aspect of nature cannot be accepted until it has been somehow "confirmed" by experiment. It does not follow that theory and experiment climb hand in hand up Jacob's ladder. It does not follow that experiments as an end in themselves are necessarily beneficial to anyone except, it may be hoped, those who perform them. It does not even follow that the theorist should scrupulously respect all such experimental data as may bear upon the branch of natural science he is trying to develop.

I offer these remarks not as lemmas of a philosophy of science, not in regard to ideal programs, but as inductions from the history of science as I have read it in the old way, searching and probing the sources in detail, case by case, problem by problem, line by line, equation by equation. Some of the data that our tragicomical history affords for these inductions I draw up here.

α. *Early experiments.* Here there is no disaster. It is plain enough that without the experiments of the seventeenth and eighteenth centuries on thermometers and calorimetry, the basis on which it is possible to think about relations between heat and work had wanted altogether. Experience with steam engines was central to the creation of thermodynamics, though the value of the numerous experiments performed upon them, often with scant comprehension of the physical processes that went on within them, is debatable.

β. *DELAROCHE & BÉRARD* (1812). Few experiments have had so great an influence on the history of physics as DELAROCHE & BÉRARD's regarding the specific heats of gases. FOX[1] tells the story:

[1] Pp. 139–140 of his *Caloric Theory*, cited above in Footnote 2 to §2A.

> In order to investigate [the dependence of the specific heat of a gas upon its density], Delaroche and Bérard modified their first experiment so as to allow air to pass through the apparatus at a pressure of 100·58 cm of mercury, as well as at atmospheric pressure, the initial temperature being very nearly the same in both cases. The result, that the volume specific heat at this higher pressure was to that at the ordinary pressure of 74·05 cm as 1·2396 to 1, was decisively, though not seriously, in error. Since the ratio of the pressures was 1·3583:1·0000, it followed that the specific heat by weight had decreased in the ratio 0·9126:1·000 as a result of the pressure increase. We know now, of course, that no variation at all should have been observed in the specific heat by weight, but in 1812 a decrease in specific heat with increasing pressure was expected and the quite unfounded confidence which Delaroche and Bérard placed in their result almost certainly owed a great deal to this fact. They based their conclusion on only two experiments conducted on air at the single higher pressure of 100·58 cm, the steady variation of pressure being impossible with their apparatus. The discrepancy between the volume specific heats deduced from the two experiments (1·2127 and 1·2665, of which 1·2396 was the mean) should in itself have made them suspicious, but without further examination they proceeded to extrapolate the results to other pressures and confidently assumed that they applied equally well to all gases. Their error, although of less than 10 per cent, was to prove one of the most influential in the whole history of the study of heat. Backed by the prestige associated with victory in the Institute's competition, the result quickly became standard and ... was to mislead many calorists.

The greatest of the calorists so misled was CARNOT, as we have seen in §§5S–5T. He was misled to the point that he not only set aside the lead offered by LAPLACE's theory of the speed of sound but also turned his back upon the whole corpus of experiment related to it. In §3F we have seen that LAPLACE himself was equally misled. While he stood fast by his theory and the experiments on the speed of sound, he let DELAROCHE & BÉRARD's result lull him into accepting the preposterous theory of specific heats that the Caloric Theory of heat delivers for an ideal gas when their ratio is assumed constant.

γ. *Steam tables.* The practical importance of steam caused experimenters to concentrate upon it. It is not a typical fluid, and its special properties offer no helpful guidance. We have seen in §§7H and 9B that since the most abundant data were for steam, KELVIN naturally had recourse to them; that in all their exuberance the data were insufficient; that as a result, KELVIN replaced wanting data by extrapolation; and that in consequence he failed to confirm the HELMHOLTZ-JOULE Determination. Had good data been available for air instead of steam, KELVIN's extrapolation would have been accurate enough, and by just the same method he would have confirmed the HELMHOLTZ-JOULE Determination. As it was, KELVIN's conclusions from steam tables helped make physicists reluctant to accept JOULE's work.

δ. *REGNAULT*. REGNAULT's experiments excited the admiration of all who studied them in their day, especially KELVIN. Comparison with data accepted now indicates that his results were very accurate. His ponderous, subsidized program of "determining the principal laws and the numerical data that enter the calculation of steam engines" began in 1842; just about then appeared the first experiments of JOULE and MAYER's Assertion; HOLTZMANN's Assertion, which was to play a key role in the thermodynamics of CLAUSIUS, lay three years in the future. The time was perfect for experiment to take the lead. CARNOT's and CLAPEYRON's publications had shown the nature of the specific heat of gases to be *of the essence* to the motive power of heat engines. REGNAULT, not mentioning anything shown by CARNOT or CLAPEYRON, set about making the most accurate thermometers and determining deviations from the ideal gas laws!

FOX[2] tells the story:

> In a history of the caloric theory of gases . . . it must surely be the sterile experimenting of Victor Regnault that has pride of place as evidence of the decline; indeed, by any standards Regnault's failure to play a significant part in the development of thermodynamics deserves more than the passing comment it has received in earlier studies of our problem.
>
> Regnault, we should recall, was a man of outstanding ability, and by the early 1840s he had the familiarity with steam-engine operation that seems to have been so important to most of the pioneers of energy conservation. Moreover, thanks to the French Government, he had been provided with assistants, equipment, and a laboratory that would have been the envy of his contemporaries both in France and elsewhere. With all this, how could he have failed?

REGNAULT did finally get to the specific heats of gases, but only in 1853, three years after CLAUSIUS had conjectured, on the basis of theory alone, that they would be found constant. REGNAULT did find, indeed, that K_p for several gases was very nearly constant. Those who accepted LAPLACE's explanation of the speed of sound had to agree that K_p/K_V was sensibly constant. Thus REGNAULT's work and the data on the speed of sound could be taken as confirming CLAUSIUS' conjecture. CLAUSIUS, unlike KELVIN, had not waited for new data. *He had dared to go ahead with the theory*, so by the

[2] P. 315 of his *Caloric Theory*, cited above in Footnote 2 to §2A.

time experiment got going, thermodynamics had been standing in print for nearly three years. Not only that, also a good value of K_p at ordinary conditions had been obtained. MAXWELL[3] tells the story, not without a tinge of gentle sarcasm:

> Hence the determinations of the specific heat of gases were generally very inaccurate, till M. Regnault brought all the resources of his experimental skill to bear on the investigation, and, by making the gas pass in a continuous current and in large quantities through the tube of his calorimeter, deduced results which cannot be far from the truth.
>
> These results, however, were not published till 1853, but in the meantime Rankine, by the application of the principles of thermodynamics to facts already known, determined theoretically a value of the specific heat of air, which he published in 1850. The value which he obtained differed from that which was then received as the best result of direct experiment, but when Regnault's result was published it agreed exactly with Rankine's calculation.

ϵ. *JOULE*. The hero of experimentation on the Interconvertibility of Heat and Work is surely JOULE. Surely he had the idea that heat could always be specified in units of work, and he claimed that his experiments supported it. His results published before 1850, however, were so inaccurate as to cast doubt upon what he claimed for them, not only among foreign critics like HELMHOLTZ but even in his friend KELVIN. *Cf.* above, §§7H and 7Iα.

True, by the end of 1850 two great theorists accepted JOULE's results: CLAUSIUS and RANKINE. Both of these theorists, however, were predisposed to them; each was guided by a molecular picture, in which he seems to have

[3] MAXWELL [1871, pp. 177–178] [1891, pp. 179–180]. Indeed, RANKINE [1851, *2*] accepted one of JOULE's values of J and used MAYER's Assertion (7B.2) to calculate K_p. He read the paper on December 2, 1850. In his earlier paper, read on February 4 of the same year, RANKINE [1850, ¶¶2 and 14] had refused to accept any of JOULE's values of J. For want of anything better he had then used DELAROCHE & BÉRARD's value of K_p to calculate J from (7B.3), as had MAYER (*cf.* §7B, above). In both of RANKINE's calculations, as in MAYER's, γ is given the value 1.4 on the basis of experiments on the speed of sound.

In fact RANKINE was incautious in being won over by a "multitude of experiments" on "substances so different as water, mercury, and air". The range of temperatures reported by JOULE [1850] was too small. KELVIN's relation (7H.12), published a year earlier, should have shown everyone that CARNOT's theory too, required heat and work to be universally interconvertible *in isothermal processes*, providing the factors of interconversion were *different at different temperatures*. The crucial question was whether or not $\theta\mu$ were independent of θ, and JOULE's experiments offered no test of that. The work of JOULE through 1850 could not even be interpreted as providing decisive experimental support for the principle of uniform and universal Interconvertibility of Heat and Work in cyclic processes.

believed firmly[4], and each sought from JOULE's work not direction toward a theory but experimental evaluation of a constant, the existence of which he had already assumed. The decisive position that both RANKINE and CLAUSIUS took was *to reject*, at last, the experimental data provided by DELAROCHE & BÈRARD.

ζ. *Fantasy*. We may imagine how different the history of thermodynamics would have been, had DELAROCHE & BÉRARD concluded that K_p was nearly constant for most gases, air being one. Rough confirmation of LAPLACE's formula for the speed of sound would have suggested that K_p/K_V = const. for air. Thus any competent theorist would have taken *the ideal gas with constant specific heats*, not as the embodiment of nature but as the most natural special case to consider. Attempting to apply the Caloric Theory, he would have derived one or both of the relations

$$K_p - K_V + p\frac{\partial K_p}{\partial p} + \rho\frac{\partial K_V}{\partial \rho} = 0 , \qquad (3F.5)_r$$

$$K_V = \frac{M}{\gamma\theta}\left[\frac{p^{1/\gamma}}{\rho}\psi'\left(\frac{p^{1/\gamma}}{\rho}\right)\right] , \qquad (3F.3)_r$$

for both were latent in LAPLACE's work, and *our imaginary theorist would have had to reject the Caloric Theory*! The date is any time from 1812 onward. The theorist in question could easily have been CARNOT, who had all the necessary apparatus. In his special case of the Caloric Theory the same thing follows from (5R.1).

That is not all. The same theorist could have calculated the motive power of a Carnot cycle *without use of any theory relating heat and work*. For an ideal gas with constant specific heats CARNOT's General Axiom is unnecessary. There is no reason why this hypothetical theorist—again, he could well have been CARNOT—would have failed then to get the entirely elementary theorem HOPPE was to publish in 1856 (above, Footnote 4 to §9D): *An ideal gas with constant specific heats interconverts heat and work uniformly in Carnot cycles*. This special case points the way to the general idea. If, as CARNOT claimed, "the motive power of heat is independent of the agents used to realize it," then *all bodies* must interconvert heat and work uniformly in Carnot cycles because the ideal gas with constant specific heats does so! CLAUSIUS' Axiom would have lain in CARNOT's hands, had DELAROCHE & BÉRARD's work been done well, or had CARNOT simply rejected

[4] RANKINE's position is made clear by our analysis of his papers (above, §§8G and 9A). CLAUSIUS withheld publication of his kinetic theory for some years, but in his paper of 1850 he alluded to it immediately after his reference to "the careful investigations of Joule" at the beginning: "To this it must be added that other facts have lately become known which support the view, that heat is not a substance, but consists in a motion of the least parts of bodies."

it. Again the fact was too much or too little experiment, for DELAROCHE & BÉRARD's data forbade those who accepted them from making the most natural first guess about the specific heats of gases.

Well then, suppose we forget DELAROCHE & BÉRARD, accept (as in historical fact we must) CARNOT's dilemma as inevitable in 1824. What about REGNAULT? Had he chosen to determine K_p in 1840, he would certainly have done so correctly, and, had he published his result, KELVIN could scarcely have failed then to accept JOULE's conclusions, inaccurate as was the experimentation from which JOULE had drawn them.

The tragicomedy of classical thermodynamics has an experimental counterpart: the tragicomedy of early experiment on the specific heats of gases.

In following the involutions of this long interlude we have reached the terminal year of our tragicomedy, 1854. We must now step backward one year. Doing so, we shall see that in 1853 the concepts of internal energy, entropy, enthalpy, and thermodynamic potential had all appeared in print —all clearly understood—all demonstrated and interrelated through explicit concepts and mathematics which while crude and long-winded are compelling in comparison with RANKINE's—and all published in such a way as to invite the oblivion which posterity has bestowed upon their author until today.

10. Schismatic Act V. Antiplot in a Dark and Empty Theatre: REECH's Discovery and Burial of a Too General Theory, and His Failure to Reduce It.

Surge ai mortali per diverse foci
  la lucerna del mondo....
...Le cose tutte quante
  hanno ordine tra loro....
DANTE, *Paradiso* I,
37–38, 103–104.

10A. REECH Discovers the Pro-entropy

[In §9C the spectators have been warned of the impending deluge[1] from the pen of REECH. They have been told also that REECH was setting about to determine the consequences of the first principles of the subject.] REECH, mentioning the works of CARNOT, CLAPEYRON, JOULE, THOMSON, RANKINE, MAYER, and CLAUSIUS, expresses the opinion (p. 357) that "too much importance has been given to pure hypotheses, losing sight of the logical train of reasoning of Mr. Carnot, which has not been broken, I think, by Mr. Regnault's objection, and which needs only to be completed from a new point of view." REECH himself adopts "the mother idea or fundamental axiom of the reasonings of Messrs. Carnot and Clapeyron" (p. 364) but refuses to accept either their assumption that the heat in a body is a function of V and θ or the new assumptions connecting heat with work which MAYER, JOULE, CLAUSIUS, RANKINE, and KELVIN had espoused. For REECH, the first principles are these:

[1] REECH [1853]. In reading this remarkable memoir I could not avoid the suspicion that I was the first to do so. Apart from the exposition by TRUESDELL & TOUPIN [1960, Chapter EII] the literature of thermodynamics has ignored it and its author except for a few references by French writers to the theorem expressed by (10D.4), below.

1. For a given gas, through each point of the V–p quadrant passes one and only one isotherm and one and only one adiabat. [He assumes tacitly that they decussate.]

2. CARNOT's General Axiom: The work $L(\mathscr{C})$ done by a fluid body in undergoing a Carnot cycle $\mathscr{C}$ is determined by its operating temperatures θ^+ and θ^- and by the heat absorbed $C^+(\mathscr{C})$ on the isotherm at the higher temperature θ^+. That work is the same in all Carnot cycles that can correspond to the three quantities θ^+, θ^-, and $C^+(\mathscr{C})$.

REECH is the first thermodynamic author, [and also probably the last,] to state openly that the theory is limited to gases and vapors. He takes for granted the thermal equation of state $p = \varpi(V, \theta)$ and the definition of work:

$$L \equiv \int_{t_1}^{t_2} p(t)\dot{V}(t)dt \ . \tag{2C.19$_r$}$$

We have shown in §9C that his reasoning suffices to prove with modest conviction the existence of a function $\Gamma(\theta)$ such that for any Carnot cycle $\mathscr{C}$

$$L(\mathscr{C}) = K(\theta^+, \theta^-)C^+(\mathscr{C}) = \Gamma(\theta^+)C^+(\mathscr{C}) - \Gamma(\theta^-)C^-(\mathscr{C}) \ , \tag{9C.10$_r$}$$

and we have given a tight proof of this result on the basis of the reversal theorem

$$C(-\mathscr{P}) = -C(\mathscr{P}) \ . \tag{2C.7$_r$}$$

[REECH's first assumption is clearly insufficient by itself to calculate anything about the heat added, because it refers only to curves on which $Q = 0$, telling us nothing about non-zero values of Q. It does not even suffice to prove the truth of the reversal theorem (2C.7), which REECH applies tacitly again and again (pp. 361, 364, 368, *et passim*). In effect REECH uses the Doctrine of Latent and Specific Heats, as had CARNOT, CLAPEYRON, KELVIN, CLAUSIUS, and all the rest. His reluctance to say so openly and in consequence avail himself of the calorimetric relations

$$Q = \Lambda_p(V, \theta)\dot{p} + \mathrm{K}_p(V, \theta)\dot{\theta} \tag{2C.8$_r$}$$

and

$$\Lambda_p = \Lambda_V \Big/ \frac{\partial p}{\partial V} \ , \qquad \mathrm{K}_p - \mathrm{K}_V = -\Lambda_V \frac{\partial p}{\partial \theta} \Big/ \frac{\partial p}{\partial V} \ , \tag{2C.9$_r$}$$

and various consequences of them we have set forth in §2C, complicates and lengthens his arguments. His memoir is so diffuse that for fear of putting the spectators altogether to sleep I report only the essence of it.]

Chapter I (pp. 358–367) recapitulates the general arguments of CARNOT, expressing them in terms of the isotherms and adiabats. REECH supposes the latter to be the curves $u =$ const. for some constitutive function ψ of the gas (pp. 358–360):

$$u = \psi(V, p) \ . \tag{10A.1}$$

The function ψ, [which of course is not defined uniquely by the property REECH assigns to it,] is to play a central part in the whole memoir. Because changes of kinetic energy are neglected, the speeds of the processes considered must be negligible (p. 359).

Chapter II (pp. 368–378) begins with the proof of "the absolute generality and total exactness" (p. 374) of his "fundamental pivot of the theory of the dynamical effects of heat" (p. 371):

$$L(\mathscr{C}) = \Gamma(\theta^+)C^+(\mathscr{C}) - \Gamma(\theta^-)C^-(\mathscr{C}) \ . \qquad (9C.1)_{1r}$$

To close the chapter (pp. 377–378) REECH takes up "what can be said for or against the relation $\Gamma(\theta) = \text{const.}$" [He does not remind the reader that this case corresponds to the basic assumption of CLAUSIUS.] He gives in favor of it the following physical argument (p. 377). Regarding $C\Gamma(\theta)$ as the mechanical equivalent of C units of heat [in all circumstances, not just in Carnot cycles], "I suppose that between two sources A and A' at different temperatures θ^- and θ^+ we renounce use of an elastic fluid as vehicle of heat from A' to A. Then the transmission will proceed freely, by radiation or by contact, from the hotter source to the colder source, and we must suppose naturally that all the heat which leaves the one will pass into the other. If we write in this case $C' = C$ then the source A' would lose a quantity of *vis viva* equal to $C\Gamma(\theta^+)$, while the source A would gain a quantity of *vis viva* equal to $C\Gamma(\theta^-)$. There would then be a loss of work or *vis viva* equal to the difference:

$$C'\Gamma(\theta^+) - C\Gamma(\theta^-) = C[\Gamma(\theta^+) - \Gamma(\theta^-)] \ , \qquad (10A.2)$$

which would singularly shock our common sense, since we have supposed that no impediment hinders the free transmission of heat from A' to A. To avoid such a difficulty" we ought to suppose both sides of (2) equal to 0. "From this point of view, the discussion would be closed," and

$$\Gamma = \text{const.} = J \ . \qquad (10A.3)$$

But then in a Carnot cycle

$$L(\mathscr{C}) = JC(\mathscr{C}) \ , \qquad (7A.1)_{1r}$$

and "there would seem to be no difference between having heat at a high temperature or a low temperature, while there is every evidence that a high temperature is a valuable thing which cannot be gotten back for nothing, once lost, without the necessary precautions.... [F]or lack of absolute evidence, I will retain the function $\Gamma(\theta)$ in all that follows...," such being "the very natural continuation of the researches of Messrs. Carnot and Clapeyron, now that Mr. Regnault has called in doubt the equation $C' = C$." [Spectator, let the logical blunder which REECH here commits serve as a paradigm of the confusion physico-philosophical reasoning can produce when applied to a mathematical question!]

Chapter III (pp. 378–410) presents the statement and proof of ***Reech's Third Theorem***: *There is a constitutive function f such that*

$$Q = f(\theta, u)\dot{u} \ , \tag{10A.4}$$

[The modern reader sees at once that in u REECH has introduced *a function that generalizes what CLAUSIUS was later to call the entropy*[2]. In §8G we have seen that in 1850 RANKINE, basing his work largely upon a molecular model, had already published results from which a conclusion of this kind can be drawn, and that in RANKINE's theory $f = \theta - \theta_0$. REECH gives his more general function u no name. To shorten our descriptions of it, we shall call it a *pro-entropy*.

[Perhaps sensing that his first proof (pp. 378–380) of the major result (4) is no more than reaffirmation], REECH gives another (p. 380, notation conformed with ours): "So that there may rest no doubt in this regard, I remark that in every state of affairs the quantity of heat necessary to go from a point θ, u to another point infinitely nearby $\theta + d\theta$, $u + du$ ought to be an infinitely small quantity, and that consequently it ought to be possible to represent such a quantity by an expression of the form

$$Qdt = a(\theta, u)d\theta + b(\theta, u)du \ . \tag{10A.5}$$

But when we put $du = 0$ in such an expression, we ought to find that $Qdt = 0$, and hence it will be necessary that $a(\theta, u) = 0 \ldots$" [That is, REECH assumes that

$$Q = a(\theta, u)\dot{\theta} + b(\theta, u)\dot{u} \ . \tag{10A.6}$$

Writing $\partial u/\partial V$ and $\partial u/\partial p$ for the partial derivatives of ψ in REECH's assumption (1), we see that

$$\begin{aligned} Q &= a\dot{\theta} + b\left[\frac{\partial u}{\partial V}\dot{V} + \frac{\partial u}{\partial p}\dot{p}\right] , \\ &= a\dot{\theta} + b\left\{\frac{\partial u}{\partial V}\dot{V} + \frac{\partial u}{\partial p}\left[\frac{\partial p}{\partial V}\dot{V} + \frac{\partial p}{\partial \theta}\dot{\theta}\right]\right\} . \end{aligned} \tag{10A.7}$$

Defining Λ_V and K_V as follows:

$$\begin{aligned} \Lambda_V &\equiv b\left(\frac{\partial u}{\partial V} + \frac{\partial u}{\partial p}\frac{\partial p}{\partial V}\right) , \\ K_V &\equiv a + b\frac{\partial u}{\partial p}\frac{\partial p}{\partial \theta} , \end{aligned} \tag{10A.8}$$

---

[2] In lectures believed to be of 1854/5 F. NEUMANN [1950] attempted to base thermodynamics upon the existence of adiabats ("kalorische Kurven"). REECH's (4) is NEUMANN's Theorem 1 in his §1; even the notations are almost the same. NEUMANN claims that many of CARNOT's considerations may be "in die neue Theorie hinübergerettet". His basic assumption is (7A.1); he arrives at (8B.2). Theorem 2 in his §5 asserts that f is the product of a universal function of θ by a possibly constitutive function of u. Thus NEUMANN, like REECH, fails to reach the central conclusion that f must be a function of θ alone (*Concepts and Logic*, Theorem 7).

we convert REECH's (6) into the statement

$$Q = \Lambda_V(V, \theta)\dot{V} + \mathrm{K}_V(V, \theta)\dot{\theta} \ . \qquad (2\text{C}.4)_r$$

Consequently REECH's starting assumptions are subsumed by the theory of calorimetry. It is possible to prove, conversely, that that theory suffices more or less for REECH's assumption (4) to hold[3]. Likewise, his assumption that through each point of the V–p quadrant runs exactly one isotherm and one adiabat, is also a consequence of (2C.4) and the constitutive inequalities

$$\Lambda_V > 0 \ , \qquad \mathrm{K}_V > 0 \ . \qquad (2\text{C}.5)_r$$

In §5C we have seen that such is the case for the V–θ quadrant, and of course we assume that the equation of state $p = \varpi(V, \theta)$ maps the one quadrant smoothly onto the other. Thus we may say justly that REECH's "first principles" are neither more nor less than

1. The theory of calorimetry.
2. CARNOT's General Axiom.

To obtain (4), REECH has not used the second of these.]

10B. REECH Generalizes the Internal Energy

Because

$$C(\mathscr{P}) = \int_{\mathscr{P}} [\Lambda_V(V, \theta)dV + \mathrm{K}_V(V, \theta)d\theta] \ , \qquad (2\text{C}.6)_2$$

from (10A.4) it follows at once that (p. 380)

$$C(\mathscr{P}) = \int_{\mathscr{P}} f(\theta(u), u)du \ , \qquad (10\text{B}.1)$$

in which θ is a given function of u on the path $\mathscr{P}$. In particular (pp. 380–383), for a Carnot cycle labelled as in Figure 3 of §5C

$$C^+ = \int_a^b f(\theta^+, u)du \ , \qquad C^- = -\int_c^d f(\theta^-, u)du \ ; \qquad (10\text{B}.2)$$

[3] By EULER's theorem on the integrating factor, there are functions g and h such that

$$Q = \Lambda_V\dot{V} + \mathrm{K}_V\dot{\theta} = g\dot{h} \ , \qquad (\text{I})$$

the argument of Λ_V, K_V, g, and h being V, θ.

REECH assumes in effect that the mapping $(V, p) \mapsto (\theta, u)$ can be inverted so as to yield, say, $V = V(\theta, u)$. Then we can set

$$f(\theta, u) \equiv g(V(\theta, u), \theta(V(\theta, u), p(V(\theta, u), \theta))) \ ,$$

and (I) becomes

$$Q = f(\theta, u)\dot{u} \ ,$$

as REECH asserts.

here a stands for u evaluated at a, *etc.* Now applying REECH's second theorem, namely

$$L(\mathscr{C}) = \Gamma(\theta^+)C^+(\mathscr{C}) - \Gamma(\theta^-)C^-(\mathscr{C}) \,, \qquad (9\text{C}.1)_{1\text{r}}$$

we see that

$$\begin{aligned} L(\mathscr{C}) &= \int_a^b \Gamma(\theta^+)f(\theta^+, u)du + \int_c^d \Gamma(\theta^-)f(\theta^-, u)du \,, \\ &= \int_{\mathscr{C}} \Gamma(\theta(u))f(\theta(u), u)du \,. \end{aligned} \qquad (10\text{B}.3)$$

But

$$L(\mathscr{P}) = \int_{\mathscr{P}} \varpi(V, \theta)dV \,. \qquad (2\text{C}.20)_{3\text{r}}$$

Thus

$$\int_{\mathscr{C}} [\Gamma(\theta(u))f(\theta(u), u)du - \varpi dV] = 0 \qquad (10\text{B}.4)$$

for every Carnot cycle $\mathscr{C}$. Now by differentiating (10A.1) we can convert (4) into an integral over a cycle in the V–p quadrant. The fact that through every point of that quadrant passes one and only one adiabat and one and only one isotherm shows that Carnot cycles are dense in that quadrant. Therefore, (4) holds for all cycles $\mathscr{C}$, not merely for Carnot cycles[1]. Hence the integrand in (4) is the differential of some function E of V and u (p. 384):

$$\dot{\text{E}} = \Gamma f \dot{u} - p\dot{V} \,. \qquad (10\text{B}.5)$$

This result is REECH's ***fundamental relation*** (his Equation (10), his notation for E being Q). [The reader will recognize this equation as being a generalized form of what was later to be called "the Gibbs relation". It is equivalent to

$$\Gamma f = \frac{\partial \text{E}}{\partial u} \,, \qquad p = -\frac{\partial \text{E}}{\partial V} \,, \qquad (10\text{B}.6)$$

p being written for the function of u and V whose value equals $\varpi(V, \theta)$, but REECH does not remark this fact. E generalizes the internal energy of CLAUSIUS, which we have discussed above in §8B. In *Concepts and Logic* E is called the *internal pro-energy*. We shall discuss it further below, at the end of the following section.

[We have stated above that if Γ = const. and $f \propto \theta$, then u is the entropy; in that case REECH's relations (6) reduce to equations having the same form

---

[1] The textbooks often appeal to the fact that Carnot cycles are dense in one or another quadrant or plane. In all cases I know but this one, the appeal is superfluous and is made only so as to avoid use of AMPÈRE's transformation (a result of "pure" mathematics which authors of thermodynamics texts seem to regard as being so modern as to destroy "physical intuition"). In the case of REECH's analysis, however, the appeal cannot be avoided, because his second and all-important fundamental principle refers to Carnot cycles alone.

as those that were to be the starting point of GIBBS' thermostatics. More precisely, REECH's equation (5) is the formal analogue for processes of GIBBS' variational statement regarding equilibrium. We shall come back to this analogy in the discussion leading to (10E.11), below.

REECH easily converts these results into counterparts when θ rather than V is taken as an independent variable alongside u. We continue to write p now for ϖ and other functions whose value is the pressure. Thus

$$\dot{\mathrm{E}} = \Gamma f \dot{u} - p\left(\frac{\partial V}{\partial u}\dot{u} + \frac{\partial V}{\partial \theta}\dot{\theta}\right) , \tag{10B.7}$$

so

$$\begin{aligned} \Gamma f - p\frac{\partial V}{\partial u} &= \frac{\partial \mathrm{E}}{\partial u} , \\ -p\frac{\partial V}{\partial \theta} &= \frac{\partial \mathrm{E}}{\partial \theta} , \end{aligned} \tag{10B.8}$$

whence it follows that

$$\frac{\partial}{\partial \theta}\left(\Gamma f - p\frac{\partial V}{\partial u}\right) = -\frac{\partial}{\partial u}\left(p\frac{\partial V}{\partial \theta}\right) . \tag{10B.9}$$

this being REECH's Equation (11). [Like (3F.4) and (5M.3), (9) is what the tradition calls a "Maxwell relation". So is

$$\frac{\partial(\Gamma f)}{\partial V} = -\frac{\partial p}{\partial u} , \tag{10B.10}$$

which is an obvious consequence of (6).]

10C. REECH Introduces and Analyses the Thermodynamic Potentials

According to REECH (p. 392) "it is one and the same thing, from the abstract standpoint, to make into an exact differential the second member of whichever ... of the equations you please," and (p. 390) "we can take as independent variables (p, θ), (V, u), (p, u) ...." REECH carries out the details in many cases. He assumes tacitly that every implicit functional relation $f(x, y, z) = 0$ is soluble for each of its arguments. He is the first author to invoke this principle, which might be called "thoughtless invertibility"; its use is to become a generic trait of thermodynamicists. The results follow by routine differential calculus. We shall list some but not all of these.

First, for the variables V, p (p. 387, misprinted) we need only calculate $\dot{u}$ from the relation

$$u = \psi(V, p) , \tag{10A.1$_r$}$$

put the result into REECH's fundamental relation

$$\dot{\mathrm{E}} = \Gamma f \dot{u} - p\dot{V} \ , \tag{10B.5$_r$}$$

and interpret the terms. The result is

$$\begin{aligned} \Gamma f \frac{\partial u}{\partial V} - p &= \frac{\partial \mathrm{E}}{\partial V} \ , \\ \Gamma f \frac{\partial u}{\partial p} &= \frac{\partial E}{\partial p} \ . \end{aligned} \tag{10C.1}$$

Here the internal pro-energy is taken as a function of V and p. Now the reciprocal of (10B.9) is (p. 387)[1]

$$\frac{\partial(\Gamma f)}{\partial \theta} \frac{\partial(\theta, u)}{\partial(p, V)} = 1 \ , \tag{10C.2}$$

this being REECH's Equation (12). Writing out the determinant and substituting for $\partial u/\partial V$ and $\partial u/\partial p$ from (1) yields (p. 387)

$$\frac{\Gamma f}{\dfrac{\partial}{\partial \theta}(\Gamma f)} = p \frac{\partial \theta}{\partial p} + \frac{\partial(\mathrm{E}, \theta)}{\partial(V, p)} \ , \tag{10C.3}$$

this being REECH's Equation (12*bis*).

To use the variables V, θ, we need only substitute (10A.4) into REECH's fundamental relation (10B.5) and so obtain (p. 389)

$$\dot{\mathrm{E}} = \Gamma Q - p\dot{V} \ . \tag{10C.4}$$

Hence

$$\begin{aligned} \Gamma \Lambda_V - p &= \frac{\partial \mathrm{E}}{\partial V} \ , \\ \Gamma \mathrm{K}_V &= \frac{\partial \mathrm{E}}{\partial \theta} \ , \end{aligned} \tag{10C.5}$$

and so

$$\frac{\partial p}{\partial \theta} = \Gamma \left(\frac{\partial \Lambda_V}{\partial \theta} - \frac{\partial \mathrm{K}_V}{\partial V} \right) + \Gamma' \Lambda_V \ . \tag{10C.6}$$

[We should expect REECH to remark here that these results reduce when $\Gamma(\theta) = J$ to CLAUSIUS' theorems:

$$\frac{\partial p}{\partial \theta} = J \left(\frac{\partial \Lambda_V}{\partial \theta} - \frac{\partial \mathrm{K}_V}{\partial V} \right) \ , \tag{8B.1$_r$}$$

$$J\Lambda_V - p = \frac{\partial \mathrm{E}}{\partial V} \ , \qquad J\mathrm{K}_V = \frac{\partial \mathrm{E}}{\partial \theta} \ , \tag{8D.3$_r$}$$]

---

[1] Of course REECH does not use the notations or formal calculus of Jacobian determinants, but such determinants had occurred frequently in the literature of hydrodynamics for 100 years. REECH's long but straightforward analysis is correct.

but he does not, though he mentions KELVIN's works and translates (4) (5), and (6) into KELVIN's notation. [Possibly he finds KELVIN's treatment clearer than CLAUSIUS'.

[A *thermodynamic potential* is a single function which determines all three constitutive functions ϖ, Λ_V, and K_V uniquely. REECH seems to sense this concept, though he does not make it explicit. He provides us with differential formulae from which we may perceive at a glance *thermodynamic potentials corresponding to the pairs of independent variables* (V, θ), (θ, p), (p, u), (u, V). One of these formulae is

$$\dot{\mathrm{E}} = \Gamma f \dot{u} - p\dot{V} \ , \qquad (10\mathrm{B}.5)_\mathrm{r}$$

which, as we have remarked, is equivalent to the relations

$$\Gamma f = \frac{\partial \mathrm{E}}{\partial u} \ , \qquad p = -\frac{\partial \mathrm{E}}{\partial V} \ . \qquad (10\mathrm{B}.6)_\mathrm{r}$$

On the presumption that Γ and f are known and invertible functions, (10B.6) delivers θ and p as functions of V and u; elimination of u between the two yields ϖ. Since θ is a function of V and u, we may eliminate u as an argument of E and so obtain E as a function of V and θ. Then (10B.5) and (10A.4) deliver Q as a linear function of $\dot{V}$ and $\dot{\theta}$, whence Λ_V and K_V are delivered by comparison with the old formula

$$Q = \Lambda_V(V, \theta)\dot{V} + \mathrm{K}_V(V, \theta)\dot{\theta} \ . \qquad (2\mathrm{C}.4)_\mathrm{r}$$

Thus *the function of u and V whose value is* E *is a thermodynamic potential.*]

For other choices of variables REECH does not write out all details. [We shall state his results in notations more compact than his.] First we set (p. 379, Equation (5), and p. 407, Equation (22))

$$F(\theta, u) \equiv \int_{u_0}^{u} f(\theta, w)dw \ ,$$

$$R \equiv \Gamma F \ , \qquad (10\mathrm{C}.7)$$

so that (10A.4) and (10B.5) may be written in the forms (his Equations $(22)_2$ and $(23)_1$),

$$\Gamma Q = \frac{\partial R}{\partial u}\dot{u} \ ,$$

$$\dot{\mathrm{E}} = \frac{\partial R}{\partial u}\dot{u} - p\dot{V} \ , \qquad (10\mathrm{C}.8)$$

respectively. Then we set [2]

$$\mathrm{X} \equiv \mathrm{E} + pV \ ,$$

$$-\mathrm{A} \equiv \mathrm{E} - R \ , \qquad (10\mathrm{C}.9)$$

$$\mathrm{Z} \equiv \mathrm{E} - R + pV \ ,$$

[2] The notations are REECH's except that he does not introduce single letters for what we denote by X and Z, and he uses as always the inevitable differentials.

so

$$\mathrm{E} + \mathrm{Z} = \mathrm{X} - \mathrm{A} \; . \tag{10C.10}$$

Then the basic equations REECH numbers (10 *bis*), (10 *ter*), and (10 *quater*) are, respectively (pp. 391–392)

$$\begin{aligned} \dot{\mathrm{X}} &= \Gamma f \dot{u} + V \dot{p} \\ \dot{\mathrm{A}} &= \frac{\partial R}{\partial \theta} \dot{\theta} + p \dot{V} \; , \\ -\dot{\mathrm{Z}} &= \frac{\partial R}{\partial \theta} \dot{\theta} - V \dot{p} \; . \end{aligned} \tag{10C.11}$$

These results, joined with

$$\dot{\mathrm{E}} = \Gamma f \dot{u} - p \dot{V} \; , \tag{10B.5$_{\mathrm{r}}$}$$

we may call ***REECH's Fourth Theorem.*** [REECH does not write out the formulae in partial derivatives that follow at once from (11): *If* X *is taken as a function of u and p*, then

$$\Gamma f = \frac{\partial \mathrm{X}}{\partial u} \, , \qquad V = \frac{\partial \mathrm{X}}{\partial p} \; ; \tag{10C.12}$$

if A *is taken as a function of θ and V*, then

$$\frac{\partial R}{\partial \theta} = \frac{\partial \mathrm{A}}{\partial \theta} \, , \qquad p = \frac{\partial \mathrm{A}}{\partial V} \; ; \tag{10C.13}$$

if Z *is taken as a function of θ and p*; then

$$\frac{\partial R}{\partial \theta} = -\frac{\partial \mathrm{Z}}{\partial \theta} \, , \qquad V = \frac{\partial \mathrm{Z}}{\partial p} \; . \tag{10C.14}$$

The corresponding conditions of integrability are

$$\begin{aligned} \frac{\partial}{\partial p}(\Gamma f) &= \frac{\partial V}{\partial u} \, , \\ \frac{\partial^2 R}{\partial V \partial \theta} &= \frac{\partial p}{\partial \theta} \, , \\ \frac{\partial^2 R}{\partial p \partial \theta} &= -\frac{\partial V}{\partial \theta} \; . \end{aligned} \tag{10C.15}$$

[Presuming Γ and f known, and hence also R known, we invoke the principle of thoughtless invertibility and so from (12) see easily that X as a function of u and p is a thermodynamic potential; from (13), so is A as a function of θ and V; from (14) so is Z as a function of θ and p.

[From these results the reader sees that REECH, *in his general theory, has introduced all four of the classical thermodynamic potentials and has found all the differential relations among them*[3]. Common names now for X, $-\mathrm{A}$, and Z are *enthalpy*, *free energy*, and *free enthalpy*. For reflections of REECH's Fourth Theorem upon the work of MASSIEU and GIBBS see the discussion surrounding (10E.11), below.]

Most of the rest of the chapter (pp. 392–407) is spent in interpretation of the four potentials. REECH appeals to various areas under curves in the V–p quadrant, and most of his interpretations, [while of course they are correct, do not seem enlightening. One, however, is important, namely,] that for E itself. In an adiabatic process (4) reduces to $\dot{\mathrm{E}} = -p\dot{V}$. Thus (pp. 392–393)

> the quantity E will increase by exactly the quantity of work put out so as to produce a physical effect of this kind; this quantity of work will remain stored in the gas as long as the variables V, p, and θ do not change. . . .

On the other hand (p. 393), in an isochoric process (4) reduces to $\dot{\mathrm{E}} = \Gamma Q$, so "a purely caloriferous change will make the quantity E increase . . . ." Thus in general (p. 394)

> the function E will represent the quantity of work or live force susceptible of being produced by the totality of the heat presently contained in a gas or elastic fluid . . . ,

that is (p. 395),

> the perfect mechanical equivalent of the heat contained in an elastic fluid.

For REECH, as for CARNOT before him, the conversion of heating into mechanical energy is proportional to a function of the temperature alone. REECH allows the possibility that that function may reduce to a constant, as in CLAUSIUS' theory, but he does not require it. His term "totality of the heat" describes well what now is called internal energy.

[REECH has every reason to be proud of his result.] He does not apply it to cases, [but he easily could]. He has proved that the constitutive function E exists locally. [As soon as it and the universal function Γ are specified, the work done by a fluid body on any path in a simply connected domain can be calculated:

$$L(\mathscr{P}) = \int_{\mathscr{P}} [\Gamma(\Lambda_V dV + \mathrm{K}_V d\theta) - d\mathrm{E}] \ . \tag{10C.16}$$

---

[3] To verify this fact, put $\Gamma(\theta) = J = 1$, $f(\theta, u) = \theta$, and denote the resulting pro-entropy, namely, what CLAUSIUS was later to call "entropy", by H. Then $R = \theta\mathrm{H}$.

REECH does not write the result so simply,] but his geometrical analysis yields equivalent statements. [Never before had anyone shown how to calculate the motive power of heat in a general cycle within CARNOT's conceptual frame without assuming the existence of a heat function, as in KELVIN's analysis (above, §7H; *cf.* also (5M.5)). REECH's fundamental relation (16) enables us to calculate the motive power of heat for any fluid body on any path, according to any theory consistent with the Doctrine of Latent and Specific Heats.]

At the end of the chapter (pp. 407–408) REECH obtains the brief expressions (7), (8), and $(9)_2$, which we, so as to shorten the arguments, have introduced earlier. These expressions, he says, are (p. 408) "*the necessary and sufficient equations of the theory of the dynamical effects of heat.*" [Here he is wrong, fatally wrong! We refrain from enodating this matter until §10E.]

10D. REECH's General Theory of Specific Heats

[Any spectator of this tragicomedy will know what to expect next, and he will not be disappointed.] In Chapter IV (pp. 410–427) REECH proceeds to construct the theory of specific heats. [He does so in generality never before achieved and rarely thereafter equalled.]

The principle of thoughtless invertibility allows us to regard V and p as functions of θ and u:

$$\dot{V} = \frac{\partial V}{\partial \theta}\dot{\theta} + \frac{\partial V}{\partial u}\dot{u} \, , \qquad \dot{p} = \frac{\partial p}{\partial \theta}\dot{\theta} + \frac{\partial p}{\partial u}\dot{u} \, . \tag{10D.1}$$

REECH's Third Theorem,

$$Q = f(\theta, u)\dot{u}, \tag*{(10A.4)$_r$}$$

in terms of these variables takes the forms

$$Q = f\dot{u} = f\frac{\dot{V} - \frac{\partial V}{\partial \theta}\dot{\theta}}{\frac{\partial V}{\partial u}} = f\frac{\dot{p} - \frac{\partial p}{\partial \theta}\dot{\theta}}{\frac{\partial p}{\partial u}} \, , \tag{10D.2}$$

which are REECH's Equations (F), p. 414. Comparison with the formal statements of the Doctrine of Latent and Specific Heats, namely

$$Q = \Lambda_V(V, \theta)\dot{V} + \mathrm{K}_V(V, \theta)\dot{\theta} = \Lambda_p(V, \theta)\dot{p} + \mathrm{K}_p(V, \theta)\dot{\theta}, \quad \text{(2C.4)}_r, \text{(2C.8)}_r$$

yields expressions for the latent and specific heats when u and θ are taken as the independent variables:

$$\Lambda_V = \frac{f}{\dfrac{\partial V}{\partial u}}\,, \qquad K_V = -\frac{f\dfrac{\partial V}{\partial \theta}}{\dfrac{\partial V}{\partial u}}\,, \tag{10D.3}$$

$$\Lambda_p = \frac{f}{\dfrac{\partial p}{\partial u}}\,, \qquad K_p = -f\frac{\dfrac{\partial p}{\partial \theta}}{\dfrac{\partial p}{\partial u}}\,,$$

these being REECH's Equations (E), (F′), and (F″), p. 414.

[If we divide $(3)_4$ by $(3)_2$ and apply a standard theorem of calculus to the result, we obtain

$$\gamma \equiv \frac{K_p}{K_V} = \frac{\dfrac{\partial p}{\partial \theta}\Big/\dfrac{\partial V}{\partial \theta}}{\dfrac{\partial p}{\partial u}\Big/\dfrac{\partial V}{\partial u}} = \frac{\dfrac{\partial p(V,u)}{\partial V}}{\dfrac{\partial p(V,\theta)}{\partial V}}\,. \tag{10D.4}$$

Although the well-known specialization of this formula to an ideal gas is sometimes called "Reech's theorem", REECH did not record it here; certainly it lay in his hands. The fact that f cancels out shows the concept of entropy to be unnecessary for the result; the pro-entropy does just as well, since in order to calculate $\partial p(V,u)/\partial V$, which is a derivative along an adiabat ($u =$ const.), we need not know how the value of the entropy varies from one adiabat to another. Indeed, from calorimetric theory alone we have shown that in an adiabatic process

$$\frac{dp}{d\rho} = \gamma\frac{\partial p}{\partial \rho}\,, \tag{3F.7$_r$}$$

which expresses the real content of (4). In my opinion REECH missed the point here; he was to see it clearly later[1].]

REECH next (pp. 414–415) introduces the concept of *specific heat along a curve*[2] $\mathscr{P}$ in the V, p quadrant. If an equation of $\mathscr{P}$ is $\phi(V,p) =$ const., upon $\mathscr{P}$

$$A(V,p)\dot{p} + B(V,p)\dot{V} = 0\,, \tag{10D.5}$$

---

[1] REECH [1868, §27].

[2] This is one of the few aspects of the work of REECH that have entered the general literature. TRUESDELL & TOUPIN [1960, §249], duly attributing to REECH the concept of specific heat along a path for the case in which there are but two independent variables, based upon it their exposition of the theory of specific heats when V is replaced by an arbitrary finite number of variables.

the functions A and B being proportional to $\partial\phi/\partial p$ and $\partial\phi/\partial V$, respectively. Then $K_{\mathscr{P}}$, *the specific heat on* $\mathscr{P}$, is defined and calculated as follows:

$$K_{\mathscr{P}} \equiv \frac{Q}{\dot{\theta}}\bigg|_{\mathscr{P}} = \frac{f\dot{u}_{\mathscr{P}}}{\dot{\theta}_{\mathscr{P}}} = f\frac{\dfrac{\partial u}{\partial V}\dot{V} + \dfrac{\partial u}{\partial p}\dot{p}}{\dfrac{\partial \theta}{\partial V}\dot{V} + \dfrac{\partial \theta}{\partial p}\dot{p}}\Bigg|_{\mathscr{P}},$$
$$= f\frac{A\dfrac{\partial u}{\partial V} - B\dfrac{\partial u}{\partial p}}{A\dfrac{\partial \theta}{\partial V} - B\dfrac{\partial \theta}{\partial p}}. \tag{10D.6}$$

[REECH's manipulations are more elaborate, and he does not record his results in just this form, but the idea of the proof is the same, and (6) is equivalent to the last of his Equations (G) on p. 416. Putting first $B = 0$ and then $A = 0$ yields

$$K_p = f\frac{\dfrac{\partial u}{\partial V}(V,p)}{\dfrac{\partial \theta}{\partial V}(V,p)}, \qquad K_V = f\frac{\dfrac{\partial u}{\partial p}(V,p)}{\dfrac{\partial \theta}{\partial p}(V,p)}, \tag{10D.7}$$

equivalent to $(3)_3$ and $(3)_4$.]

Most of the rest of the chapter concerns geometric interpretations of the properties of differential forms in two variables. REECH also sets about to express various thermodynamic quantities in terms more easily accessible to experiment, with the idea of applying the results so as to determine from experimental data the crucial functions Γ and f. In particular, he seeks to eliminate the pro-entropy u, and in some cases he succeeds in doing so.

For example, if we take V and p as independent variables, recall that $u = \psi(V,p)$, and substitute (10C.1) into (7), we obtain

$$\Gamma K_p = \frac{\dfrac{\partial E}{\partial V} + p}{\dfrac{\partial \theta}{\partial V}} = \frac{\left(\dfrac{\partial E}{\partial V} + p\right)\dfrac{\partial \theta}{\partial p}}{\dfrac{\partial \theta}{\partial V}\dfrac{\partial \theta}{\partial p}},$$
$$\Gamma K_V = \frac{\dfrac{\partial E}{\partial p}}{\dfrac{\partial \theta}{\partial p}} = \frac{\dfrac{\partial E}{\partial p}\dfrac{\partial \theta}{\partial V}}{\dfrac{\partial \theta}{\partial p}\dfrac{\partial \theta}{\partial V}}. \tag{10D.8}$$

Subtraction followed by use of (10C.3) yields

$$\Gamma(K_p - K_V) = \frac{\dfrac{\partial(E,\theta)}{\partial(V,p)} + p\dfrac{\partial \theta}{\partial p}(V,p)}{\dfrac{\partial \theta}{\partial p}(V,p)\dfrac{\partial \theta}{\partial V}(V,p)} = \frac{\Gamma f}{\dfrac{\partial \theta}{\partial p}(V,p)\dfrac{\partial \theta}{\partial V}(V,p)\dfrac{\partial}{\partial \theta}(\Gamma f)}. \tag{10D.9}$$

(included in REECH's Equations (O), p. 424). If $\Gamma(K_p - K_V)$ and the thermal equation of state are known, as from experiment they may be, this relation determines $\partial[\log(\Gamma f)]/\partial\theta$.

REECH obtains also equations which deliver ΓK_p and ΓK_V in terms of the functions A, E, and R alone (Equations (O′), p. 426). "These different formulae will acquire great importance in what follows, when it will be our concern to find the algebraic expressions of the functions A, E, R, ... on the basis of certain experimental results."

To conclude the chapter (p. 427), REECH recalls that CARNOT's theory in effect supposed f to be a function of u alone, in which case f cancels out of many of the relations, yielding "a crowd of very remarkable theorems.... But as soon as $f(\theta, u)$ depends upon the variable θ and can differ from one elastic fluid to another, all these theorems disappear, and the main thing will be to find a system of experimentation by means of which, some day, we shall know the universal function $\Gamma(\theta)$ and the special function $f(\theta, u)$ for each kind of elastic fluids."

[We note that REECH regards Γ as universal and f as constitutive. This passage may be the earliest in which the generic principles of the subject are explicitly separated from constitutive relations. However, we shall soon see that the next step, indeed the only remaining step required to complete the classical theory, was to proclaim f a universal function of θ alone.

[We have outlined only the first four of REECH's nine chapters, only the first seventy of his nearly 200 pages. The rest consider the effects of different possible choices of f, applications to perfect gases, and the theory of vapors. The following section makes it plain why we give no account of all that.]

10E. Critique: the Fatal Failure of REECH's Analysis

In §10C we have seen that REECH claimed his basic differential relations to be necessary and sufficient. Necessary they are, but sufficient they are not! Had REECH understood CARNOT's analysis fully, he might have discerned and corrected his own error.

To see this, let us write his relations $(10D.3)_{1,2}$ in the forms

$$\frac{\Lambda_V}{f} = \frac{\partial u}{\partial V}, \qquad \frac{K_V}{f} = \frac{\partial u}{\partial \theta}. \tag{10E.1}$$

It follows that

$$\frac{\partial}{\partial \theta}\left(\frac{\Lambda_V}{f}\right) = \frac{\partial}{\partial V}\left(\frac{K_V}{f}\right). \tag{10E.2}$$

Hence

$$\Lambda_V \frac{\partial f}{\partial \theta} = f\left(\frac{\partial \Lambda_V}{\partial \theta} - \frac{\partial K_V}{\partial V}\right) + K_V \frac{\partial f}{\partial V}. \tag{10E.3}$$

Comparison with REECH's formula

$$\frac{\partial p}{\partial \theta} = \Gamma\left(\frac{\partial \Lambda_V}{\partial \theta} - \frac{\partial \mathrm{K}_V}{\partial V}\right) + \Gamma' \Lambda_V \tag{10C.6$_r$}$$

yields

$$\Lambda_V = \frac{f}{\frac{\partial}{\partial \theta}(f\Gamma)}\frac{\partial p}{\partial \theta} + \frac{\Gamma \mathrm{K}_V}{\frac{\partial}{\partial \theta}(f\Gamma)}\frac{\partial f}{\partial V}\,. \tag{10E.4}$$

What a strangely complicated result! CARNOT's own reasoning, checked and endorsed meanwhile by CLAPEYRON, KELVIN, and CLAUSIUS, from *just the same assumptions* had led to the *General CARNOT-CLAPEYRON Theorem*:

$$\mu \Lambda_V = \frac{\partial p}{\partial \theta}\,. \tag{5L.4$_r$}$$

The mathematics being correct in both cases, *the two different evaluations of Λ_V must agree*! A glance shows that they do agree if

$$\frac{\partial f}{\partial V} = 0\,, \qquad \mu = \frac{(f\Gamma)'}{f}\,. \tag{10E.5}$$

Moreover, if f and Γ are both universal functions, these conditions are necessary as well as sufficient. This agreement does not prove but does strongly suggest that it is possible for any material to take f as a function of θ alone, and that since both μ and Γ are universal functions, so also is f.

Such, indeed, is the case. Over a hundred years were to pass before anyone again took up the logical implications of CARNOT's axioms and developed them to the point of *sufficient* as well as necessary local relations. My discovery in 1971 that REECH had failed in his program impelled me to undertake to complete it; the product is *Concepts and Logic*, to which the reader must refer if he would learn the details. Here I report only the contents of Theorem 11 in §10:

Let the theory of calorimetry be assumed. Then either of the following two sets of local restrictions is equivalent to CARNOT's General Axiom:

1. *There are continuously differentiable functions f and g of θ alone, f being positive and g increasing, such that*

$$\frac{\partial}{\partial \theta}\left(\frac{\Lambda_V}{f}\right) = \frac{\partial}{\partial V}\left(\frac{\mathrm{K}_V}{f}\right), \qquad \frac{g'}{f}\Lambda_V = \frac{\partial p}{\partial \theta}\,. \tag{10E.6}$$

That is, REECH's f *must* be a function of θ alone, and REECH's $f\Gamma$, here denoted by g, is a monotonically increasing function. The function f is unique to within a constant multiplicative factor; when f has been specified, g is unique to within an additive constant.

2. *In addition to g and f as previously specified there are functions $u(V, \theta)$ and* $\mathrm{E}(V, \theta)$ such that

$$\Lambda_V = f\frac{\partial u}{\partial V}, \qquad \mathrm{K}_V = f\frac{\partial u}{\partial \theta},$$
$$\frac{g}{f}\Lambda_V = \varpi + \frac{\partial \mathrm{E}}{\partial V}, \qquad \frac{g}{f}\mathrm{K}_V = \frac{\partial \mathrm{E}}{\partial \theta}. \tag{10E.7}$$

The reader will recognize u as being REECH's pro-entropy and E as being REECH's generalization of CLAUSIUS' internal energy. Also he will easily obtain the following formula for the work done in a Carnot cycle:

$$L(\mathscr{C}) = \frac{g(\theta^+) - g(\theta^-)}{f(\theta^+)} C^+(\mathscr{C}),$$
$$= \frac{g(\theta^+)}{f(\theta^+)} C^+(\mathscr{C}) - \frac{g(\theta^-)}{f(\theta^-)} C^-(\mathscr{C}), \tag{10E.8}$$

so $[g - g(\theta_0)]/f$ replaces REECH's Γ in (9C.1), his "fundamental pivot of the theory of the dynamical effects of heat". That is, Γ must be the quotient of an increasing function by a positive function. Thus REECH's results are merely necessary, *not sufficient* for the truth of CARNOT's General Axiom when the theory of calorimetry is assumed. Moreover, the defect of REECH's function Γ from uniqueness is not trivial. REECH seems not to have noticed this fact, perhaps because his proof of (9C.1) is so obscure. The simple proof we have given above in §9C rests upon the definition (9C.9) and so makes this failure of uniqueness obvious because θ_0 is arbitrary. The choice of θ_0 does not affect the left-hand side of (8) because

$$\frac{C^+(\mathscr{C})}{f(\theta^+)} = \frac{C^-(\mathscr{C})}{f(\theta^-)}. \tag{10E.9}$$

The developments in *Concepts and Logic* provide a common framework upon which CARNOT's theory, CLAUSIUS' theory, and other possibilities may be discussed. The mathematics used is no more than was available in CARNOT's day. The program is REECH's, and REECH had the tools with which to effect it as well as the will to do so. Had he succeeded, classical thermodynamics would have been set upon a sound base of experience and reason conjoined. History was not to be that way. The tragicomic muse of thermodynamics would not allow something so simple. REECH failed. Even his attempt was forgotten.

Appendix: The Later Work of REECH

The terminal date set for this tragicomedy is 1854, and I leave to others the task of tracing the later history of the field. For REECH, however, I make an exception, because it is little likely that anyone else will take the necessary trouble. While Historians of Science are ready enough to dismiss the standard folklore about the methods and ideas of the giants, and even more ready to reject outright any absolute standard of truth in science, they usually accept uncritically the party line of today's profession of physics as to what the various disciplines are and what must come out triumphant in the end. Since I regard this latter folklore as more dangerous, because less obviously hieratic, than the former, and because I regard criticism just as important as discovery, I append here an outline of the work of the only substantial and constructive critic classical thermodynamics excited in its formative period.

REECH's work was set aside, unread. HELMHOLTZ[1] in the course of a general and in most instances fair and precise summary of researches on the theory of heat through 1851 devotes a paragraph to the first note of REECH[2] but, mentioning only a subsidiary numerical detail, bestows oblivion upon its program and the beautiful theorem it announces.

When REECH's great memoir appeared, HELMHOLTZ[3] dismissed it as being too general to deserve study:

> Mr. REECH's basic equation [*i.e.* $(9C.1)_1$] reflects only an incomplete acceptance of the principle of conservation of force. He assumes that force cannot be generated from nothing, but he leaves open the possibility that it can disappear with no effect.

HELMHOLTZ refrains from further summary of REECH's results because "they contain more undetermined functions than do those of Clausius and W. Thomson." HELMHOLTZ is fair and accurate as far as he goes, but he gives no idea either of REECH's aims or of the abundance of new results in REECH's

[1] HELMHOLTZ [1855, p. 590].
[2] REECH [1851].
[3] HELMHOLTZ [1856, *1*].

memoir. He judges REECH's work strictly on the basis of what he regards as physically correct at the moment, disregarding not only the past but also the future of the subject. Perhaps HELMHOLTZ's review is responsible for the tradition's failure to notice REECH's pro-entropy and four thermodynamic potentials.

Next REECH[4] published a "very succinct recapitulation" of recent researches in thermodynamics. He points out that his result (10B.5) shows "by simple inspection" that E [in his notation, Q] "represents the quantity of work that can be produced by all the heat that is in an elastic fluid" and is thus "the mechanical equivalent of that quantity of heat." With extraordinary self-control, REECH does not remind the reader that he had published all this in 1853 (above, §10C). While "it is generally accepted today" that $\Gamma = \text{const.}$, so (8A.1) holds, "My aim was to prepare in advance formulae for the greatest number if not for all experiments to be done by the physicists, and to do so in the greatest generality possible...." Below in §11I we shall discuss the improvement of CLAUSIUS' work of 1854 that REECH supplies in this same note. He accepts there CLAUSIUS' conclusion that the integrating factor is a function of θ alone. In addition he develops in detail the special forms his numerous results assume when CLAUSIUS' "first principle", which in his notation amounts to the statement $\Gamma = \text{const.}$, is adopted. Of course he is right; CLAUSIUS' results are all there, special cases.

In reviewing this paper HELMHOLTZ[5] justly remarks that it is neither very succinct nor very understandable. He now has even less patience with REECH and rejects the work out of hand: "... Mr. REECH by use of a lot of undetermined functions attempts to retain in his formulae a generality that here has no purpose at all but makes the essay extraordinarily difficult to study." He seems to be giving excuses for not studying it. In fact it is discursive and vague, harder to read than REECH's long memoir, and is not at all a recapitulation. On the other hand, HELMHOLTZ is most unfair in speaking of "a lot" ("eine Menge") of undetermined functions. The theory of CLAUSIUS and KELVIN has in principle the one undetermined function μ, which those authors determined by subsidiary hypotheses or appeal to experiment. REECH's theory has precisely two undetermined functions, f and Γ, no more; he expressly eschews further hypotheses, and he regards the experimental data, with which he was thoroughly familiar, as being not accurate enough to be accepted. HELMHOLTZ in superficially rejecting REECH's ideas misses the essential flaw of REECH's analysis: insufficiency. As the reader of *Concepts and Logic* will know, REECH's mathematical assumptions *require* that f be a non-vanishing function of θ alone and that Γ be reduced as follows in terms of f and a monotone function g of θ alone:

$$\Gamma = g/f \,. \tag{1}$$

[4] REECH [1856].
[5] HELMHOLTZ [1859, *1*].

Hence

$$\mu = g'/f \cdot \tag{2}$$

Theorem 1 in §7I, above, asserts that CLAUSIUS' "obvious subsidiary hypothesis" regarding an ideal gas, namely HOLTZMANN's Assertion

$$J\Lambda_V = p \ , \tag{7D.1)$_r$ (}$$

is *equivalent* to the HELMHOLTZ-JOULE Determination

$$\mu = J/\theta \ . \qquad (5N.7)_r$$

As REECH's starting point is more general than CLAUSIUS', his development, had he only seen how to push through the mathematical analysis, would have led to all of CLAUSIUS' results *from assumptions weaker than CLAUSIUS'*. It would have delivered among other things the implication

$$\left.\begin{array}{c}\text{CARNOT's General Axiom}\\ \text{and}\\ \text{CLAUSIUS' "subsidiary hypothesis"}\end{array}\right\} \Rightarrow \text{CLAUSIUS' "first principle".}$$

In symbols which will be thoroughly familiar to such spectators as remain in the theater,

$$\left.\begin{array}{l} L(\mathscr{C}) = G(\theta^+, \theta^-, C^+(\mathscr{C})) \quad \text{for all Carnot cycles } \mathscr{C}\\ \text{and}\\ J\Lambda_V = p \quad \text{for ideal gases}\end{array}\right. \Rightarrow L(\mathscr{C}) = JC(\mathscr{C}) \quad \text{for all cycles } \mathscr{C} \ .$$

Such was not to be the course of history.

In 1858 REECH[6] announced and summarized a new memoir, in which he cites "facts of experiment that tend to make us assume that the function $\Gamma(\theta)$ should be a constant...." Also "I cite and prove, in my own way, the theorem of Mr. Clausius", according to which (10A.4) is "too general" and should be replaced by $Q = \gamma(\theta)\dot{u}$, "the function $\gamma(\theta)$ being supposed the same for all kinds of elastic fluids, be they gases or be they vapors."

Then REECH's relation (10B.5) reduces to $\dot{\mathrm{E}} = \Gamma\gamma\dot{u} - p\dot{V}$, "simple inspection" of which shows that "the thing denoted by E will have to be considered as being the quantity of heat proper to an elastic fluid." Moreover, the experiments of REGNAULT show that for air K_p = const., "which obliges me to make $\gamma(\theta) = \theta$ + const." This memoir seems not to have been published. As is mentioned in the text above, some years ago I found a strict local proof that REECH's f must reduce to a function of θ alone, and a proof in the large may be found in *Concepts and Logic*, §10. Judging from REECH's attempts to demonstrate other things, I think it unlikely he found any mathematical proof; probably he appealed to some sort of "absurdity".

[6] REECH [1858].

With this powerful theorem in hand, REECH held fast to the remainder of his original position: CARNOT's point of view was correct and needed only mathematical development. His opinion was sound; it is substantiated in detail by *Concepts and Logic*. Logic is logic. Its consequences can be obtained without the logician's having to know whether they were obtained by someone else, hundreds or thousands of years ago. In illustration of that fact I mention that in outlining *Concepts and Logic* I had only REECH's first two papers before me and did not know his assertion of 1858.

The first textbook of thermodynamics, written by ZEUNER, appeared in 1860. ZEUNER[7] attributed to REECH[8] an argument leading to CLAUSIUS' "Second Principal Theorem" of 1854. We present this argument below, at the end of §11H; as we remark there, it is really due to RANKINE.

In 1863 CLAUSIUS[9] with his usual generosity mentioned REECH's paper of 1856 as having "rendered" or "reproduced" (wiedergegeben) his proof of the Second Law. As was his wont, he repeated this statement again and again in subsequent publications and republications: "Reech's version of my proof". I find no evidence that CLAUSIUS ever took the trouble to read REECH's work carefully enough to understand any part of it that did not "reproduce" something of his own.

Aroused by a controversy between DUPRÉ and CLAUSIUS regarding vapors, in 1863 REECH[10] showed that "the equations found by Messrs. Dupré and Clausius can be established without our having to know whether there is or is not a mechanical equivalent of heat". Replying to new objections by DUPRÉ and CLAUSIUS, REECH[11] stated, "My aim was to show what equations could have been used at any period without knowing that a given amount of heat was equivalent to a given amount of work." Seeming to relinquish his position of 1858, he called for more data:

> For a long time I have been waiting for Mr. Regnault to publish his last experiments, from which to exhibit [the integrating factors] in different cases.... In order for there to be a mechanical equivalent of heat that does not depend upon the nature of the elastic fluid..., it suffices rigorously that [the ratio of integrating factors] be a function of θ, the same for all kinds of fluids.

Again it is the mortifying influence of REGNAULT's fancy apparatus and subsidized precision. REECH fails to take advantage of the experimental facts already known. He seems not to trust any datum that does not come from REGNAULT's laboratory.

---

[7] ZEUNER [1860, §10].
[8] REECH [1856].
[9] CLAUSIUS [1863, §3].
[10] REECH [1863, *1*].
[11] REECH [1863, *2*].

Attacked next by MOIGNO, REECH[12] once more showed CLAUSIUS' results to be special cases of his own. "In summary: He who can do more can do less. That is why my theory contains as a special case that which is called the mechanical theory of heat." REECH concludes in disgust, "If these explanations are not considered sufficient, I must give up using the French language."

In these late notes REECH makes no claim of priority. He does not refer to his earlier papers but derives afresh everything he wishes to assert.

Finally REECH wrote a textbook[13]. In it he developed the Doctrine of Latent and Specific Heats as a basis and then adopted CLAUSIUS' axiom (7A.1) as the foundation of thermodynamics. The treatment suffers from REECH's preference for differential arguments drawn from appeal to graphs. It would have been helped by greater fluency in integral calculus, especially by use of AMPÈRE's transformation. Otherwise it is the clearest presentation of the subject I have ever seen. It was published obscurely, and it has never been noticed.

When MASSIEU's note on characteristic functions[14] appeared, REECH replied[15] by calling attention to similar considerations in the book he had published in the preceding year. He remarked with full justice that what CLAUSIUS called the entropy stood already, unnamed, in his book, and that all of MASSIEU's results were either in that book or easy consequences of equations there. For example, he had emphasized (10C.4), suitably specialized. REECH says not a word about the priority in the whole subject of thermodynamic potentials which his memoir of 1853 in fact establishes. He merely remarks that he regards his theory "as simple and not less general than that of most of the savants who have written about thermodynamics."

In 1873 GIBBS' magnificent researches on the theory of heat began to appear. Despite the titles of his papers, his work concerns thermostatics, not thermodynamics: It does not treat of processes but rather compares putative equilibria. However, GIBBS' most immediate results are just the same *in form* as certain relations in the thermodynamics of fluids governed by the Doctrine of Latent and Specific Heats. In this sense we may compare GIBBS' equations with counterparts in thermodynamics. Among the most famous is the "GIBBS relation"[16]

$$d\mathrm{E} = \theta d\mathrm{H} - p dV, \tag{3}$$

which connects the internal energy E, taken as being a function of H and V, with the elementary heat added and the elementary work done. As we have stated in §10B, the "GIBBS relation" is nothing more nor less than REECH's

[12] REECH [1863, *3*].
[13] REECH [1868]. See especially §90ff.
[14] MASSIEU [1869, *1* and *2*].
[15] REECH [1869].
[16] GIBBS [1873, *2*].

(10B.5) and (10B.6), specialized as they must be if they are to respect the uniform and universal Interconvertibility of Heat and Work. *Cf.* also REECH's (10C.4).

That is not all. Not only does (10B.6) show that E taken as a function of u and V is a thermodynamic potential, we have seen in §10C that REECH had introduced GIBBS' three further thermodynamic potentials. This work of REECH had been published in 1853. By GIBBS' day, 1873, REECH's too general relations would not have looked interesting, had GIBBS seen them. But again the tragicomic muse played one of her pretty tricks. When, two years later, GIBBS[17] remarked that MASSIEU had found "fundamental equations" different from those that he was then introducing, he cited and described MASSIEU's notes appearing *in the same volume* as REECH's note of 1869. In that note, which was prompted by MASSIEU's first, REECH had merely indicated how easy it was to obtain a "fundamental equation" using any pair of independent variables we might please. He then cited only his textbook of 1868; he did not mention his paper of 1853 or its contents regarding thermodynamic potentials. Of course, MASSIEU's functions do not appear there, and that GIBBS in 1875 would introduce for the thermostatics based on the Interconvertibility of Heat and Work the very same "*fundamental equations*" as those REECH had derived for any theory compatible with CARNOT's General Axiom, REECH could not have known in 1868! The muse was malicious indeed!

By 1873 REECH, still alive, had fallen silent. Perhaps he knew that it was futile for him to write. Nobody had ever paid attention to him when his works were new and original; why should anyone pay attention now, when they were largely superseded?

There is a parallel between REECH's fate and WATERSTON's. Both were a generation too early in what they sought to do; the works of both were defective in one way or another; the defects were all that their contemporaries noticed; and both were thereafter dismissed, ignored, and then forgotten. There is a difference, too. WATERSTON wrote excellently well, but he could not get his papers published. REECH, a mediocre writer, published easily and too much. For different reasons, the works of both went unread.

REECH's publications are patient and dispassionate; the later ones are clearer and more explicit than any other works on thermodynamics at the time. REECH shows none of CLAUSIUS' greed for priority and recognition, and he never attacks any person or his work. It is hard to account for the oblivion in which the tradition has buried him. I hesitantly conjecture three reasons:

1. His great memoir of 1853 was far too long; it was awkward, and it claimed a generality in part illusory. Readers tend to dismiss unread

[17] GIBBS [1875, pp. 86–87 of the reprint in *Collected Papers*].

any subsequent publication by such an author: "more of the same".

2. Critical logic in physics was going out of fashion.
3. REECH was neither a famous physicist nor a member of any group of physicists.

I incline toward the third as the strongest reason. Otherwise his work would have influenced the textbooks, one of the principal aims of which is to make things simple and clear for beginners.

NOTE ADDED IN PROOF. The statement on p. 299 to the effect that REECH's textbook has never been noticed is not correct. JULES MOUTIER in Chapter III, §3 of his book, *La thermodynamique et ses principales applications*, Paris, Gauthier-Villars, 1885, cited it for the proof of (3F.7) as a theorem of calorimetry and proposed and justified naming it "Théorème de Reech".

Postscript on Maximum Efficiency

The formal structure of the classical thermodynamics of reversible processes in fluids is now complete, though only if we include the results of REECH appropriately specialized. One point is left dangling. Are Carnot cycles truly the most efficient for given operating temperatures? In all these years, through the thousands of pages of print we have analysed, nothing better than the negations of CARNOT (§§5D–5E) was produced. CARNOT's argument or some substitute even worse is all that is found in textbooks today. The proposition is mathematical; it can be regarded as true only if mathematically proved.

The special features of the Caloric Theory makes such a proof easy, and in this tragicomedy I have provided two proofs, one in §5M and another in §7H; the former has not been published before, and the latter is one I found in 1973.

What of CLAUSIUS' theory? The proposition is still true, but the proof is not so immediate. So far as I know, the first proof is by corollary from a more general result[1] I published in 1973. Later[2] I found a better proof. The matter is discussed from a pro-historical standpoint in Chapter 13 of *Concepts and Logic*. A counterexample is adduced there to show that CARNOT's claim is *not* a consequence of his General Axiom alone. Therefore CARNOT's argument, since it makes no reference to any particular theory of heat, cannot be correct.

---

[1] TRUESDELL [1973, *2*].

[2] TRUESDELL [1976].

11. Orthodox Act V. KELVIN's Absolute Temperatures. CLAUSIUS' Second Paper: Irreversibility and Oracling

Lo nostro scender conviene esser tardo,
sì che s'ausi un poco in prima il senso
al tristo fiato; e poi no i fia riguardo.
DANTE, *Inferno* XI, 10–12.

11A. KELVIN's Remarks on Dissipation

In 1852 KELVIN[1] had presented his ideas on irreversible processes, which he claimed to be "necessary consequences" of his "axiom" about "inanimate material agency" (above, §9B).

> I. When heat is created by a reversible process (so that the mechanical energy thus spent may be *restored* to its primitive condition), there is also a transference from a cold body to a hot body of a quantity of heat bearing to the quantity created a definite proportion depending on the temperatures of the two bodies.
>
> II. When heat is created by any unreversible process (such as friction), there is a *dissipation* of mechanical energy, and a full *restoration* of it to its primitive condition is impossible.
>
> III. When heat is diffused by *conduction*, there is a *dissipation* of mechanical energy, and perfect *restoration* is impossible.

---

[1] THOMSON [1852, *2*].

THOMSON [1853, *2*] later tried to calculate the total amount of mechanical energy dissipated when a body whose temperature varies from point to point in an arbitrary way is confined in a vessel with adiabatic walls and allowed to come to thermal equilibrium. He does so by supposing each element of the body to act like a Carnot engine. This is the earliest paper on thermodynamics in which a non-constant temperature field appears.

> IV. When radiant heat or light is absorbed, otherwise than in vegetation, or in chemical action, there is a *dissipation* of mechanical energy, and perfect *restoration* is impossible.

KELVIN does not disclose how he has derived these statements from his axiom. The remainder of his note concerns the question, "*How far is the loss of power experienced by steam in rushing through narrow steam-pipes compensated, as regards the economy of the engine, by the heat* (containing an exact equivalent of mechanical energy) *created by the friction*?" His tool is his formula for the motive power of a Carnot cycle:

$$\frac{L(\mathscr{C})}{JC^{+}(\mathscr{C})} = 1 - \exp\left(-\frac{1}{J}\int_{\theta-}^{\theta+} \mu\, d\theta\right) . \qquad (9B.9)_r$$

He concludes from an experimental datum that in the best steam engines "at least three-fourths of the work spent in any kind of friction is utterly wasted." He draws some "general conclusions...from the propositions stated above, and known facts with reference to the mechanics of animal and vegetable bodies":—

> 1. There is at present in the material world a universal tendency to the dissipation of mechanical energy.
>
> 2. Any *restoration* of mechanical energy, without more than an equivalent of dissipation, is impossible in inanimate material processes, and is probably never effected by means of organized matter, either endowed with vegetable life or subjected to the will of an animated creature.
>
> 3. Within a finite period of time past, the earth must have been, and within a finite period of time to come the earth must again be, unfit for the habitation of man as at present constituted, unless operations have been, or are to be performed, which are impossible under the laws to which the known operations going on at present in the material world are subject.

11B. KELVIN's Absolute Temperatures

In Footnote 6 to §7H we have mentioned the "absolute temperature" KELVIN introduced in 1848, using the framework of the Caloric Theory. [That temperature was not the result of any new development in thermodynamics then, nor was it used in any of the later works of the pioneers.] In 1854 KELVIN turned back to his early work, "wholly founded", as he wrote, "on CARNOT's uncorrected theory", and replaced it by a new and slightly different development.

The paper[1] of 1848 opens as follows:

> THE determination of temperature has long been recognized as a problem of the greatest importance in physical science. It has accordingly been made a subject of most careful attention, and, especially in late years, of very elaborate and refined experimental researches; and we are thus at present in possession of as complete a practical solution of the problem as can be desired, even for the most accurate investigations. The theory of thermometry is however as yet far from being in so satisfactory a state. The principle to be followed in constructing a thermometric scale might at first sight seem to be obvious, as it might appear that a perfect thermometer would indicate equal additions of heat, as corresponding to equal elevations of temperature, estimated by the numbered divisions of its scale. It is however now recognized (from the variations in the specific heat of bodies) as an experimentally demonstrated fact that thermometry under this condition is impossible, and we are left without any principle on which to found an absolute thermometric scale.
>
> Next in importance to the primary establishment of an absolute scale, independently of the properties of any particular kind of matter, is the fixing upon an arbitrary system of thermometry, according to which results of observations made by different experimenters, in various positions and circumstances, may be exactly compared.

For this purpose the air thermometer does very well, as KELVIN goes on to explain in detail.

> Now it is found by Regnault that various thermometers, constructed with air under different pressures, or with different gases, give indications which coincide so closely, that, unless when certain gases, such as sulphurous acid, which approach the physical condition of vapours at saturation, are made use of, the variations are inappreciable. This remarkable circumstance enhances very much the practical value of the air-thermometer; but still a rigorous standard can only be defined by fixing upon a certain gas at a determinate pressure, as the thermometric substance. Although we have thus a strict principle for constructing a *definite* system for the estimation of temperature, yet as reference is essentially made to a specific body as the standard thermometric substance, we cannot consider that we have arrived at an *absolute* scale, and we can only regard, in strictness, the scale actually adopted

[1] THOMSON [1848].

> as *an arbitrary series of numbered points of reference sufficiently close for the requirements of practical thermometry.*
>
> In the present state of physical science, therefore, a question of extreme interest arises: *Is there any principle on which an absolute thermometric scale can be founded:* It appears to me that Carnot's theory of the motive power of heat enables us to give an affirmative answer.

That theory allows us to estimate "*the value of a degree* . . . by the [maximum] mechanical effect to be obtained from the descent of a unit of heat through it . . .":

> The characteristic property of the scale which I now propose is, that all degrees have the same value; that is, that a unit of heat descending from a body A at the temperature T^0 of this scale, to a body B at the temperature $(T - 1)^0$, would give out the same mechanical effect, whatever be the number T. This may justly be termed an absolute scale, since its characteristic is quite independent of the physical properties of any specific substance.

KELVIN then turns to REGNAULT's data on the latent heat of steam; he states that because the densities of saturated vapor at different temperatures are not yet determined accurately, he has no choice but to use the laws of ideal gases. He appends a table for interconverting degrees of air thermometer with these new degrees.

[This paper, which contains no equations and no mathematical reasoning, can be understood only by a person thoroughly familiar with the Caloric Theory as CARNOT and CLAPEYRON presented it. KELVIN bases his scale upon the CARNOT-CLAPEYRON Theorem (5L.4), which we may write in the form

$$\mu = \frac{\partial p}{\partial \theta} \Big/ \Lambda_V \,, \tag{11B.1}$$

provided we know that $\Lambda_V \neq 0$. His definition of "absolute temperature" τ is

$$\tau = F(\theta) \,, \qquad F \equiv \int \mu(\theta) d\theta \,. \tag{11B.2}$$

According to CARNOT's Special Axiom, the work done by a Carnot cycle $\mathscr{C}$ is given by

$$L(\mathscr{C}) = \left(\int_{\theta^-}^{\theta^+} \mu d\theta \right) C^+(\mathscr{C}) \,. \tag{5L.7$_{2r}$}$$

Using the definition (2), we write this statement in terms of the "absolute" operating temperatures τ^+ and τ^-:

$$L(\mathscr{C}) = (\tau^+ - \tau^-) C^+(\mathscr{C}) \,. \tag{11B.3}$$

Just as KELVIN claims, a unit difference of this "absolute temperature" equals precisely the maximum work that can be done by a unit of heat let down through 1° in a cyclic process, no matter what be the temperature of the furnace.

[KELVIN's second development[2], appropriate to the "dynamical theory", rejects (5L.7) and (3) as being peculiar to the Caloric Theory. On the contrary, (1) remains valid, and the definition (2) still makes sense, though what the numbers τ so obtained are good for remains to be discovered.]

In his earlier paper, KELVIN now states, he showed

> that CARNOT's function (derivable from the properties of any substance whatever, but the same for all bodies at the same temperature), or any arbitrary function of CARNOT's function, may be defined as temperature, and is therefore the foundation of an absolute system of thermometry. We may now adopt this suggestion with great advantage, since we have found that CARNOT's function varies very nearly in the inverse ratio of what has been called "temperature from the zero of the air-thermometer," that is, Centigrade temperature by the air-thermometer increased by the reciprocal of the coefficient of expansion; and we may define temperature simply as the reciprocal of CARNOT's function. When we take into account what has been proved regarding the mechanical action of heat*, and consider what is meant by CARNOT's function, we see that the following explicit definition may be substituted:—
>
> *If any substance whatever, subjected to a perfectly reversible cycle of operations, takes in heat only in a locality kept at a uniform temperature, and emits heat only in another locality kept at a uniform temperature, the temperatures of these localities are proportional to the quantities of heat taken in or emitted at them in a complete cycle of the operations.*

Again there are no equations, [but an informed reader might have grasped KELVIN's intentions]. First, the new "absolute temperature" T is defined as follows:

$$\mathrm{T} \equiv \frac{J}{\mu} . \tag{11B.4}$$

KELVIN had made the same proposal a year earlier[3] except that then he subtracted a constant "which might have any value, but ought to have for its value the reciprocal of the coefficient of expansion of air in order that the

[2] JOULE & THOMSON [1854, Theoretical Deductions, Section IV].

* [KELVIN's Footnote:] Dynamical Theory of Heat [*i.e.* THOMSON [1851]], §§42, 43.

[3] THOMSON [1853, *2*, Equation (6)].

system of measuring temperature here adopted may agree approximately with that of the air thermometer." Then he did not explain why he chose this particular new absolute temperature. [The statement about Carnot cycles KELVIN now has set in italics can only mean that T is to be so defined that for a Carnot cycle $\mathscr{C}$

$$\frac{C^{+}(\mathscr{C})}{\mathrm{T}^{+}} = \frac{C^{-}(\mathscr{C})}{\mathrm{T}^{-}} \,.\Big] \tag{11B.5}$$

KELVIN provides no proof that (5) does follow from (4). [He may well have inferred it from the argument given above at the beginning of §9B. If we are justified in replacing θ by T in (2), we conclude that KELVIN's old and new absolute temperatures are related as follows:]

$$\tau = J \log \mathrm{T} + \text{const.} \tag{11B.6}$$

KELVIN was later[4] to publish a formula of this kind with specific numerical constants appropriate to particular choices of a degree and a zero.

The section concludes with a table comparing θ as determined by an air thermometer with T. In most circumstances the differences are small. For details, see the end of the appendix to §9D, above. [That is, while KELVIN's old absolute temperature was entirely different from any empirical temperature in common use, the degrees according to the air thermometer are very nearly proportional to absolute degrees according to his new definition.]

11C. CLAUSIUS' Two "Laws" of Thermodynamics

In 1854 CLAUSIUS[1] took up thermodynamics at the point he had left it in 1850. He mentions that "other authors" had treated the subject, but he takes no account of their work.

CLAUSIUS here remarks that thermodynamics may be based on a "First Law" and a "Second Law". In the title of the paper he uses the vague German word "Hauptsatz", and in the text the still vaguer word "Satz"; his first translator renders the former "Fundamental Theorem", the latter, "theorem". The "Second Law", then, is "Carnot's theorem", which CLAUSIUS says he had used in 1850 in "its original form", although at that

---

[4] Annotation to the reprint of THOMSON [1848] in KELVIN's *Papers*, 1882.

[1] CLAUSIUS [1854]. As I cannot understand parts of the argument in this paper, I depart from my usual practice of translating afresh everything I quote, adopting instead the safer course of quoting from an accepted English translation, first published in 1856. If that is nonsense, at least it is not my fault. Also the typically British fused participles and misplaced adverbs are not mine.

In 1851 CLAUSIUS had published a paper, the first of several, on that fetish of thermodynamics, the steam engine, and a note about vapors. I find in them nothing that contributes to the general theory.

time he admitted that he knew CARNOT's work only through CLAPEYRON's and KELVIN's versions of it. [The English terms "First Law" and "Second Law" are soon to be introduced by RANKINE[2].

[In this paper CLAUSIUS abandons CLAPEYRON's lead and strikes out on his own. His mathematics becomes more primitive, his expression vaguer.]

For "the first theorem" CLAUSIUS presents a new "demonstration..., because it is at once more general and more concise" than his former one. This "demonstration" is a statement:

> *Mechanical work may be transformed into heat, and conversely heat into work, the magnitude of the one being always proportional to that of the other.*

He adds, "I have divided the work done by heat into *internal* and *external* work, which are subjected to essentially different laws." The "internal" work is that which contributes to the change of the internal energy E. Thus the statement of the "first theorem", which CLAUSIUS numbers (I), is

$$\dot{\mathrm{E}} = JQ + P \, . \qquad (8\mathrm{B}.2)_{1\mathrm{r}}$$

> In order to give special forms to equation (I), in which it shall express definite properties of bodies, we must make special assumptions with respect to the foreign influences to which the body is exposed. For instance, we will assume that the only active external force, or at least the only one requiring consideration in the determination of work, is an external pressure everywhere normal to the surface, and equally intense at every point of the same, which is always the case with liquid and gaseous bodies when other foreign forces are absent, and might at least be the case with solid bodies. It will be seen that under this condition it is not necessary, in determining the external work, to consider the variations in form experienced by the body, and its expansion or contraction in different directions, but only the total change in its volume. We will further assume that the pressure always changes very gradually, so that at any moment it shall differ so little from the opposite expansive force of the body, that both may be counted as equal. Thus the pressure constitutes a property of the body itself, which can be determined from its other contemporaneous properties.

That is,

$$P \equiv -p\dot{V} \, . \qquad (8\mathrm{B}.2)_{2\mathrm{r}}$$

---

[2] RANKINE [1859, §§236, 241, 243]; of course RANKINE's "Laws" themselves differ from CLAUSIUS'.

Next, CLAUSIUS considers both p and E to be functions of V and θ. He concludes from (8B.2) that

$$J\Lambda_V - p = \frac{\partial \mathrm{E}}{\partial V}, \qquad J\mathrm{K}_V = \frac{\partial \mathrm{E}}{\partial \theta}, \tag{8D.3)$_r$}$$

and hence [as KELVIN had shown in 1851 (*cf.* §§9B and 9D, above)] follows the local form of CLAUSIUS' "first principle" of 1850, freed of the restriction to ideal gases:

$$\frac{\partial p}{\partial \theta} = J\left(\frac{\partial \Lambda_V}{\partial \theta} - \frac{\partial \mathrm{K}_V}{\partial V}\right). \qquad (8B.1)_r$$

Next comes the "second theorem".

> Carnot's theorem, when brought into agreement with the first fundamental theorem, expresses a relation between two kinds of transformations, the transformation of heat into work, and the passage of heat from a warmer to a colder body, which may be regarded as the transformation of heat at a higher into heat at a lower temperature. In its original form it may be enunciated in some such manner as the following:—*In all cases where a quantity of heat is converted into work, and where the body effecting this transformation ultimately returns to its original condition, another quantity of heat must necessarily be transferred from a warmer to a colder body; and the magnitude of the last quantity of heat, in relation to the first, depends only upon the temperatures of the bodies between which heat passes, and not upon the nature of the body effecting the transformation.*

CLAUSIUS regards this "theorem" as "incomplete" because its "demonstration . . . is based upon too simple a process, in which only two bodies losing or receiving heat are employed . . . ." He replaces CARNOT's "theorem" by a new "principle":

> *Heat can never pass from a colder to a warmer body without some other change, connected therewith, occurring at the same time.*

As he writes, "Everything we know . . . confirms this . . . . Without further explanation, therefore, the truth of the principle will be granted."

11D. CLAUSIUS' Equivalence-Value of a Transformation

To exploit his new principle, CLAUSIUS regards a Carnot cycle as too simple, so he introduces a more complicated cyclic process and applies it to an ideal gas placed successively in communication with three different bodies. He applies the "first theorem" in the form

$$L(\mathscr{C}) = JC(\mathscr{C}) \qquad (7A.1)_{1r}$$

for all cycles $\mathscr{C}$.

> These several processes can be either reversible ... or not, and the law which governs the transformations will vary accordingly. Nevertheless, the modification which the law for non-reversible processes suffers may be easily applied afterwards, so that at present we will confine ourselves to the consideration of *reversible* circular processes.

Clausius convinces himself that a transformation of heat from work has a "mathematical value", which "may be called its equivalence-value":

> With respect to the magnitude of the equivalence-value, it is ... clear that the value of a transformation from work into heat must be proportional to the quantity of heat produced, and besides this it can only depend upon the temperature.

Thus, he claims, "the equivalence-value of the transformation of the quantity of heat C, of the temperature θ, from work, may be represented generally by

$$Cf(\theta) \,, \qquad (11D.1)$$

wherein $f(\theta)$ is a [positive] function of the temperature, which is the same for all cases." [To avoid confusion, I replace Clausius' Q here by C.]

> In a similar manner the value of the transmission of the quantity of heat C from the temperature θ_1 to the temperature θ_2, must be proportional to the quantity transmitted, and besides this, can only depend upon the two temperatures. In general, therefore, it may be expressed by
>
> $$CF(\theta_1, \theta_2) \qquad (11D.2)$$
>
> wherein $F(\theta_1, \theta_2)$ is a function of both temperatures, which is the same for all cases, and of which we at present only know that, without changing its numerical value, it must change its sign when the two temperatures are interchanged; so that
>
> $$F(\theta_2, \theta_1) = -F(\theta_1, \theta_2) \,. \qquad (11D.3)$$
>
> In order to institute a relation between these two expressions, we have the condition, that in every reversible circular process of the above kind, the two transformations which are involved must be equal in magnitude, but opposite in sign; so that their algebraical sum must be zero. For instance, in the process for a gas, so fully

described above, the quantity of heat C, at the temperature θ, was converted into work; that gives $-Cf(\theta)$ as its equivalence-value, and that of the quantity of heat C_1, transferred from the temperature θ_1 to θ_2, will be $C_1F(\theta_1, \theta_2)$, so that we have the equation

$$-Cf(\theta) + C_1F(\theta_1, \theta_2) = 0 \, . \qquad (11D.4)$$

Let us now conceive a similar process executed in an opposite manner, so that the bodies K_1 and K_2, and the quantity of heat C_1, passing between them, remain the same as before; but that instead of the body K of the temperature θ, another body K' of the temperature θ' be employed; and let us call the quantity of heat produced by the work in this case C',—then, analogous to the last, we shall have the equation

$$C'f(\theta') + C_1F(\theta_2, \theta_1) = 0 \, . \qquad (11D.5)$$

Adding these two equations, and applying (3), we have

$$-Cf(\theta) + C'f(\theta') = 0 \, . \qquad (11D.6)$$

If now we regard these two circular processes together as one circular process, which is of course allowable, then in the latter the transmissions of heat between K_1 and K_2 will no longer enter into consideration, for they precisely cancel one another, and there remain only the quantity of heat C taken from K and transformed into work, and the quantity C' generated by work and given to K'. These two transformations of the *same* kind, however, may be so divided and combined as again to appear as transformations of different kinds. If we hold simply to the fact that a body K has lost the quantity of heat C, and another body K' has received the quantity C', we may without hesitation consider the part common to both as transferred from K to K', and regard only the other part, the excess of one quantity over the other, as a transformation from work into heat, or *vice versa*. For example, let the temperature θ' be greater than θ, so that the above, being a transmission from the colder to the warmer body, will be negative. Then the other transformation must be positive, that is, a transformation from work into heat, whence it follows that the quantity of heat C' imparted to K' must be greater than the quantity C lost by K. If we divide C' into the two parts

$$C \quad \text{and} \quad C' - C \, ,$$

the first will be the quantity of heat transferred from K to K_1, and the second the quantity generated from work.

According to this view the double process appears as a process of the same kind as the two simple ones of which it consists, for the circumstances that the generated heat is not imparted to a third body, but to one of the two between which the transmission of heat takes place, makes no essential difference, because the temperature of the generated heat is arbitrary, and may therefore have the same value as the temperature of one of the two bodies, in which case a third body would be superfluous. Consequently, for the two quantities of heat C and $C' - C$, an equation of the same form as (5) must hold, *i.e.*

$$(C' - C)f(\theta') + CF(\theta, \theta') = 0 . \tag{11D.7}$$

Eliminating C' by (6) yields

$$F(\theta, \theta') = f(\theta') - f(\theta) . \tag{11D.8}$$

CLAUSIUS introduces a function T as follows:

$$f(\theta) = \frac{1}{\mathrm{T}} , \tag{11D.9}$$

and T_1 denotes $1/f(\theta_1)$, *etc.* In this notation CLAUSIUS enunciates thus

> the second fundamental theorem..., which in this form might appropriately be called *the theorem of the equivalence of transformations...*:
>
> *If two transformations which, without necessitating any other permanent change, can mutually replace one another, be called equivalent, then the generation of the quantity of heat C at the temperature θ from work, has the equivalence-value*
>
> $$\frac{C}{\mathrm{T}} , \tag{11D.10}$$
>
> *and the passage of the quantity of heat C from the temperature θ_1 to the temperature θ_2, has the value*
>
> $$C\left(\frac{1}{\mathrm{T}_2} - \frac{1}{\mathrm{T}_1}\right) , \tag{11D.11}$$
>
> *wherein* T *is a function of the temperature, independent of the nature of the process by which the transformation is effected.*

CLAUSIUS supposes that

> the several bodies K_1, K_2, K_3, &c., serving as reservoirs of heat at the temperatures θ_1, θ_2, θ_3, &c., have received during the process the quantities of heat C_1, C_2, C_3, &c., whereby the loss of a quantity of heat will be counted as the gain of a negative quantity of heat; then the total value N of all the transformations will be
>
> $$N = \frac{C_1}{\mathrm{T}_1} + \frac{C_2}{\mathrm{T}_2} + \frac{C_3}{\mathrm{T}_3} + \text{\&c.} = \sum \frac{C}{\mathrm{T}} . \tag{11D.12}$$

It is here assumed that the temperatures of the bodies K_1, K_2, K_3, &c. are constant, or at least so nearly constant, that their variations may be neglected. When one of the bodies, however, either by the reception of the quantity of heat C itself, or through some other cause, changes its temperature during the process so considerably, that the variation demands consideration, then for each element of heat dC we must employ that temperature which the body possessed at the time it received it, whereby an integration will be necessary. For the sake of generality, let us assume that this is the case with all the bodies; then the foregoing equation will assume the form

$$N = \int \frac{dC}{\mathrm{T}}, \tag{11D.13}$$

wherein the integral extends over all the quantities of heat received by the several bodies.

If the process is *reversible*, then, however complicated it may be, we can prove, as in the simple process before considered, *that the transformations which occur must exactly cancel each other, so that their algebraical sum is zero.*

For were this not the case, then we might conceive all the transformations divided into two parts, of which the first gives the algebraical sum zero, and the second consists entirely of transformations having the same sign. By means of a finite or infinite number of simple circular processes, the transformations of the first part must admit of being made in an opposite manner, so that the transformations of the second part would alone remain without any other change. Were these transformations *negative*, *i.e.* from heat into work, and the transmission of heat from a lower to a higher temperature, then of the two the first could be replaced by transformations of the latter kind, and ultimately transmissions of heat from a lower to a higher temperature would alone remain, which would be compensated by nothing, and therefore contrary to the above principle. Further, were those transformations *positive*, it would only be necessary to execute the operations in an inverse manner to render them negative, and thus obtain the foregoing impossible case again. Hence we conclude that the second part of the transformations can have no existence.

Consequently the equation

$$\int \frac{dC}{\mathrm{T}} = 0 \tag{11D.14}$$

is the analytical expression of the second fundamental theorem in the mechanical theory of heat.

This "analytical expression" is his Equation (II).

11E. CLAUSIUS' Application to the Doctrine of Latent and Specific Heats

CLAUSIUS applies these assertions to

a circular process consisting of a series of changes of condition made by a body which ultimately returns to its original state, and for simplicity, let us assume that all parts of the body have the same temperature; then in order that the process may be reversible, the changing body when imparting or receiving heat can only be placed in communication with such bodies as have the same temperature as itself, for only in this case can the heat pass in an opposite direction. Strictly speaking, this condition can never be fulfilled if a motion of heat at all occurs; but we may assume it to be so nearly fulfilled, that the small differences of temperature still existing may be neglected in the calculation. In this case it is of course of no importance whether θ, in the equation (II), represents the temperature of the reservoir of heat just employed, or the momentary temperature of the changing body, inasmuch as both are equal. The latter signification being once adopted, however, it is easy to see that any other temperatures may be attributed to the reservoirs of heat without producing thereby any change in the expression $\int dC/\mathrm{T}$ which shall be prejudicial to the validity of the foregoing equation. As with this signification of θ the several reservoirs of heat need no longer enter into consideration, it is customary to refer the quantities of heat, not to them, but to the changing body itself, by stating what quantities of heat this body successively receives or imparts during its modifications.... [I]t follows, therefore, that when for every quantity of heat dC which the body receives or, if negative, imparts during its modifications the temperature of the body at the moment be taken into calculation, the equation (II) may be applied without further considering whence the heat comes or whither it goes, provided always that the process is reversible.

CLAUSIUS now specializes these ideas and assertions to fluid bodies such as had been considered heretofore. He asserts as "the condition of the body being defined by its temperature θ and volume V" the Doctrine of Latent and Specific Heats:

$$Q = \Lambda_V(V, \theta)\dot{V} + \mathrm{K}_V(V, \theta)\dot{\theta} \,, \tag{2C.4$_r$}$$

and he then interprets dC in (11D.14) as being $\Lambda_V dV + \mathrm{K}_V d\theta$. Then $(\Lambda_V dV + \mathrm{K}_V d\theta)/\mathrm{T}$ is an exact differential; therefore

$$\frac{\partial}{\partial\theta}\left(\frac{\Lambda_V}{\mathrm{T}}\right) = \frac{\partial}{\partial V}\left(\frac{\mathrm{K}_V}{\mathrm{T}}\right) . \tag{11E.1}$$

That is,

$$\frac{\mathrm{T}'}{\mathrm{T}}\Lambda_V = \frac{\partial \Lambda_V}{\partial \theta} - \frac{\partial \mathrm{K}_V}{\partial V}, \tag{11E.2}$$

whence by use of the relation

$$\frac{\partial p}{\partial \theta} = J\left(\frac{\partial \Lambda_V}{\partial \theta} - \frac{\partial \mathrm{K}_V}{\partial V}\right) \tag*{(8B.1)$_\mathrm{r}$}$$

it follows that

$$J\frac{\mathrm{T}'}{\mathrm{T}}\Lambda_V = \frac{\partial p}{\partial \theta}, \tag{11E.3}$$

and so

$$J\Lambda_V = \mathrm{T}\frac{\partial p}{\partial \mathrm{T}}. \tag{11E.4}$$

The former of these results CLAUSIUS recognizes as being "the before-mentioned equation established by Clapeyron"; [that is, to within the qualifications explained in §§5J and 5L, it is the General CARNOT-CLAPEYRON Theorem

$$\mu\Lambda_V = \frac{\partial p}{\partial \theta}, \tag*{(5L.4)$_\mathrm{r}$}$$

specialized in such a way as to make it compatible with the Interconvertibility of Heat and Work,] and

$$\mu = J\frac{\mathrm{T}'}{\mathrm{T}}. \tag{11E.5}$$

[The result (4) shows that CLAUSIUS in his formalistic way notices that the value of his function T might be taken as an independent variable in place of θ. The Λ_V delivered by (4) is a function of V and T.] In annotating this passage for the reprint of 1864 he remarked that "in the same way" one could derive from (1) the relation

$$\frac{\partial}{\partial \mathrm{T}}\left(\frac{\Lambda_V}{\mathrm{T}}\right) = \frac{\partial}{\partial V}\left(\frac{\mathrm{K}_V}{\mathrm{T}}\right), \tag{11E.6}$$

[in which both Λ_V and K_V are regarded as functions of V and T.]

11F. CLAUSIUS' Remarks on Irreversible Processes

We proceed now to the consideration of *non-reversible* circular processes.

In the proof of the previous theorem, that in any compound reversible process the algebraical sum of all the transformations must be zero, it was first shown that the sum could not be *negative*, and afterwards that it could not be *positive*, for if so it would only be necessary to reverse the process in order to obtain a negative sum. The first part of this proof remains unchanged even when the process is not reversible; the second part, however, cannot be applied in such a case. Hence we obtain the following theorem, which applies generally to all circular processes, those that are reversible forming the limit:—

The algebraical sum of all transformations occurring in a circular process can only be positive.

A transformation which thus remains at the conclusion of a circular process without another opposite one, and which according to this theorem can only be positive, we shall, for brevity, call an *uncompensated* transformation.

The different kinds of operations giving rise to uncompensated transformations are, as far as external appearances are concerned, rather numerous, even though they may not differ very essentially. One of the most frequently occurring examples is that of the transmission of heat by mere conduction, when two bodies of different temperatures are brought into immediate contact; other cases are the production of heat by friction, and by an electric current when overcoming the resistance due to imperfect conductibility, together with all cases where a force, in doing mechanical work, has not to overcome an equal resistance, and therefore produces a perceptible external motion, with more or less velocity, the *vis viva* of which afterwards passes into heat. An instance of the last kind may be seen when a vessel filled with air is suddenly connected with an empty one; a portion of air is then propelled with great velocity into the empty vessel and again comes to rest there. It is well known that in this case just as much heat is present in the whole mass of air after expansion as before, even if differences have arisen in the several parts, and therefore there is no heat permanently converted into work. On the other hand, however, the air cannot again be compressed into its former volume without a simultaneous conversion of work into heat.

The principle according to which the equivalence-values of the uncompensated transformations thus produced are to be determined, is evident from what has gone before, and I will not here enter further into the treatment of particular cases.

Later CLAUSIUS[1] was to use the "analytical expression" (11D.13) to express the "Second Law" for "every cyclic process possible at all" in the form

$$\int \frac{dC}{\mathrm{T}} \leqq 0 \; . \tag{11F.1}$$

[1] CLAUSIUS [1862, Eq. (Ia)]. Our dC here is what CLAUSIUS denoted by $-dQ$ in this paper, by dQ in his later work.

11G. Clausius' Determination of His Universal Function T

CLAUSIUS cannot determine T "entirely without hypothesis", so he returns to the "accessory assumption" made in his former paper, that is, to [HOLTZMANN's Assertion]

$$J\Lambda_V = p \ , \qquad (7D.1)_r$$

(*cf.* §8B, above). CLAUSIUS states that it "has been verified by the later experiments of REGNAULT, and in all probability is accurate for all gases to the same degree" as is the thermal equation of state $pV = R\theta$. Of course, putting (7D.1) into the relation

$$J\frac{T'}{T}\Lambda_V = \frac{\partial p}{\partial \theta} \qquad (11E.3)_r$$

shows that at once

$$T = K\theta \ . \qquad (11G.1)$$

For K "the simplest value ... is unity". Thus "T is nothing more than the temperature counted from ... about −273°C below the freezing-point, and, considering the point thus determined as the absolute zero of temperature, T is simply *the absolute temperature.*"

11H. Critique: Empirical and Absolute Temperatures

In this same year, 1854, KELVIN and CLAUSIUS introduced independently what might seem to be the same concept of absolute temperature. In fact their ideas, while consistent with each other, were somewhat different. KELVIN, though his treatment is so lacunary, indeed careless, as to leave the reader in doubt as to what he meant and did, operated fully within the Doctrine of Latent and Specific Heats. CLAUSIUS, considering circumstances not necessarily inquadrable by that Doctrine, used a sort of physical divination to hale up his function T as a daemon of an incipient general thermodynamics. For him the formal manipulations show only what T can do when Q is determined by the Doctrine of Latent and Specific Heats. He does not exhibit anything definite for more general circumstances, and I cannot supply any logical structure to replace his bare claims.

I can, however, fill all the gaps in KELVIN's and CLAUSIUS' specific arguments regarding the quantities of classical thermodynamics. These gaps are

1. CLAUSIUS does not demonstrate that there is such a function as he claims T to be.
2. While KELVIN gives an explicit definition of his function T, it does not make sense unless $\mu > 0$; furthermore, KELVIN does not demonstrate that his T has the properties he claims for it.

This is not all. The critical reader will have noticed that KELVIN's second definition shifts ground, as the vague phrase "any arbitrary function of CARNOT's function" suggests. KELVIN's requirement of 1848 was that the ratio of the work done to heat absorbed in a Carnot cycle should depend only upon the difference of its "absolute" operating temperatures. If T as defined by KELVIN, namely

$$\mathrm{T} \equiv \frac{J}{\mu}, \tag{11B.4$_\mathrm{r}$}$$

is in fact an empirical-temperature scale, we can use it in place of θ in KELVIN's determination of the motive power of a Carnot cycle:

$$\frac{L(\mathscr{C})}{JC^+(\mathscr{C})} = 1 - \exp\left(-\frac{1}{J}\int_{\theta^-}^{\theta^+} \mu d\theta\right). \tag{9B.9$_\mathrm{r}$}$$

The result is

$$\frac{L(\mathscr{C})}{JC^+(\mathscr{C})} = 1 - \frac{\mathrm{T}^-}{\mathrm{T}^+}. \tag{11H.1}$$

Because $\mathrm{T}^-/\mathrm{T}^+$ is not a function of $\mathrm{T}^+ - \mathrm{T}^-$, KELVIN's second "absolute temperature" T *does not satisfy the requirement* he had laid down in 1848. Moreover, if we apply KELVIN's first definition, namely

$$\tau = F(\theta)\,, F \equiv \int \mu(\theta)d\theta\,, \tag{11.B2$_\mathrm{r}$}$$

(9B.9) assumes the form

$$\frac{L(\mathscr{C})}{JC^+(\mathscr{C})} = 1 - e^{-(\tau^+ - \tau^-)/J}. \tag{11H.2}$$

Thus KELVIN's first "absolute temperature" τ *continues even in the new, corrected theory to satisfy both of the requirements he laid down originally.* Of course the linear dependence

$$L(\mathscr{C}) = (\tau^+ - \tau^-)C^+(\mathscr{C})\,, \tag{11B.3$_\mathrm{r}$}$$

which made the definition of τ particularly attractive in the Caloric Theory, is irretrievably lost in the "dynamical theory".

Why KELVIN should have abandoned his original idea and chosen subsequently to base his absolute temperature upon heat exchange rather than work done, I cannot explain in terms of the aspects of the theory he himself published. Of course for other reasons, especially those drawn from the kinetic theory of gases, it is easy enough to see why the second definition is preferable.

There are deeper failings in KELVIN's hasty work on this subject. He tells us that his two "absolute temperatures" τ and T are independent of the thermometer used to specify the temperature θ that enters the determination of $\mu(\theta)$ through the General CARNOT-CLAPEYRON Theorem (11B.1). True, for a given scale of empirical temperature CARNOT's General Axiom makes the function μ universal in the sense that we shall get the same quotient $(\partial p/\partial\theta) \div \Lambda_V$ *for all bodies*, and it will be a function of θ only. That Axiom does not tell us that *for all scales* we shall get the same function μ this way. Indeed, we shall not, nor should we.

To clarify this matter, we first observe that KELVIN has used the term "absolute" in three distinct senses:

1. Temperature independent of the choice of thermometric fluid.
2. Temperature whose "value of a degree" is a function of the ratio of maximum work done to heat absorbed in a cyclic process whose greatest and least temperatures differ by one degree, no matter in what part of the scale those temperatures lie.
3. Temperature the ratio of whose values at the furnace and the refrigerator of a Carnot engine equals the ratio of heat absorbed to heat emitted in a cycle of that engine.

Meanings 2 and 3 are incompatible and so must define different absolute temperatures. Hence in what follows I shall use the unqualified term "absolute" *in the first sense only.*

α. *Empirical temperature.* To investigate KELVIN's ideas in terms of the first sense and so to correct his reference to "any arbitrary function of CARNOT's function", we must first explain "empirical temperature". Here I endeavor to do so in terms of the concepts available in 1854. Necessarily the treatment cannot be explicit[1].

As we have seen, early students of heat appealed often to an in effect primitive concept of an *ideal gas*, to the behavior of which nearly all real gases approximated very closely in most circumstances then available to experiment. By reference to such a gas the letter θ as it has been used in this tragicomical history could have been *defined* by the thermal equation of state $pV = R\theta$, augmented by AVOGADRO's hypothesis:

$$R \propto \mathbb{M}\ , \qquad \text{(11H.3)}$$

[1] An explicit treatment starts from the abstract concept of hotness, which was introduced by MACH [1896, Historische Uebersicht der Entwicklung der Thermometrie, §1, and Kritik des Temperaturbegriffes, §5]. MACH assumes explicitly that the set of all hotnesses is a 1-dimensional continuous manifold in the sense RIEMANN had introduced in his *Habilitationsvortrag* of 1854, published in 1868. Most of the few modern authors who treat the matter explicitly assume or prove that the set of hotnesses is a 1-dimensional differentiable manifold diffeomorphic to the real line, totally ordered under an intrinsic ordering to be interpreted as "not cooler than". The work of SERRIN, at this writing largely unpublished, is the clearest and best. An empirical-temperature scale is a local co-ordinate system or "chart" upon the hotness manifold which is order-preserving, say, with the convention that a hotter hotness is mapped onto a larger real number.

Elsewhere I have tried to explain the matter in terms of MACH's concepts and others available to theorists working in the 1870s (TRUESDELL [1979]).

$\mathbb{M}$ being the molecular weight or some mass characteristic of the substance. That is, if subscripts 1 and 2 indicate measured values for two such gases in thermal equilibrium with each other, then

$$\frac{p_1/\rho_2}{p_2/\rho_2} = \frac{\mathbb{M}_2}{\mathbb{M}_1} . \tag{11H.4}$$

If we choose some arbitrary positive scale-factor k and set

$$\theta \equiv k^{-1}\mathbb{M}p/\rho , \tag{11H.5}$$

the resulting θ will be an *ideal-gas temperature* on some scale such as "Fahrenheit" or "Réaumur" or "Celsius", determined by k. "Absolute cold" corresponds to $p/\rho = 0$ for all ideal-gas scales. Over large ranges of hotness many real gases were found to agree excellently well among themselves in the sense made specific by (3). Hence the volumes of a body of such a gas maintained at constant pressure would be nearly constant multiples of θ as defined by (5). In other words, such a gas would be used as a *thermometric fluid* to provide a *thermometer* which would read *empirical temperatures* very close to the ideal-gas temperature. This fact made and makes the concept "ideal gas" a natural one, much as the nearly rigid deportment of many real bodies in many circumstances abundantly justifies the imaginary "rigid body".

As the reader of this tragicomical history will have seen, an "ideal-gas temperature", which up to this point we have denoted by θ, did perfectly well for the early thermodynamics.

On the other hand, there was in principle no objection to using as a thermometer any body susceptible of appreciable changes of volume when heated at constant pressure. The volumes so obtained are also empirical temperatures. While other methods of determining empirical temperature are not excluded, the scales obtained by measured volumes suffice to make the concept concrete as well as clear.

KELVIN seems to have perceived that *the basic ideas of the theory of heat should be independent of the choice of thermometer*. He seems also to have wished to excise the concept of "ideal-gas temperature", which in fact he had never used. Writing a quarter of a century later than the closing year of our tragicomedy, he fulminated against the ideal gas[2]:

> ... intelligence in thermodynamics has been hitherto much retarded, and the student unnecessarily perplexed, and a mere quicksand has been given as a foundation for thermometry, by building from the beginning on an ideal substance called perfect gas, with none of its properties realized rigorously by any real substance, and with some of them... unknown, and utterly unassignable, even by guess.

[2] THOMSON [1878, §46].

MACH, writing in 1896, certainly had no better word for ideal-gas temperature[3]:

> Until very recently those who worked in this field seem to have looked more or less unconsciously for a natural measure of temperature, for a real temperature, for a sort of Platonic idea of temperature, of which the temperature read on a thermometer would be only an incomplete, inexact expression.

Among the proponents of this to him reprehensible idea MACH could cite even CLAUSIUS. Neither KELVIN nor MACH provided any rules of transformation for quantities under change of empirical-temperature scale.

If we are to banish the concept of ideal-gas temperature, we must go back over all developments in thermodynamics up to 1854 and discard every reference to ideal gases. In particular, *all axioms must be phrased in terms of an arbitrary empirical-temperature scale.* For that we may continue to use the letter θ, but we may not attribute to any one scale any property that distinguishes it from others. On the contrary, we must show that *all axioms are invariant under change of empirical-temperature scale.*

β. Change of scale. Let the empirical temperatures of one and the same body be measured by two thermometers. The two empirical temperatures so found, say θ and θ^*, must be invertible functions of one another:

$$\theta^* = f(\theta) \ , \qquad \theta = f^{-1}(\theta^*) \ . \tag{11H.6}$$

The example of water just above its freezing point shows that such is not always the case. If we subject a body of water at atmospheric pressure to conditions which make its temperature according to the air thermometer decrease from 6°C to 2°C, we shall find that its volume decreases to a minimum at about 4°C and then increases again. Therefore, in those conditions air and water cannot both provide empirical-temperature scales. *One or the other must be rejected* as a thermometer in the range including 4°C.

That is not all. Even in the range below 4°C, should we try to use as a thermometer a body of water at atmospheric pressure, we should find that if a Body 1 was *hotter* than Body 2 according to the air thermometer, it would be *colder* than Body 2 according to the water thermometer. This fact has important bearing on CARNOT's General Axiom (5I.1) when that is regarded as referring to an empirical temperature. A Carnot cycle absorbs heat at the *higher* of the two operating temperatures. CARNOT's General Axiom is invariant under change of empirical-temperature scale if and only if the function f in the transformations (6) is an *increasing* function. Thus in the circumstances just mentioned, air and water cannot both be thermometric fluids: One must be rejected. Experience teaches us that the engine

[3] MACH [1896, Kritik des Temperaturbegriffes, §14].

which absorbs heat at the higher temperature according to the air thermometer does positive work and so conforms with CARNOT's General Axiom. Thus it is the water thermometer that we reject in the range including 4°C and air temperatures lower than that.

Turning back now to the fundamental axioms of the theory of calorimetry, namely

$$p = \varpi(V, \theta) , \qquad \frac{\partial p}{\partial V} < 0 , \qquad (2A.2)_r, (2A.5)_{1r}$$

$$Q = \Lambda_V(V, \theta)\dot{V} + K_V(V, \theta)\dot{\theta} , \qquad K_V > 0 , \quad (2C.4)_r, (2C.5)_{2r}$$

we see that they involve differentiation. They cannot be invariant under change of empirical-temperature scale unless that change carries differentiable functions into differentiable functions. Therefore we require both f and f^{-1} in (6) to be *differentiable*. Taking account of the fact that f is an increasing function, we see that *a change of scale f must satisfy the additional condition*

$$f' > 0 . \qquad (11H.7)$$

For convenience we shall write

$$d\theta^* = f'(\theta)d\theta . \qquad (11H.8)$$

Of course we require that p and Q, the meanings of which as functions of time in a process have nothing to do with the way we might choose to measure temperature, shall be independent of the choice of empirical-temperature scale. Then the inequality $(2A.5)_1$ is likewise independent. By use of the chain rule of differential calculus we see that

$$\frac{\partial p}{\partial \theta^*} d\theta^* = \frac{\partial p}{\partial \theta} d\theta . \qquad (11H.9)$$

Here, as usual in works of physics,

$$\frac{\partial p}{\partial \theta^*} \equiv \frac{\partial}{\partial \theta^*} \varpi(V, f^{-1}(\theta^*)) . \qquad (11H.10)$$

To consider the invariance of the quantities appearing in (2C.4), we let subscript θ and θ^* indicate the scale used and so obtain as the condition that Q be independent of scale

$$\Lambda_{V,\theta^*}(V, \theta^*)\dot{V} + K_{V,\theta^*}(V, \theta^*)\dot{\theta}^* = \Lambda_{V,\theta}(V, \theta)\dot{V} + K_{V,\theta}(K, \theta)\dot{\theta} . \quad (11H.11)$$

Hence Λ_V and K_V transform as follows under change of scale:

$$\begin{aligned} \Lambda_{V,\theta^*} &= \Lambda_{V,\theta} , \\ K_{V,\theta^*} d\theta^* &= K_{V,\theta} d\theta , \end{aligned} \qquad (11H.12)$$

the left-hand sides being evaluated at V, θ^* when the right-hand sides are evaluated at V, θ.

With these rules not only are *all axioms invariant* under change of empirical-temperature scales, but *so also are the signs of $\partial p/\partial V$, $\partial p/\partial \theta$, Λ_V,* *and* K_V. Thus the various conditions and classifications dependent upon those signs are invariant. For example, since a differential equation for determining the adiabats is $d\theta/dV = -\Lambda_V/K_V$, the qualitative behavior of the adiabats near a point is the same for all scales. Consequently Carnot cycles exist according to one scale if and only if they exist according to all scales.

γ. *Invariance of* CARNOT's* function.* The proof of the General CARNOT-CLAPEYRON Theorem:

$$\mu\Lambda_V = \frac{\partial p}{\partial\theta}\,, \tag{5L.4$_r$}$$

is valid for an arbitrary empirical-temperature scale. Hence

$$\mu_\theta\Lambda_{V,\theta} = \frac{\partial p}{\partial\theta}\,, \qquad \mu_{\theta*}\Lambda_{V,\theta*} = \frac{\partial p}{\partial\theta^*}\,. \tag{11H.13}$$

Comparison with $(12)_1$ and (9) shows that

$$\mu_\theta d\theta = \mu_{\theta*}d\theta^*\,. \tag{11H.14}$$

KELVIN's first "absolute temperature" τ is defined by (11B.2), which we may write as follows:

$$\tau \equiv \int^\theta \mu_\theta(x)dx\,. \tag{11H.15}$$

From (14) we conclude that

$$\tau_{\theta*}(\theta^*) = \tau_\theta(\theta) \quad \text{if} \quad \theta^* = f(\theta)\,: \tag{11H.16}$$

The value of τ is independent of the choice of empirical-temperature scale used to calculate it.

But is τ an empirical-temperature scale? Certainly τ is a differentiable function of θ, but it satisfies the essential requirement (7) if and only if

$$\mu > 0\,. \tag{5J.8$_r$}$$

CARNOT's own argument shows that

$$\mu \geqq 0\,, \tag{5J.9$_r$}$$

and the pioneer thermodynamicists seem to have assumed the stronger inequality (5J.8). In CARNOT's own work (5J.8) holds for an entirely different reason, namely, CARNOT treated only of an ideal gas, so for him (5L.6) held, and he assumed explicitly that $\Lambda_V > 0$, so for him $\mu = R/(V\Lambda_V) > 0$. In later work we encounter frequently HOLTZMANN's Assertion $J\Lambda_V = p$, whence again $\Lambda_V > 0$, so it follows again that $\mu > 0$ for ideal gases. Since μ is a universal function, the same for all fluid bodies, to determine its sign for ideal gases determines its sign for all fluid bodies. Well and good. But if we are to construct a thermodynamics without mention of ideal gases, these arguments collapse! Can μ vanish? From (14) we see that if $\theta^* = f(\theta_0)$, then $\mu_\theta(\theta_0) = 0$ if and only if $\mu_{\theta*}(\theta^*) = 0$, so a zero of μ is an intrinsic property, independent of scale, but that is all. As we have mentioned in §5J, CARNOT's General Axiom does *not* suffice to ensure that (5J.8) shall hold; it allows μ to vanish on a set with empty interior.

δ. The missing Thermometric Axiom. KELVIN seems to have taken it for granted that $\mu > 0$. Indeed, by 1854 he knew that for θ on the air-temperature scale the HELMHOLTZ-JOULE Determination $\mu = J/\theta$ conformed closely with measurements, whence of course $\mu > 0$. Possibly, therefore, KELVIN regarded (5J.8) as established by experiment, but as he did not mention the matter, perhaps he simply overlooked the problem here. Perhaps, on the contrary, he recalled a calorimetric formula he had published in 1851, namely

$$K_p - K_V = -\Lambda_V \frac{\partial p}{\partial \theta} \Big/ \frac{\partial p}{\partial V} \ . \qquad (2C.9)_{2r}$$

If for each θ_0 there is *some* body such that on the isotherm $\theta = \theta_0$ there is *one* point where

$$K_p > K_V \ , \qquad (2C.10)_{2r}$$

then $(2C.9)_2$ shows that at that point

$$\Lambda_V \frac{\partial p}{\partial \theta} > 0 \ . \qquad (9F.3)_r$$

At that point, then, neither Λ_V nor $\partial p/\partial \theta$ vanishes, and both have the same sign. From the General CARNOT-CLAPEYRON Theorem,

$$\mu \Lambda_V = \frac{\partial p}{\partial \theta} \ , \qquad (5L.4)_r$$

we now determine $\mu(\theta_0)$ uniquely, and $\mu(\theta_0) > 0$. By assumption, for each θ_0 there is an appropriate body by means of which the foregoing determination may be effected[3].

While KELVIN may have regarded $(2C.10)_2$ or its consequence (5J.8) as being established directly by experiment, a careful theorist cannot do so. Thermodynamics is designed not only to correlate experimental data already

---

[3] CARNOT's argument to derive (5L.4) fails at a point where $\Lambda_V = 0$, because the adiabat, if it exists, is tangent to the isotherm there. An argument based on continuity may be used to circumvent this difficulty on isotherms where the body in question has one point at which $\Lambda_V \neq 0$ (*Concepts and Logic*, Theorem 7_{ext} in Chapter 9), but there is no way to prove the existence of $\mu(\theta_0)$ for a body such that $\Lambda_V = 0$ on all points on the isotherm $\theta = \theta_0$.

The difficulty is greatly diminished as soon as we invoke the fact that μ is a universal function, the same for all bodies, so for each value of θ *one body suffices*. The Thermometric Axiom provides one way to exploit this fact.

at hand but also, and even more, to predict the results of experiments yet unperformed. To this end the theorist must lay down explicit axioms. Only on such a basis may he draw specific conclusions which, in time, some experiment may *contradict*. He who stays on the safe side by assuming only what is *already* known saves himself trivially from risk but at the same time adds nothing to theoretical science, for he can predict *nothing*—false or true.

The foregoing makes it clear that in order to ensure the truth of (5J.8) the theorist must lay down a *Thermometric Axiom*, need for which seems to have escaped the pioneers[4]. The discussion above has revealed *two formulations of such an axiom*. Henceforth we shall assume that the conclusion (5J.8) holds. We shall interpret the term "thermometric fluid", which we have used several times in the preceding pages, as referring only to fluids from which μ may be determined by use of the General CARNOT-CLAPEYRON Theorem. For example, a fluid such that $p = \varpi(V)$ is not a thermometric fluid.

It follows immediately that τ *as defined by* (15) *is an empirical-temperature scale which is absolute in that its value depends neither on the empirical-temperature scale used to specify* μ *nor on the choice of body used to determine* μ.

But any function of τ with positive derivative has the properties just stated. KELVIN should have written, not "any function of μ", but "any function of τ which has a positive derivative". *Thus there are infinitely many empirical-temperature scales that are absolute* in KELVIN's first sense. (It is possible to show that all of these are functions of τ, but only on the basis of a more explicit approach to the whole problem.) An example is provided by KELVIN's second absolute temperature T, which may be defined as follows in terms of some constant J having the dimensions of work $\div$ heat:

$$\mathrm{T} \equiv e^{\tau/J} \tag{11H.17}$$

in place of KELVIN's

$$\mathrm{T} \equiv \frac{J}{\mu}\,. \tag*{(11B.4)$_r$}$$

This much follows from CARNOT's General Axiom and the Thermometric Axiom alone. It may be applied to any particular theory of the relations between heat and work. *Inherent in CARNOT's general frame of ideas if those are supplemented by a Thermometric Axiom is the existence of infinitely many* absolute *scales of empirical temperature*, scales *independent* of the choice of empirical-temperature scale used to state the axioms, independent of the fluid used to determine μ.

Use of any absolute scale enables us to dispense with the ideal gas.

---

[4] That such an axiom is necessary, has been emphasized by Mr. SERRIN in his recent writings and lectures. References and discussion in historical contexts have been provided by TRUESDELL [1979].

ϵ. *Constitutive restrictions expressed in terms of* τ *and* T. As τ and T are particular empirical-temperature scales, we may use either of them in place of θ in all relations derived so far. From (14), (15), and (17) we see at once that

$$\mu_\tau = \mu_\theta \frac{d\theta}{d\tau} = \mu_\theta / \mu_\theta = 1 \ , \qquad \mu_T = \mu_\tau \frac{d\tau}{dT} = J/T \ . \tag{11H.18}$$

The last of these is equivalent KELVIN's definition (11B.4).

Because of (11H.17) we may write the General CARNOT-CLAPEYRON Theorem (5L.4) in the following forms:

$$\Lambda_{V,\tau} = \frac{\partial p}{\partial \tau} \ , \qquad J\Lambda_{V,T} = T \frac{\partial p}{\partial T} \ . \tag{11H.19}$$

The second is familiar in the now accepted thermodynamics, but it is in no way restricted thereto. We recall that for each choice of the positive constant J a different scale T results. All the remaining constitutive restrictions and auxiliary formulae may be expressed in terms of τ and T if we please[5], but we already have enough to clarify the second and third senses in which KELVIN used the term "absolute".

ζ. *The integrating factor: proof of Clausius' statements.* Using T as an empirical-temperature scale, we can rewrite $(19)_2$ as follows:

$$J\left[\frac{\partial}{\partial T}\left(\frac{\Lambda_{V,T}}{T}\right) - \frac{\partial}{\partial V}\left(\frac{K_{V,T}}{T}\right)\right] = \frac{1}{T}\left[-\frac{\partial p}{\partial T} + J\left(\frac{\partial \Lambda_{V,T}}{\partial T} - \frac{\partial K_{V,T}}{\partial V}\right)\right] \ . \tag{11H.20}$$

So far, J is any positive constant having the dimensions of work $\div$ heat. Now let us adopt CLAUSIUS' principle of the uniform and universal Interconvertibility of Heat and Work in cycles, and for the J in (20) take the same J as appears in CLAUSIUS' basic relation

$$\frac{\partial p}{\partial \theta} = J\left(\frac{\partial \Lambda_V}{\partial \theta} - \frac{\partial K_V}{\partial V}\right) , \tag{8B.1$_r$}$$

a relation which is invariant under change of scale and hence must hold in particular when T is used. Then (20) reduces to

$$\frac{\partial}{\partial T}\left(\frac{\Lambda_{V,T}}{T}\right) - \frac{\partial}{\partial V}\left(\frac{K_{V,T}}{T}\right) = 0 \ . \tag{11H.21}$$

Integrating the left-hand side of this relation over the area included by a simple cycle $\mathscr{C}$, by use of AMPÈRE's transformation and (2C.4) we conclude that

$$\int_{t_1}^{t_2} \frac{Q}{T}\, dt = 0 \ . \tag{11H.22}$$

[5] TRUESDELL [1979].

For a simple Carnot cycle $\mathscr{C}$ this statement reduces to

$$\frac{C^+(\mathscr{C})}{\mathrm{T}^+} = \frac{C^-(\mathscr{C})}{\mathrm{T}^-}, \qquad (11\mathrm{H}.23)$$

so

$$\frac{L(\mathscr{C})}{JC^+(\mathscr{C})} = 1 - \frac{\mathrm{T}^-}{\mathrm{T}^+}. \qquad (11\mathrm{H}.1)_\mathrm{r}$$

The spectators will recognize $(19)_2$, (21), and (22) as having exactly the same forms as CLAUSIUS' (11E.4), (11E.6), and (11D.14). The difference is that while CLAUSIUS' function T is divined, here T is KELVIN's second absolute temperature, a specific function of θ which exists *because of the General CARNOT-CLAPEYRON Theorem and the Thermometric Axiom*, on which the definition (17) is based. While CLAUSIUS has to resort to a metaphysical "equivalence-value" in order to conclude that T "can only depend upon the temperature", the definition (17) makes T outright an empirical-temperature function, provided, of course, that $\mu > 0$. If there is such a thing as CLAUSIUS' T, the properties he asserts for it show that it must be a constant multiple of the T we have exhibited. That KELVIN's T^{-1} is an integrating factor for *Qdt*, we have proved by the straightforward argument leading to (21) and (22).

While the foregoing proof is supremely easy and calls upon only mathematics which was in common use in the 1850s, I have found no trace of it in the sources. The reason may be that the authors of textbooks have not perceived that KELVIN's second absolute scale exists whether or not the "First Law" holds, and that the relation $(19)_2$, familiar as it is in the context of the "First Law", in no way requires that law.

The constitutive restrictions $(19)_2$ and (21) provide the basis of the classical thermodynamics of reversible processes in fluids as that doctrine is understood today. The former subsists in consequence of the existence of T, the constant J being arbitrary. The latter subsists only when heat and work are uniformly and universally interconvertible in Carnot cycles and when J is chosen as the corresponding mechanical equivalent of a unit of heat.

η. *Proof of KELVIN's statements.* The formal aspects of KELVIN's work on absolute temperatures have been handled at the beginning of this section. In regard to them all that remains is to note the neat proof of KELVIN's elegant formula

$$\frac{L(\mathscr{C})}{JC^+(\mathscr{C})} = 1 - \exp\left(-\frac{1}{J}\int_{\theta^-}^{\theta^+} \mu d\theta\right) \qquad (9\mathrm{B.g})_\mathrm{r}$$

we get if we use the definitions (17) and (11B.2) to conclude that

$$\mathrm{T} = \exp\left(\frac{1}{J}\int \mu(\theta)d\theta\right), \qquad (11\mathrm{H}.25)$$

then substitute the result into (21). This reverses KELVIN's path: He started from (9B.9), which he had proved by the argument we have reproduced above in §9B. That proof did not cast much light upon the result. Failure to uncover its conceptual basis may have cost KELVIN the independent discovery of CLAUSIUS' assertions, proof of whose concrete aspects we have given just above.

The relation (23) shows that for CLAUSIUS' theory T is "absolute" in KELVIN's third sense.

The essential gap in KELVIN's presentation was not formal but conceptual: He did not prove that T was in fact an empirical-temperature scale. We have filled that gap in Part δ of this section.

θ. *Summary.* In regard to absolute temperature all of CLAUSIUS' statements and all but one of KELVIN's are correct within the framework of classical thermodynamics. By assembling and interlocking them through mathematical proof we have provided and justified the formal structure of classical thermodynamics of fluid bodies described by two variables: a geometric quantity, such as volume V, and the particular absolute temperature T.

Whatever may have been KELVIN's reason for abandoning τ and moving to T, the advantage in doing so is plain: *Without ever mentioning the existence of such a thing as an ideal gas, we find in* T *a temperature which is absolute in the sense that it is independent of the choice of thermometer, yet it gives to all thermodynamic equations the same simple form that the existence of an ideal gas with constant specific heats would imply.* All this follows from just three assumptions restricting the Doctrine of Latent and Specific Heats:

1. The General CARNOT-CLAPEYRON Theorem (itself a consequence of CARNOT's General Axiom).
2. A Thermometric Axiom sufficient to conclude that μ never vanishes.
3. CLAUSIUS' First Law in its earliest form: The work done by a Carnot cycle is the uniform and universal equivalent of the net gain of heat.

The loss in using T is that it does *not* satisfy the requirement KELVIN laid down first: Degrees of heat do *not* have "the same value" in terms of the maximum work their differences can produce in operating a reversible heat engine. On the other hand, degrees according to τ are entirely different from the degrees read upon any common thermometer, while the difference between T and centigrade temperature according to the air thermometer with its zero set at absolute cold is negligible in most circumstances.

Although CLAUSIUS obtained the formal relations of thermodynamics in terms of his absolute temperature as an independent variable, he shows no evidence of having grasped its intrinsic meaning, the meaning that KELVIN's definition, namely

$$\mathrm{T} \equiv \frac{J}{\mu}, \qquad (11\mathrm{B}.4)_r$$

gives it right away. CLAUSIUS' argument for the existence of his T is so vague and mysterious that we cannot subject it to scrutiny. KELVIN's definition is clear: If μ exists and is positive, KELVIN's T exists. By using data from their experiment on the porous plug JOULE & KELVIN were able to determine T as a function of θ. In this way they calculated the deviations of θ according to the air-thermometer from the absolute temperature T. All CLAUSIUS could do in this regard was fall back on the old expedient he had used in 1850: Appealing to the properties of ideal gases, he reaffirmed HOLTZMANN's Assertion. I do not think that thermodynamics has ever been developed correctly without some appeal to a particular material. The thermometric axiom discussed above in Part δ of this section calls upon such a material in order to ensure that $\mu > 0$. With this addition, KELVIN's development becomes sound. CLAUSIUS, on the contrary, never could free himself of an *a priori* concept of an ideal gas. Neither KELVIN nor CLAUSIUS felt the need to prove that what he called T was in fact an empirical-temperature scale. We have seen that a Thermometric Axiom is necessary if we are to prove that the temperature measured on a common thermometer is a differentiable function of KELVIN's T. Then T is an empirical-temperature scale which is absolute in KELVIN's first and third senses.

It would be far too much to say that either KELVIN or CLAUSIUS established all this. Their physical principles were abundantly sufficient and had been so for at least three years before they undertook the work we have been discussing. Their treatments of absolute temperature—KELVIN's having the great merit of concreteness but hasty and fragmentary, CLAUSIUS' riddling and magical—add up to another near miss, obscure from failure to apply mathematical criticism to a mathematical question.

11I. Critique: CLAUSIUS' "Laws" of Thermodynamics

The message of CLAUSIUS' paper is not in the simple, concrete statements we have been at some pains to justify in the preceding section but rather in ideas inchoate.

In his paper of 1850 CLAUSIUS had given reasoning to prove that an internal energy exists. Now, like KELVIN before him (§9D, above), he regards internal energy as something directly motivated in physics. He states the "First Law" as an axiom for all processes, not just cyclic ones:

$$\dot{\mathrm{E}} = JQ + P \,. \qquad (8B.2)_{1r}$$

The main properties of internal energy follow almost by inspection if E is assumed to be a function of V and θ and if P has the classical form $(8B.2)_2$. *However, these specializing assumptions are not inherent to the idea.*

Ever since the appearance of this paper it has been customary to introduce E as a primitive concept in the mathematical structure.

CLAUSIUS' verbal statement of the "Second Law" makes no sense, for "some other change, connected therewith" introduces two new and unexplained concepts: "other change" and "connection" of changes. Neither of these finds any place in CLAUSIUS' formal structure. All that remains is a Mosaic prohibition. A century of philosophers and journalists have acclaimed this commandment; a century of mathematicians have shuddered and averted their eyes from the unclean.

The brief remarks on irreversible processes make no sense either, since a "process" has not been defined or illustrated except within a structure that provides only bodies susceptible of no processes but reversible ones. CLAUSIUS tells us that he could easily calculate "the equivalence-values of the uncompensated transformations" but gives no illustration and no idea what we could do with such a quantity if we had it.

Even the definition of "circular process" is vague: "the series of changes are such that through them the body returns to its original condition", but what is that? CLAUSIUS states at once that $\Delta \mathrm{E} = 0$ in a circular process, but in fact for bodies susceptible of irreversible processes there is no reason to think that a cycle in such variables as V and θ will leave E unchanged.

Like KELVIN, CLAUSIUS seems to see that direct postulation of the two basic "Laws" makes it unnecessary to use the Doctrine of Latent and Specific Heats, which had been the common starting point of all attempts to construct a thermodynamics. Here is a tiny glimmer of light toward a theory of bodies described by more than two variables; far more, toward a theory that allows processes to be irreversible.

CLAUSIUS does nothing with his potentially greater generality. All his specific calculations fall back on the Doctrine of Latent and Specific Heats as an added assumption.

While KELVIN's operative statement of the "Second Law" (§9B, above) is the General CARNOT-CLAPEYRON Theorem, CLAUSIUS' is

$$\int \frac{dC}{\mathrm{T}} = 0 \,. \tag{11D.14}_{\mathrm{r}}$$

Within the framework of the theory of calorimetry this statement is obviously synonymous with *the global existence of entropy*. CLAUSIUS was to wait fourteen years to see that! Here is another of our muse's little jokes. While she had put conditions of integrability into CLAUSIUS' tool box, she seems not to have taught him what to do with them. Four years after the appearance of the work we are discussing, he is to publish a mathematical paper explaining (to himself?) how to use them, and in annotating the reprint of his papers in 1864 he is to say a good deal about them. Then there was the entropy already standing in print in the papers of RANKINE, as we have seen in §§8G and 9A, but in them it was entwined in the obscurity of RANKINE's molecular model. Finally there was REECH's pro-entropy, speciously general but nevertheless enjoying the full panoply of formal relations we today associate with the entropy. But our muse gave REECH the gift of Cassandra: Nobody paid any attention to him.

Indeed, REECH[1] was quick to remark in effect that if his integrating factor f were assumed to be a function of θ alone, say γ, then his formula (10A.4) applied to a Carnot cycle $\mathscr{C}$ would yield

$$\begin{aligned} C^{+}(\mathscr{C}) &= \gamma(\theta^{+})\Delta u \ , \\ C^{-}(\mathscr{C}) &= \gamma(\theta^{-})\Delta u \ . \end{aligned} \tag{11I.1}$$

Thus

$$L(\mathscr{C}) = J[\gamma(\theta^{+}) - \gamma(\theta^{-})]\Delta u \ , \tag{11I.2}$$

and

$$\frac{C^{+}(\mathscr{C})}{\gamma(\theta^{+})} = \frac{C^{-}(\mathscr{C})}{\gamma(\theta^{-})} \ . \tag{11I.3}$$

This is CLAUSIUS' Second Principal Theorem; as γ is universal, CLAUSIUS' conclusions about an absolute temperature T follow, and $\mathrm{T} = \gamma(\theta)$.

That is not all. If $f(\theta, u) = \gamma(\theta)$, *REECH's function u is a function of what* CLAUSIUS, twelve years later, *was to introduce afresh and call the entropy*! REECH's analysis of the Carnot cycle by use of it is elegant and complete. Many modern textbooks present the Carnot cycle in just this way. But that argument is not REECH's discovery either, for it already stood in RANKINE's paper of 1851, which we have analysed in §9A; *cf.* (9A.11)–(9A.14).

For this, none of the inspired assertions about "equivalence-values" and no compositions and resolutions of processes are necessary.

11J. Critique: Irreversible Processes

The now familiar term "dissipation" seems to derive from KELVIN's note of 1852, the contents of which we have presented in §11A. Everyone today accepts KELVIN's statements there. They express something that is become part of the way we look at nature.

---

[1] REECH [1856, pp. 61–66] was impeded by the insufficiency and incompleteness of his own earlier mathematical analysis (above, §10E). He did not know that his integrating factor f could always be taken as a function of θ alone, but "certain properties of vapors" had led him in 1853 to suppose so for those substances. Then he regarded f as constitutive. Now he uses CARNOT's construction (§5F) to conclude that it would be "absurd" for γ to vary from one substance to another. His pro-entropy u remains for him a function of p and θ.

In a later note REECH [1858] makes his position clear: "I cite and prove in my own way the theorem of Mr. CLAUSIUS, according to which my expressions for C^{+} and C^{-} are too general, and should be replaced by [(11I.1)], the function $\gamma(\theta)$ being supposed the same for all kinds of elastic fluids...."

It would be difficult to find any clear connection between KELVIN's pronouncements in this note and the extremely limited body of concrete thermodynamics by then constructed. The overtones of belief in an omnipotent divine will capable of altering divine laws are apparent: "inanimate material agency", "material world", "inanimate material processes", "the will of an animated creature", "man as at present constituted", "at present in the material world". Just what there was about the modest if respectable little budget of specific statements which thermodynamics in its early years delivered that made its creators bold to apply them to "the earth" and "the universe", is not clear. I suppose the tragicomic muse was schadenfroh.

KELVIN's note reads as if it were a program for future research toward a general theory of thermomechanics in deformable bodies that can accept conduction and radiation of heat. Unfortunately such was not to be the course of history.

KELVIN himself was to study dissipation explicitly and efficiently only in the context of linear conductors of electricity—a branch, that is, of the theory of circuits, in which purely mechanical dissipation had long been understood and could serve as a model. In the thermodynamics of dissipation in general nothing concrete was to be achieved during the period of this tragicomedy, and very little thereafter until the 1960s.

More attention was to be paid to the oracles of CLAUSIUS, which have been repeated, embroidered, and glossed in all the textbooks. What they are to mean is another matter.

Seven times in the past thirty years have I tried to follow the argument CLAUSIUS offers to conclude that the integrating factor T exists *in general*, is a function of θ alone, and is the same for all bodies, and seven times has it blanked and gravelled me. For that reason I have reproduced the central passages in §§11D and 11F. Let the reader judge of this. I cannot explain what I cannot understand.

I cannot even see what CLAUSIUS' "analytical expression"

$$\int \frac{dC}{\mathrm{T}} = 0 \qquad (11\mathrm{D}.14)_r$$

means. I should like to think that dC here stands *in general* for Qdt, and that the integral is a double one, Q itself being the value of some kind of integral that "extends over all the quantities of heat received by the several bodies".

CLAUSIUS does not give the reader any further explanation or example that would clarify any case except the one already abundantly treated in earlier papers. He has shown himself able to correct CARNOT's work, but he cannot produce anything concrete in a domain of thermodynamics through which CARNOT had not cut a trail.

Epilogue: Götterdämmerung

... Coscïenza fusca
o de la propria o de l'altrui vergogna
pur sentirà la tua parola brusca.
Ma nondimen, rimossa ogne menzogna,
tutta tua visïon fa manifesta;
e lascia pur grattar dov' è la rogna.
Ché se la voce tua sarà molesta
nel primo gusto, vital nodrimento
lascerà poi, quando sarà digesta.
Questo tuo grido farà come vento,
che le più alte cime più percuote;
e ciò non fa d'onor poco argomento.
DANTE, *Paradiso* XVII, 124–135

CLAUSIUS' first paper, while entangled and slack, was in aim and result constructive. From his second paper, on the contrary, through the murk and gloom emerges a growing aura of retreat and impending failure. While in all work analysed up to now there was no hint that conditions were any more specific than the equations themselves suggested, in this paper CLAUSIUS assumes that "the pressure always changes very gradually," though he specifies no time scale sufficient to give meaning to the term "gradual". Here the tergiverse "quasistatic process", hinted at by REECH, first slithers onto the scene. It joins the "state" as a principal engine of the mystic double-talk that makes thermodynamics different in kind from all the rest of classical physics.

But that is far from the worst. Hitherto thermodynamics had been, like any other theory in mathematical physics, pretty largely a model for the way things are. In CLAUSIUS' hands it now begins to change into a model for the way things are not. The old theory, based on the Doctrine of Latent and Specific Heats, makes all processes "reversible". CLAUSIUS seems suddenly to see that in nature we cannot run engines backward. To the great geometers of the previous century this idea (which, if I may be permitted an unhistorical conjecture, would not have seemed the least bit startling or new) would have been a challenge to construct a general and inclusive theory. To CLAUSIUS it was sufficient reason to confine the circumstances so as to fit the existing theory.

And if the new principles are to hold outside the constitutive class specified by the Doctrine of Latent and Specific Heats—indeed, that Doctrine *cannot* apply to irreversible processes—what is the more general class that CLAUSIUS considers? As BRIDGMAN wrote[1],

> ... why is thermodynamics restricted to the formulation of necessary conditions, and why is it so impotent in its endeavor to frame sufficient conditions? Other branches of physics are not thus restricted. Or why is it that it is so impotent to deal with irreversible processes? There are certain irreversible processes that are of a patent simplicity, and that can be completely measured by the instruments which give us our thermodynamic information, such as thermal conductivity. Why should not the physicist be able to deal with the thermodynamic implications of thermal conduction?

Indeed. The thread of the plot that began to spin out in Act I hangs. It will continue to hang for nearly 150 years, despite brilliant particular exceptions: MAXWELL's derivation of the field equation for balance of energy according to his kinetic theory of gases, 1866; KIRCHHOFF's consequent analysis of the absorption and dispersion of sound in a viscous, heat-conducting gas, 1868; MAXWELL's prophetic study of thermo-elastic, dissipative interaction in rarefied gases, 1876; the work of DUHEM, HADAMARD, and others on thermo-elasticity in general, including thermal effects on shock waves and acceleration waves. But these did not unite thermodynamics and mechanics.

At similar stages in mechanics and electromagnetism, recognition of a major gap of this kind led to a clear concept and exploitation of the role of constitutive relations. In those subjects a key problem well solved was father to a new key problem posed; growth in concept matched the growth of special cases mastered; when physics called for new tools, it was the creating geometer himself who set about to forge them. Not so in thermodynamics. CLAUSIUS, instead of enlarging his theory enough to include the processes he chose to call "uncompensated" (irreversible) contented himself with describing them in physical terms. We have quoted his remarks in full in §11D. The "particular cases" into which CLAUSIUS chose not to enter include not only FOURIER's theory of heat conduction but also most phenomena described by mechanics and many described by electromagnetism. With this decree, thermodynamics turned its back on the real world. Henceforth, relinquishing steam engines, it would treat mainly a "universe"—an infinite space filled with some gas, the constitutive relation of which was specified

[1] BRIDGMAN [1941, p. 5 of the 1961 edition].

only for the case when it was at rest with uniform density and temperature. Hence grew the thermodynamicists' notorious disregard for constitutive relations altogether, their illusion that thermodynamics can do without them and deal only in extremely general principles. The defeat of CARNOT's program had been formally acknowledged by a treaty of neutrality supervised by mutual blindness.

The thermodynamics of the nineteenth century began a new style, in which the physicist applied what mathematics he happened to know. If that mathematics did not suffice, he cut down the problem to its size. Such a physicist did not think it necessary for his students to learn any mathematics beyond what he himself had been taught when he was a student. Mathematical research meanwhile advanced swiftly, but little of it was learnt by physicists. It became purer and purer. Physicists finally began to think that all this new mathematics was useless; that mathematicians neglected their duty to teach the good old mathematics already used in physics and hence (obviously!) destined to suffice it forever; and that they themselves should teach "physical" mathematics to their students: mathematical tools for the physicist!

Se Dio ti lasci, lettor, prender frutto
di tua lezione....
DANTE, *Inferno* XX, 19–20.

Sources

listed according to the *dates of their earliest publication*. If no date is supplied in the course of a reference, the date of publication is that under which the item is listed. The collection is intended to include all works on mathematical thermodynamics in general from its beginnings through 1854, except for reviews, of which only those of real importance are listed. Later publications and works on related subjects such as application to steam or other vapors and experimental determination of quantities are included only if they are referred to in the text. Omitted are textbooks, works on continuum mechanics or kinetic theory, purely historical writings, and purely critical studies; all these I have cited at the points where reference is made to them.

Italic numbers following an entry direct the reader to the sections of the book which mention the entry.

1779 LAMBERT, J. H.: *Pyrometrie oder vom Maaße des Feuers und der Wärme*, Berlin, Haude und Spener.
(*4A*)

1784 LAVOISIER, A., & P.-S. DE LAPLACE: "Mémoire sur la chaleur", *Histoire de l'Académie Royale des Sciences* (Paris), *Année 1780, avec les Mémoires de Mathématique et Physique*, 355–408 = pp. 283–333 of *Œuvres de LAVOISIER*, Tome 2, 1878.
Transl. "Abhandlung über die Wärme", pp. 3–55 of A. L. LAVOISIER & P.-S. DE LAPLACE, *Zwei Abhandlungen über die Wärme, ed.* J. ROSENTHAL, Ostwalds Klassiker No. 40, Leipzig, Engelmann, 1892.
(*2C*)

1802 BIOT, J. B.:
1. "Sur la théorie du son", *Journal de Physique* **55**, 173–182.
(*3A*)
2. "Sur la propagation du son" [Extract of the preceding, signed by I. B.], *Bulletin des Sciences, Société Philomathique de Paris* **3**, No. 63, 116–118 (An 10 = 1801/2).
[I have seen this rare work only in what appears to be a reprinted volume published by Libraire Klostermann over the date 1811.]
(*3A*)

GAY-LUSSAC, J. L.: "Recherches sur la dilatation des gaz et des vapeurs", *Annales de Chimie* (1) **43**, 137–175.

Transl. W. W. RANDALL, "Researches upon the rate of expansion of gases and vapors", pp. 27–48 of W. W. RANDALL, *The Expansion of Gases by Heat*, New York *etc.*, American, 1902.
Transl. "Untersuchungen über die Ausdehnung der Gasarten und der Dämpfe durch die Wärme", *Annalen der Physik* **12**, 257–291 = pp. 3–25 of *Das Ausdehnungsgesetz der Gase*, ed. W. OSTWALD, Ostwalds Klassiker No. 44, Leipzig, Engelmann, 1894.
(*2A*)

1803 BLACK, J.: *Lectures on the Elements of Chemistry*, 2 vols., London, Longman & Rees, and Edinburgh, William Creech.
(*2C*)

1804 BIOT, J. B.: "Mémoire sur la propagation de la chaleur, lu à la Classe des Sciences Mathématiques et Physiques de l'Institut National", *Bibliothèque Britannique* **27** (1804/1805), 310–329 = [with modernized spelling and other slight changes] *Journal des Mines* An XIII [1804/5], 203–224.
(*4A, 4E*)

BRANDES, H. W.: "Untersuchungen über die Fortpflanzung des Schalles in der Luft von BIOT", *Annalen der Physik* **18**, 385–400.
[This paper is a paraphrase of BIOT [1802, *1* and *2*] with notes and remarks by BRANDES and references to works of AMONTONS, EULER, LAMBERT, LAGRANGE, DALTON, and CHLADNI.]
(*3A*)

1805 LAPLACE, P.-S. DE: *Traité de Mécanique Céleste*, Tome 4, Paris, Bachelier = *Œuvres Complètes de LAPLACE*, Tome 4, Paris, Gauthier-Villars, 1880.
Transl., with notes by N. BOWDITCH, *Mécanique Céleste*, Volume 4, Boston, C. C. Little and J. Brown, 1839.
(*3F*)

1806 BRANDES, H. W.: *Die Gesetze des Gleichgewichts und der Bewegung flüssiger Körper, dargestellt von Leonhard Euler, uebersetzt, mit einigen Abänderungen und Zusätzen*, Leipzig, S. L. Crusius.
(*2A, 3A*)

1808 POISSON, S.-D.: "Mémoire sur la théorie du son" (1807), *Journal de l'Ecole Polytechnique* **7**, cahier 14, 319–392.
(*3A*)

[S.-D.] P[OISSON]: "Mémoire sur la propagation de la chaleur dans les corps solides, par M. Fourier", *Nouveau Bulletin des Sciences par la Société Philomathique de Paris* **1**, 112–116 = pp. 213–221 of *Œuvres de FOURIER*, Volume 2, 1890.
(*4B*)

1816 BIOT, J. B.:
1. Traité de Physique Expérimentale et Mathématique, Tome I, Chapitre IX: "Mesure de la dilatation des gaz par la chaleur", Paris, Deterville.

Transl. W. W. RANDALL, "The determination of the rate of expansion of gases by heat", pp. 53–61 of W. W. RANDALL, *The Expansion of Gases by Heat*, New York *etc.*, American, 1902.
(*2A*)
2. Traité de Physique Expérimentale et Mathématique, Tome IV, livre 7me: "Du calorique, soit rayonnant, soit latent", Paris, Deterville.
(*4A, 6A*)

LAPLACE, P.-S. DE: "Sur la vitesse du son dans l'air et dans l'eau", *Annales de Chimie et de Physique* (2)**3**, 238–241 = pp. 297–300 of *Œuvres Complètes de LAPLACE*, Tome 14, 1912.
Transl. R. B. LINDSAY, "Velocity of sound in air and water", pp. 181–182 of *Acoustics: Historical and Philosophical Development*, ed. R. B. LINDSAY, Stroudsburg (Pa.), Dowden, Hutchinson and Ross, 1972.
(*3C, 5T*)

1818 PETIT, A. T.: "Sur l'emploi du principe des forces vives dans le calcul de l'effet des machines", *Annales de Chimie et de Physique* (2)**8**, 287–305.
(*5K, 5U*)

1821 LAPLACE, P.-S. DE: "Sur l'attraction des sphères et sur la répulsion des fluides élastiques", *Connaissance des Tems . . . pour l'An 1824*, 328–343 = pp. 273–290 of *Œuvres Complètes de LAPLACE*, Tome 13, 1904.
Partial abstract: "Sur l'attraction des corps sphériques, et sur la répulsion des fluides élastiques", *Annales de Chimie et de Physique* (2) **18**, 181–190.
(*3C*)

1822 FOURIER, J.: *Théorie Analytique de la Chaleur*, Paris = *Œuvres de FOURIER*, Tome 1, 1888.
Transl. A. FREEMAN, *The Analytical Theory of Heat*, Cambridge University Press, 1878.
(*3C, 4B–4F, 4I, 7A*)

LAPLACE, P.-S. DE:
1. "Développement de la théorie des fluides élastiques et application de cette théorie à la vitesse du son", *Connaissance des Tems . . . pour l'An 1825*, 219–227 = pp. 291–301 of *Œuvres Complètes de LAPLACE*, Tome 13, 1904.
(*3C*)
2. "Continuation du mémoire précédent sur le développement de la théorie des fluides élastiques", *Connaissance des Tems . . . pour l'An 1825*, 302–323 = pp. 133–160 of *Œuvres Complètes de LAPLACE*, Tome 5, 1882 [*i.e.*, a part of LAPLACE [1823]].
(*3C*)

3. "Sur la vitesse du son", *Connaissance des Tems . . . pour l'An 1825*, 371–372 = pp. 303–304 of *Œuvres Complètes de* LAPLACE, Tome 13, 1904.
(*3C*)

1823 LAPLACE, P.-S. DE: *Traité de Mécanique Céleste*, Tome 5, Livre XII, Paris, Bachelier = pp. 97–160 of *Œuvres Complètes de* LAPLACE, Tome 5, 1882.
(*3B, 3C, 3D, 5T*)

POISSON, S.-D.:

1. "Sur la vitesse du son", *Annales de Chimie et de Physique* (2)**23**, 5–16.
(*3D, 3E, 5T*)

2. "Sur la chaleur des gaz et des vapeurs", *Annales de Chimie et de Physique* (2)**23**, 337–352.
Transl., with notes by [J.C.] P[OGGENDORF], "Ueber die Wärme der Gase und Dämpfe", *Annalen der Physik und Chemie* **76**, 269–288 (1824).
Transl., with notes by J. HERAPATH, "On the caloric of gases and vapours", *Philosophical Magazine* **62**, 328–338 (1823).
(*3D*)

3. "Sur la vitesse du son", *Connaissance des Tems . . . pour l'An 1826*, 248–277.
[§I of this article = POISSON [1823, *1*].]
(*3D, 3E, 5T*)

1824 CARNOT, S.: *Réflexions sur la Puissance Motrice du Feu et sur les Machines Propres à développer cette Puissance*, Paris, Bachelier = *Annales Scientifiques de l'Ecole Normale Supérieure* (2)**1** (1872), 393–457. Reprinted, along with CARNOT's manuscript notes (1824–1832), Paris, 1878.
Transl. R. H. THURSTON, *Reflections on the Motive Power of Heat and on Machines fitted to develop this Power*, New York, 1890; new edition, New York, 1943.
Transl. W. F. MAGIE, "Reflections on the motive power of heat and on engines suitable for developing this power", pp. 1–60 of *The Second Law of Thermodynamics*, ed. W. F. MAGIE, New York and London, 1899.
Corrected and annotated reprint of THURSTON's translation, with selections from the manuscript notes, and provided with an introduction by E. MENDOZA, *Reflections on the Motive Power of Fire and on Machines fitted to develop that Power*, New York, Dover Publications, 1960.
Transl. W. OSTWALD, *Betrachtungen über die bewegende Kraft des Feuers und die zur Entwickelung dieser Kraft geeigneten Maschinen*, ed. W. OSTWALD, Ostwalds Klassiker No. 37, Leipzig, Engelmann, 1892.

[For the references in the text above, the first edition has been used. It is readily available in one or more photoreproductions.]
(*2C*, *5A–5D*, *5F–5I*, *5K*, *5O*, *5Q*, *5S*, *5T*, *7A*, *7H*)

1825 Ivory, J.: "On the laws of the condensation and dilatation of air and the gases, and the velocity of sound", *Philosophical Magazine* **66**, 3–13.
(*3E*)

1826 Meikle, H.:
1. "On the law of temperature", *The Annals of Philosophy* **28** = (n.s.) **12**, 366–369.
(*3E*)
2. "The theory of the air-thermometer", *The Edinburgh New Philosophical Journal* **1**, 332–341.
(*3E*)

1827 Ivory, J.:
1. "Investigation of the heat extricated from air when it undergoes a given condensation", *Philosophical Magazine* (n.s.) **1**, 89–94.
(*2C*, *3E*)
2. "Application of the variations of temperature in air that changes its volume to account for the velocity of sound", *Philosophical Magazine* (n.s.) **1**, 249–255.
(*3E*)

Poisson, S.-D.: "Observations relatives à un article de Mr. Ivory ....", *Philosophical Magazine* (n.s.) **2**, 11–16.
(*3E*)

1828 Ivory, J.: "Answer to an article by Mr. Henry Meikle ...", *Philosophical Magazine* (n.s.) **4**, 321–326.
(*3E*)

Meikle, H.:
1. "On Mr. Ivory's investigations of the velocity of sound", *Quarterly Journal of Science, Literature, and Art* **26** = (n.s.) **4**, 124–135.
(*3E*)
2. "Reply to Mr. James Ivory's answer ...", *Quarterly Journal of Science, Literature, and Art* **26** = (n.s.) **4**, 315–319.
(*3E*)

1829 Meikle, H.: "On the relation between density, pressure, and temperature of air; and on experiments regarding the theory of clouds, rain, &c.; with a conjecture about thunder and lightning", *Quarterly Journal of Science, Literature, and Art* **27** = (n.s.) **5**, 56–75.
(*3E*)

1830 Weber, W.: "Ueber die specifische Wärme fester Körper, insbesondere der Metalle", *Annalen der Physik und Chemie* **96** = (2) **20**, 177–213.
(*2C*)

1832 DUHAMEL, J.-M.-C.: "Sur les équations générales de la propagation de la chaleur dans les corps solides dont la conductibilité n'est pas la même dans tous les sens", *Journal de l'Ecole Polytechnique* **13**, Cahier 21, 356–399.
(4F)

1833 FOURIER, J.: "Mémoire d'analyse sur le mouvement de la chaleur dans les fluides", *Mémoires de l'Académie Royale des Sciences de l'Institut de France* (2) **12**, 507–530 = (with editorial changes) pp. 595–616 of *Œuvres de FOURIER*, Tome 2, 1890.
(2A, 4F, 4G)

POISSON, S.-D.: *Traité de Mécanique*, Seconde éd., Tome 2, Paris, Bachelier.
Transl., with notes by H. H. HARTE, *A Treatise of Mechanics*, Volume 2, London, Longman and Co., *etc.*, 1842.
(2C, 3F)

1834 CLAPEYRON, E.: "Mémoire sur la puissance motrice de la chaleur", *Journal de l'Ecole Polytechnique* **14**, Cahier 23, 153–190.
Transl., "Ueber die bewegende Kraft der Wärme", *Annalen der Physik und Chemie* **135** = (2)**59**, 446–451, 566–586 (1843).
Transl., "Memoir on the motive power of heat", pp. 347–376 of *Scientific Memoirs* **1**, ed. R. TAYLOR, London, 1837; reprinted, New York and London, Johnson Reprint Corp., 1966.
Transl. K. SCHREBER, *Abhandlung über die bewegende Kraft der Wärme*, ed. K. SCHREBER, Ostwalds Klassiker No. 216, Leipzig, Akademische Verlagsgesellschaft, 1926.
Transl. E. MENDOZA, "Memoir on the motive power of heat", pp. 71–105 of MENDOZA's edition of CARNOT [1824].
(2C, 5L, 5O, 5Q, 6A)

1835 POISSON, S.-D.: *Théorie Mathématique de la Chaleur*, Paris, Bachelier.
(4D)

1837 DUHAMEL, J.-M.-C.: "Second mémoire sur les phénomènes thermo-mécaniques" (1835), *Journal de l'Ecole Polytechnique* **15**, Cahier 15, 1–57.
(6B)

KELLAND, P.: *Theory of Heat*, Cambridge, for Deighton and Parker.
(2C)

1838 DUHAMEL, J.-M.-C.: "Mémoire sur le calcul des actions moléculaires développées par les changements de température dans les corps solides", *Mémoires Présentés par Divers Savants à l'Académie des Sciences de l'Institut de France* (2) **5**, 440–485.
(6B)

1842 MAYER, J. R.: "Bemerkungen über die Kräfte der unbelebten Natur", *Annalen der Chemie und Pharmacie* **42**, 233–240 = (somewhat revised) pp. 3–12 of J. R. MAYER, *Die Mechanik der Wärme*, Stuttgart, 1867 = pp. 3–12 of *ibid.*, 2nd ed., 1874 = pp. 23–30 of *ibid.*, 3rd ed., 1893 = pp. 3–8 of R. MAYER, *Die Mechanik der Wärme*, ed. and annot. A. VON OETTINGEN, Ostwalds Klassiker No. 180, Leipzig, Engelmann, 1911.

Transl. G. C. FOSTER, "Remarks on the forces of inorganic nature", *Philosophical Magazine* (4) **24** (1862), 371–377 = pp. 251–258 of *The Correlation and Conservation of Forces*, ed. E. L. YOUMANS, New York, Appleton, 1865 = pp. 27–33 of G. SARTON, "The discovery of the law of conservation of energy", *Isis* **13** (1929–30), 18–34 (1929) = pp. 197–203 of W. F. MAGIE, *A Source Book in Physics*, New York and London, 1935 = pp. 72–77 of S. G. BRUSH, *Kinetic Theory*, Volume 1, Oxford *etc.*, Pergamon Press, 1965.

Transl. R. B. LINDSAY, "On the forces of inorganic nature", pp. 68–74 of R. B. LINDSAY, *Julius Robert Mayer, Prophet of Energy*, Oxford *etc.*, Pergamon, 1973 = pp. 277–283 of *Energy: Historical Development of the Concept*, ed. R. B. LINDSAY, Stroudsburg (Pa.), Dowden, Hutchinson & Ross, 1975.

(*7B*)

REGNAULT, V.: "Recherches sur la dilatation des gaz", *Annales de Chimie et de Physique* (3) **4**, 5–63; **5**, 52–83.

Transl. W. W. RANDALL, "Researches upon the rate of expansion of gases", pp. 65–119, 123–150 of W. W. RANDALL, *The Expansion of Gases by Heat*, New York etc., American, 1902.

Transl., "Untersuchung über die Ausdehnung der Gase", *Annalen der Physik und Chemie* **131** = (2) **55**, 391–414, 557–584, **133** = (2) **57**, 115–150 = pp. 88–164 of *Das Ausdehnungsgesetz der Gase*, ed. W. OSTWALD, Ostwalds Klassiker No. 44, Leipzig, Engelmann, 1894.

(*2A*)

1843 NEUMANN, F. E.: "Die Gesetze der Doppelbrechung des Lichts in comprimirten oder ungleichförmig erwärmten unkrystallinischen Körpern", *Abhandlungen der königlichen Akademie der Wissenschaften zu Berlin* $\mathbf{1841}_2$, 1–254 = (with an introduction and notes by W. VOIGT) pp. 5–256 of *Franz Neumann's Gesammelte Werke*, Band 3, 1912.

[The introduction is printed, with small differences, in *Bericht über die zur Bekanntmachung geeigneten Verhandlungen der königlichen Preussischen Akademie der Wissenschaften zu Berlin* **1841**, 330–353 = *Annalen der Physik und Chemie* **130** = (2) **54** (1841), 449–476. The entire work was issued separately in 1842 by the press of the Academy.]

(*6B*)

1845 HOLTZMANN, C.: *Ueber die Wärme und Elasticität der Gase und Dämpfe*, Mannheim.

Transl. W. FRANCIS, "On the heat and elasticity of gases and vapours", pp. 189–217 of *Scientific Memoirs*, Volume 4, ed.

RICHARD TAYLOR, London, 1846; reprinted, New York and London, Johnson Reprint Co., 1966.
Editorial summary, *Annalen der Physik und Chemie*, Ergänzungsband **II** (1848), 183–191.
(*7D*, *7E*)

JOULE, J. P.:
1. "On the changes of temperature produced by the rarefaction and condensation of air", *Philosophical Magazine* (3) **26**, 369–383 = pp. 172–189 of JOULE's *Scientific Papers*, Volume 1, 1887.
(*7G*)
2. "On the existence of an equivalent relation between heat and the ordinary forms of mechanical power", *Philosophical Magazine* (3) **27**, 205–207 = pp. 202–205 of JOULE's *Scientific Papers*, Volume 1, 1887 = pp. 345–348 of *Energy: Historical Development of the Concept*, ed. R. B. LINDSAY, Stroudsburg (Pa), Dowden, Hutchinson & Ross, 1975.
(*7G*)

MAYER, J. R.: *Die organische Bewegung in ihrem Zusammenhange mit dem Stoffwechsel. Ein Beitrag zur Naturkunde*, Heilbronn, Dechsler = (somewhat revised) pp. 16–126 of J. R. MAYER, *Die Mechanik der Wärme*, Stuttgart, 1867 = pp. 15–126 of *ibid.*, 2nd ed., 1874 = pp. 45–128 of *ibid.*, 3rd ed., 1893 = pp. 9–79 of R. MAYER, *Die Mechanik der Wärme*, ed. and annot. A. VON OETTINGEN, Ostwalds Klassiker No. 180, Leipzig, Engelmann, 1911. Transl. R. B. LINDSAY, "The motions of organisms and their relation to metabolism. An essay in natural science", pp. 76–145 of R. B. LINDSAY, *Julius Robert Mayer, Prophet of Energy*, Oxford *etc.*, Pergamon, 1973.
(*7B*)

1847 HELMHOLTZ, H. v.: "Ueber die Erhaltung der Kraft", Berlin, G. Reimer = (with added notes) pp. 12–75 of HELMHOLTZ's *Wissenschaftliche Abhandlungen*, Band 1, 1881 = Ostwalds Klassiker No. 1, Leipzig, Engelmann, 1889.
Transl. J. TYNDALL, "On the conservation of force; a physical memoir", pp. 114–162 of *Scientific Memoirs, Natural Philosophy*, ed. J. TYNDALL & W. FRANCIS, London, Taylor & Francis, 1853; reprinted, New York and London, Johnson Reprint Corp., 1966.
(*7B*, *7E*, *7F*)

JOULE, J. P.: "On matter, living force, and heat", *Manchester Courier*, May 5 and 12 = pp. 265–276 of JOULE's *Scientific Papers*, Volume 1, 1887 = pp. 385–390 of E. C. WATSON, "Joule's only *general* exposition of the principle of conservation of energy", *American Journal of Physics* **15** (1947), 383–390 = pp. 349–360 of *Energy: Historical Development of the Concept*, ed. R. B. LINDSAY, Stroudsburg (Pa.), Dowden, Hutchinson & Ross, 1975.
(*7G*)

REGNAULT, V.: *Relation des expériences...pour déterminer les principales lois et les données numériques qui entrent dans le calcul des machines à vapeur*, Paris, Didot, Tome 1 = *Mémoires de l'Académie des Sciences de l'Institut de France* **21**, 1–767.
(*9D* App.)

1848 THOMSON, W. (later Lord KELVIN): "On an absolute thermometric scale founded on Carnot's theory of the motive power of heat, and calculated from Regnault's observations", *Proceedings of the Cambridge Philosophical Society* **1** (1843/1863), No. 5, 66–71 = *Philosophical Magazine* (3) **33**, 313–317 = (with added notes) pp. 100–106 of W. THOMSON's *Mathematical and Physical Papers*, Volume 1, 1882 = pp. 52–58 of *The Second Law of Thermodynamics*, ed. J. KESTIN, Stroudsburg (Pa.), Dowden, Hutchinson & Ross, 1976.
(*5D*, *7H*, *11B*)

1849 THOMSON, J.: "Theoretical considerations on the effect of pressure in lowering the freezing point of water", *Transactions of the Royal Society of Edinburgh* **16**, 575–580 = (with alterations) *Cambridge and Dublin Mathematical Journal* **5** (1850), 248–255 = pp. 156–164 of W. THOMSON's *Mathematical and Physical Papers*, Volume 1, 1882.
(*5C*)

THOMSON, W. (later Lord KELVIN): "An account of Carnot's theory of the motive power of heat; with numerical results deduced from Regnault's experiments on steam", *Transactions of the Royal Society of Edinburgh* **16**, 541–574 = (with annotations) pp. 113–155 of W. THOMSON's *Mathematical and Physical Papers*, Volume 1, 1882.
(*2C*, *5C*, *5J*, *5L*, *5Q*, *7H*)

1850 CLAUSIUS, R.: "Über die bewegende Kraft der Wärme und die Gesetze, welche sich daraus für die Wärmelehre selbst ableiten lassen", *Annalen der Physik und Chemie* **155** = (2) **79** = (3) **19**, 368–397, 500–524 = (annotated and augmented) pp. 16–90 of CLAUSIUS' *Abhandlungen über die mechanische Wärmetheorie*, Volume 1, 1864 = pp. 3–52 of Ostwalds Klassiker No. 99, ed. and annot. M. PLANCK, Leipzig, Engelmann, 1898.
Transl. J. TYNDALL, "On the moving force of heat, and the laws regarding the nature of heat itself which are deducible therefrom", *Philosophical Magazine* (4) **2** (1851), 1–21, 102–119.
Transl. W. F. MAGIE, "On the motive power of heat, and on the laws which can be deduced from it for the theory of heat", pp. 63–107 of *The Second Law of Thermodynamics*, ed. W. F. MAGIE, New York and London, 1899 = pp. 107–152 of MENDOZA's edition of CARNOT [1824].
Transl. of the annotated and augmented version in the *Abhandlungen*, "On the moving force of heat and the laws of heat which

may be deduced therefrom", pp. 14–80 of CLAUSIUS' *The Mechanical Theory of Heat*, ed. T. A. HIRST, London, Van Voorst, 1867. (*2C, 7A, 7B, 7D, 7E, 7H, 8A, 8B, 8E*)

JOULE, J. P.: "On the mechanical equivalent of heat", *Philosophical Transactions of the Royal Society* (London) **140**, 61–82 = pp. 298–328 of JOULE's *Scientific Papers*, Volume 1, 1887.
(*9D* App., *9G*)

MAXWELL, J. C.: "On the equilibrium of elastic solids", *Transactions of the Royal Society of Edinburgh* **20** (1850/1853), Part 1, 87–120 = pp. 30–73 of MAXWELL's *Scientific Papers*, Volume 1, 1890.
(*6B*)

RANKINE, W. J. McQ.: "On the mechanical action of heat, especially in gases and vapours", *Transactions of the Royal Society of Edinburgh* **20** (1850/1853), Part 1, 147–190 = pp. 234–284 of RANKINE's *Miscellaneous Scientific Papers*, 1881.
(*7I, 8G, 9A, 9G*)

1851 RANKINE, W. J. McQ.:

1. "On the centrifugal theory of elasticity, as applied to gases and vapours" (1850), *Philosophical Magazine* (4) **2**, 509–542 = pp. 16–48 of RANKINE's *Miscellaneous Scientific Papers*, 1881.
(*8G, 9A*)

2. "Note as to the dynamical equivalent of temperature in liquid water and the specific heat of atmospheric air and steam", *Transactions of the Royal Society of Edinburgh* **20** (1850/1853), Part 2, pp. 191–193 = pp. 285–287 of RANKINE's *Miscellaneous Scientific Papers*, 1881.
(*9G*)

3. "On the economy of heat in expansive machines: Forming the fifth section of a paper on the mechanical action of heat", *Transactions of the Royal Society of Edinburgh* **20** (1850/1853), Part 2, 205–210 = pp. 300–306 of RANKINE's *Miscellaneous Scientific Papers*, 1881.
(*7B, 9A*)

REECH, F.: "Notice sur un Mémoire intitulé: *Théorie de la force motrice du calorique*", *Comptes Rendus Hebdomadaires des Séances de l'Académie des Sciences*, Paris, **33**, 567–571.
(*9C, 10* App.)

STOKES, G. G.: "On the conduction of heat in crystals", *Cambridge and Dublin Mathematical Journal* **6**, 215–238 = pp. 203–227 of STOKES's *Mathematical and Physical Papers*, Volume 3, 1901.
(*4F*)

THOMSON, W. (later Lord KELVIN): "On the dynamical theory of heat, with numerical results deduced from Mr. Joule's equivalent of a thermal unit, and M. Regnault's observations on steam, Parts I–III", *Transactions of the Royal Society of Edinburgh* **20**

(1850/1853), Part 2, 261–268, 289–298 = (with some additions) *Philosophical Magazine* (4) **4** (1852), 8–21, 105–117, 168–176 = (with annotations) pp. 174–210 of W. THOMSON's *Mathematical and Physical Papers*, Volume 1, 1882 = pp. 109–144 of *The Second Law of Thermodynamics*, ed. W. F. MAGIE, New York and London, 1899.
Transl., "Mémoire sur la théorie dynamique de la chaleur, contenant quelques résultats numériques déduits de l'équivalent de l'unité de chaleur de M. Joule, et des observations de M. Regnault sur la vapeur d'eau", §I of W. THOMSON, "Deux mémoires sur la théorie dynamique de la chaleur", *Journal de Mathématiques Pures et Appliquées* **12** (1852), 209–241.
Transl. W. BLOCK, "Über die dynamische Theorie der Wärme mit numerischen Ergebnissen aus Herrn Joules Äquivalent einer thermischen Einheit und Regnaults Messungen an Dampf", pp. 3–42 of *Über die dynamische Theorie der Wärme*, ed. and annot. W. BLOCK, Ostwalds Klassiker No. 193, Leipzig etc., Engelmann, 1914.
(*2C*, *3C*, *5J*, *7B*, *7H*, *8B*, *8D*, *9B*, *11B*)

1852 JOULE, J. P., & W. THOMSON (later Lord KELVIN): "On the thermal effects experienced by air in rushing through small apertures", *Philosophical Magazine* (4) **4**, 481–492 = pp. 333–345 of W. THOMSON's *Mathematical and Physical Papers*, Volume 1, 1882 = pp. 216–230 of JOULE's *Scientific Papers*, Volume 2, 1887.
(*9D* App.)

RANKINE, W. J. McQ.: "On the centrifugal theory of elasticity and its connection with the theory of heat", *Transactions of the Royal Society of Edinburgh* **20** (1850/1853), Part 3, 425–440 = pp. 49–66 of RANKINE's *Miscellaneous Scientific Papers*, 1881.
(*3F*, *8G*, *9E*)

THOMSON, W. (later Lord KELVIN):
1. "On the dynamical theory of heat, Part IV. On a method of discovering experimentally the relation between the mechanical work spent, and the heat produced by the compression of a gaseous fluid", *Transactions of the Royal Society of Edinburgh* **20** (1850/1853), Part 3, 475–482 = *Philosophical Magazine* (4) **4**, 424–434 = pp. 210–222 of W. THOMSON's *Mathematical and Physical Papers*, Volume 1, 1882.
Transl., "Mémoire sur une méthode de déterminer par l'expérience la relation qui existe entre le travail dépensé et la chaleur produite dans la compression d'un gaz", §II of W. THOMSON, "Deux mémoires sur la théorie dynamique de la chaleur", *Journal de Mathématiques Pures et Appliquées* **12**, 241–252.
Transl. W. BLOCK, "Über eine Methode zur experimentellen Feststellung der Beziehung zwischen der aufgewendeten mechanischen Arbeit und der bei der Kompression einer Flüssigkeit im Gaszustand entstehenden Wärme", pp. 43–56 of *Über die dynamische*

Theorie der Wärme, ed. and annot. W. Block, Ostwalds Klassiker No. 193, Leipzig etc., Engelmann, 1914.
(*2C, 7D, 7H, 7I, 9D*)
2. "On a universal tendency in nature to the dissipation of mechanical energy", *Transactions of the Royal Society of Edinburgh* **20** (1850/1853), Part 3, 139–142 = *Philosophical Magazine* (4) **4**, 304–306 = pp. 511–514 of W. Thomson's *Mathematical and Physical Papers*, Volume 1, 1882 = pp. 194–197 of *The Second Law of Thermodynamics*, ed. J. Kestin, Stroudsburg (Pa.), Dowden, Hutchinson & Ross, 1976.
(*11A*)
3. "Synthetical investigation of the duty of a perfect thermodynamic engine founded on the expansions and condensations of a fluid, for which the gaseous laws hold and the ratio (k) of the specific heat under constant pressure to the specific heat in constant volume is constant; and modification of the result by the assumption of Mayer's hypothesis", *Philosophical Transactions of the Royal Society* (London) **142**, 78–80 = (with a postscript) *Cambridge and Dublin Mathematical Journal* **8** (1853), 250–253 = pp. 326–329 of W. Thomson's *Mathematical and Physical Papers*, Volume 1, 1882. The extract from the letter of Kelvin to Joule which appears in a footnote on pp. 327–328 of the last publication appeared originally on pp. 66–67 of *Philosophical Transactions of the Royal Society* (London) **142**, appended to a paper by Joule.
Transl. W. Block, "Synthetische Untersuchung der Leistung einer vollkommenen thermodynamischen Maschine, die auf der Verdünnung und Verdichtung einer Flüssigkeit beruht, für welche die Gasgesetze gelten, und bei der das Verhältnis (k) der spezifischen Wärme unter konstantem Druck zu der spezifischen Wärme unter konstantem Volumen konstant ist, und Modifikation des Ergebnisses durch die Annahme von *Mayers* Hypothese", pp. 175–179 of W. Thomson, *Über die dynamische Theorie der Wärme*, ed. and annot. W. Block, Ostwalds Klassiker No. 193, Leipzig etc., Engelmann, 1914.
(*9B, 9D*)

1853 Clausius, R.: "Ueber einige Stellen der Schrift von Helmholtz 'über die Erhaltung der Kraft'", *Annalen der Physik und Chemie* **165** = (2) **89** = (3) **29**, 568–579.
(*7F*)

Rankine, W. J. McQ.:
1. "On the general law of the transformation of energy", *Proceedings of the Philosophical Society of Glasgow* **3**, 276–280 = *Philosophical Magazine* (4) **5**, 106–117 = pp. 203–208 of Rankine's *Miscellaneous Scientific Papers*, 1881.
(*7B, 9E*)

2. "On the absolute zero of the perfect gas thermometer", *Transactions of the Royal Society of Edinburgh* **20** (1850/1853), Part 4, 561–563 = pp. 307–309 of RANKINE's *Miscellaneous Scientific Papers*, 1881.
(*9D* App.)

3. "On the mechanical action of heat", *Transactions of the Royal Society of Edinburgh* **20** (1850/1853), Part 4, 565–589 = pp. 310–336 of RANKINE's *Miscellaneous Scientific Papers*, 1881.
(*2C, 8G, 9A, 9E*)

REECH, M. F.: "Théorie générale des effets dynamiques de la chaleur", *Journal de Mathématiques Pures et Appliquées* **18**, 357–568.
[This memoir seems to have been published also as a separate work in 1854.]
(*2C, 9C, 10A–10D*)

REGNAULT, V.: "Recherches sur les chaleurs spécifiques des fluides élastiques", *Comptes Rendus Hebdomadaires des Séances de l'Académie des Sciences* (Paris) **36**, 676–687.
Transl., "Researches upon the specific heat of elastic fluids", *Philosophical Magazine* (4) **5**, 473–483.
Transl., "Untersuchungen über die spezifischen Wärmen der elastischen Flüssigkeiten", *Annalen der Physik und Chemie* **165** = (2) **89** = (3) **29**, 335–348.
(*8C, 9D* App.)

THOMSON, W. (later Lord KELVIN):

1. "On the dynamical theory of heat, Part V. On the quantities of mechanical energy contained in a fluid in different states, as to temperature and density", *Transactions of the Royal Society of Edinburgh* **20** (1850/1853), 475–482 = *Philosophical Magazine* (4) **9** (1855), 523–531 = (with annotations), pp. 222–232 of W. THOMSON's *Mathematical and Physical Papers*, Volume 1, 1882.
Transl. W. BLOCK, "Über die Mengen mechanischer Energie in einer Flüssigkeit in verschiedenen Zuständen von Temperatur und Dichte", pp. 56–67 of W. THOMSON, *Über die dynamische Theorie der Wärme*, ed. and annot. W. BLOCK, Ostwalds Klassiker No. 193, Leipzig etc., Engelmann, 1914.
(*5O, 7H, 8D, 9D*)

2. "On the restoration of mechanical energy from an unequally heated space", *Philosophical Magazine* (4) **5**, 102–105 = pp. 554–558 of W. THOMSON's *Mathematical and Physical Papers*, Volume 1, 1882.
(*7H, 11A, 11B*)

THOMSON, W. (later Lord KELVIN), & J. P. JOULE: "On the thermal effects of fluids in motion", *Philosophical Transactions of the Royal Society* (London) **143**, 357–365 = 346–356 of W. THOMSON's *Mathematical and Physical Papers*, Volume 1, 1882 = pp. 231–245 of JOULE's *Scientific Papers*, Volume 2, 1887.

Transl., "Ueber die Wärmewirkung bewegter Flüssigkeiten", *Annalen der Physik und Chemie* **173** = (2) **97** = (4) **7** (1856), 576–589.
(*9D* App.)

1854 CLAUSIUS, R.: "Über eine veränderte Form des zweiten Hauptsatzes der mechanischen Wärmetheorie", *Annalen der Physik und Chemie* **169** = (2) **93** = (4) **3**, 481–506 = (annotated) pp. 127–154 of CLAUSIUS' *Abhandlungen über die mechanische Wärmetheorie*, Volume 1, 1864.
Transl., "Sur une forme nouvelle du second théorème principal de la théorie mécanique de la chaleur", *Journal de Mathématiques Pures et Appliquées* **20** (1855), 63–86.
Transl., "On a modified form of the second fundamental theorem in the mechanical theory of heat", *Philosophical Magazine* (4) **12** (1856), 81–98.
Transl. of the annotated version in *Abhandlungen*, same title as preceding, pp. 111–135 of CLAUSIUS' *The Mechanical Theory of Heat*, ed. T. A. HIRST, London, Van Voorst, 1867.
(*8B, 11C–11G*)

HELMHOLTZ, H. v.: "Erwiederung auf die Bemerkungen von Herrn Clausius", *Annalen der Physik und Chemie* **167** = (2) **91** = (4) **1**, 241–260 = pp. 76–93 of HELMHOLTZ's *Wissenschaftliche Abhandlungen*, Volume 1, 1881.
(*7F*)

JOULE, J. P., & W. THOMSON (later Lord KELVIN): "On the thermal effects of fluids in motion, Part II", *Philosophical Transactions of the Royal Society* (London) **144**, 321–364 = (with pp. 357–361½ and 362½–364 excised) pp. 357–400 of W. THOMSON's *Mathematical and Physical Papers*, Volume 1, 1882 = pp. 247–299 of JOULE's *Scientific Papers*, Volume 2, 1887.
(*7H, 9B, 9D* App., *9F, 11B*)

RANKINE, W. J. McQ.: "On the geometrical representation of the expansive action of heat, and the theory of thermo-dynamic engines", *Philosophical Transactions of the Royal Society* (London) **144**, 115–175 = pp. 339–409 of RANKINE's *Miscellaneous Scientific Papers*, 1881.
(*9E*)

1855 HELMHOLTZ, H. v.: "Theorie der Wärme", *Fortschritte der Physik* **6** and **7** (1850/1), 561–598.
(*5J, 7D–7F, 8F, 10* App.)

THOMSON, W. (later Lord KELVIN): "On the thermo-elastic and thermo-magnetic properties of matter", *Quarterly Journal of Pure and Applied Mathematics* **1** (1855/1857), 55–77 = (with notes, corrections, and additions) *Philosophical Magazine* (5) **5** (1878), 4–27 = (with paragraphs numbered and with some corrections, additions,

and cross-references) Part VII of "On the dynamical theory of heat", pp. 291–313 of W. THOMSON's *Mathematical and Physical Papers*, Volume 1, 1882.
Transl. W. BLOCK, §182–206 of "Über die thermoelastischen, thermomagnetischen und pyroelektrischen Eigenschaften der Materie", pp. 135–161 of W. THOMSON, *Über die dynamische Theorie der Wärme*, ed. and annot. W. BLOCK, Ostwalds Klassiker No. 193, Leipzig etc., Engelmann, 1914.
(*6B*)

1856 CLAUSIUS, R.:
1. "On the discovery of the true form of Carnot's function", *Philosophical Magazine* (4) **11**, 388–390.
(*7F, 9B*)
2. "Ueber die Anwendung der mechanischen Wärmetheorie auf die Dampfmaschine", *Annalen der Physik und Chemie* **173** = (2) **97** = (4) **7**, 441–476, 513–558 = (annotated and augmented) pp. 155–241 of CLAUSIUS' *Abhandlungen über die mechanische Wärmetheorie*, Volume 1, 1864.
Transl., "On the application of the mechanical theory of heat to the steam-engine", *Philosophical Magazine* (4) **12**, 241–265, 338–354, 426–443 = *American Journal of Science* (2) **22**, 180–203, 364–374, and **23** (1857), 25–46.
Transl. of the annotated and augmented version in *Abhandlungen*, and same title, pp. 136–214 of CLAUSIUS' *The Mechanical Theory of Heat*, ed. T. A. HIRST, London, Van Voorst, 1867.
(*5J*)
3. "Notiz über den Zusammenhang zwischen dem Satze von der Aequivalenz von Wärme und Arbeit und dem Verhalten der permanenten Gase", *Annalen der Physik und Chemie* **174** = (2) **98** = (4) **8**, 173–178.
(*9D*)

HELMHOLTZ, H. v.:
1. Review of REECH [1853], *Fortschritte der Physik* **9** (1853), 404–405.
(*7F, 10* App.)
2. Review of RANKINE [1853, *1*], *Fortschritte der Physik* **9** (1853), 406–409.
(*7F, 9E*)

HOPPE, R.: "Ueber die Wärme als Aequivalent der Arbeit", *Annalen der Physik und Chemie* **173** = (2) **97** = 4 (**7**), 30–34.
(*9D*)

REECH, F.: "Recapitulation très-succincte des recherches algébriques sur la théorie des effets mécaniques de la chaleur par différents auteurs", *Journal de Mathématiques Pures et Appliquées* (2) **1**, 58–75.
(*1, 10* App., *11I*)

THOMSON, W. (later Lord KELVIN): "On the discovery of the true form of Carnot's function", *Philosophical Magazine* (4) **11**, 447–448 = pp. 45–46 of Volume 5 of W. THOMSON's *Mathematical and Physical Papers*, 1911.
(*9B*)

1858 REECH, F.: "Note sur un mémoire intitulé: *Théorie des propriétés calorifiques et expansives des fluides élastiques*", *Comptes Rendus Hebdomadaires des Séances de l'Académie des Sciences* (Paris) **46**, 84–89.
(*10* App., *11I*)

1859 HELMHOLTZ, H. v.:
1. Review of REECH [1856], *Fortschritte der Physik* **12** (1856), 345.
(*7F*, *10* App.)
2. Review of CLAUSIUS [1856, *2*], *Fortschritte der Physik* **12** (1856), 355–356.
(*5J*, *7F*)

RANKINE, W. J. McQ.: *A Manual of the Steam Engine and other Prime Movers*, London and Glasgow, Griffin.
(*2C*, *3C*, *8G*, *9E*, *11C*)

1860 JOCHMANN, E.: "Beiträge zur Theorie der Gase", *Zeitschrift für Mathematik und Physik* **5**, 24–39, 96–131.
(*9D* App.)

ZEUNER, G.: *Grundzüge der mechanischen Wärmetheorie*, Freiberg, J. G. Engelhardt.
(*10* App.)

1862 CLAUSIUS, R.: "Ueber die Anwendung des Satzes von der Aequivalenz der Verwandlungen auf die innere Arbeit", *Vierteljahrsschrift der naturforschenden Gesellschaft* (Zürich) **7**, 48–95 = *Annalen der Physik und Chemie* **192** = (2) **116** = (4) **26**, 73–112 = (annotated and augmented) pp. 242–296 of CLAUSIUS' *Abhandlungen uber die mechanische Wärmetheorie*, Volume 1, 1864.
Transl., "On the application of the theorem of the equivalence of transformations to the internal work of a mass of matter", *Philosophical Magazine* (4) **24**, 81–97, 201–213 = pp. 133–161 of *The Second Law of Thermodynamics*, ed. J. KESTIN, Stroudsburg (Pa.), Dowden, Hutchinson and Ross, 1976.
Transl., "Sur l'application du théorème de l'équivalence des transformations au travail interieur", *Journal de Mathématiques Pures et Appliquées* (2) **7**, 209–245.
Transl. of the annotated and augmented version in *Abhandlungen*, "On the application of the theorem of the equivalence of transformations to interior work", pp. 215–266 of CLAUSIUS' *The Mechanical Theory of Heat*, ed. T. A. HIRST, London, Van Voorst, 1867.
(*8G*, *8H*, *11F*)

REGNAULT, V.: "Mémoire sur la chaleur spécifique des fluides élastiques" (read 18 April 1853, printed at the end of 1855 but not then distributed to the public), *Mémoires de l'Académie des Sciences de l'Institut Impérial de France* **26**, iii–x, 1–333 = pp. iii–x, 1–333 of REGNAULT's *Relations des Expériences . . . pour déterminer les principales lois et les données numériques qui entrent dans le calcul des machines à vapeur*, Paris, Didot, Tome 2.
(*8C*)

1863 CLAUSIUS, R.: "Ueber einen Grundsatz der mechanischen Wärmetheorie", *Annalen der Physik und Chemie* **196** = (2) **120** = (4) **30**, 426–452 = (annotated) pp. 297–321 of CLAUSIUS' *Abhandlungen über die mechanische Wärmetheorie*, Volume 1, 1864.
Transl. of the annotated version, "On an axiom in the mechanical theory of heat", pp. 267–289 of CLAUSIUS' *The Mechanical Theory of Heat*, ed. T. A. HIRST, London, Van Voorst, 1867.
(*8H*, *10* App.)

REECH, F.:
1. "Sur les propriétés calorifiques et expansives des fluids élastiques", *Comptes Rendus Hebdomadaires des Séances de l'Académie des Sciences* (Paris) **56**, 1240–1242.
(*10* App.)
2. "Note sur les propriétés calorifiques et expansives des gaz", *Comptes Rendus Hebdomadaires des Séances de l'Académie des Sciences* (Paris) **57**, 505–509.
(*9C*, *10* App.)
3. "Réponse de M. Reech à M. l'Abbé Moigno", *Les Mondes* **3**, 713–717 (10 décembre).
(*10* App.)

STEFAN, J.: "Über die Fortpflanzung der Wärme", *Sitzungsberichte der Akademie der Wissenschaften, Wien*, $\mathbf{47}_2$, 326–344.
(*4F*)

1865 CLAUSIUS, R.: "Ueber verschiedene für die Anwendung bequeme Formen der Hauptgleichungen der mechanischen Wärmetheorie", *Vierteljahrsschrift der naturforschenden Gesellschaft* (Zürich) **10**, 1–59 = *Annalen der Physik und Chemie* **201** = (2) **125** = (5) **5**, 353–400 = (annotated and augmented) pp. 1–56 of CLAUSIUS' *Abhandlungen über die mechanische Wärmetheorie*, Volume 2, 1867.
Transl. H.-F. BESSARD, "Sur diverses formes facilement applicables, qu'on peut donner aux équations fondamentales de la théorie mécanique de la chaleur", *Journal de Mathématiques Pures et Appliquées* (2) **10**, 361–400.
Transl. R. B. LINDSAY, "On different forms of the fundamental equations of the mechanical theory of heat and their convenience for application", pp. 162–193 of *The Second Law of Thermodynamics*, ed. J. KESTIN, Stroudsburg (Pa.), Dowden, Hutchinson and Ross, 1976.

Transl. of the augmented version in *Abhandlungen*, "On several convenient forms of the fundamental equations of the mechanical theory of heat", pp. 327–376 of CLAUSIUS' *The Mechanical Theory of Heat*, ed. T. A. HIRST, London, Van Voorst, 1867.
(*9E*)

1868 REECH, F.: "Théorie des machines motrices et des effets mécaniques de la chaleur" (Leçons . . . recueillies et rédigées par E. LECLERT), *Revue Maritime et Coloniale* **24**, 790–842; **25** (1869), 355–377; **26** (1869), 318–357, 690–727.
(*2C, 3F, 9C, 10D, 10* App.)

1869 MASSIEU, F.:
1. "Sur les fonctions caractéristiques des divers fluides", *Comptes Rendus Hebdomadaires des Séances de l'Académie des Sciences* (Paris) **69**, 858–862.
(*10* App.)
2. "Addition au précédent mémoire sur les fonctions caractéristiques", *Comptes Rendus Hebdomadaires des Séances de l'Académie des Sciences* (Paris) **69**, 1057–1061.
(*10* App.)

REECH, F.: "Equations fondamentales dans la théorie mécanique de la chaleur", *Comptes Rendus Hebdomadaires des Séances de l'Académie des Sciences* (Paris) **69**, 913–916.
(*10* App.)

1871 MAXWELL, J. C.: *Theory of Heat*, London, Longmans Green and Co.
(*2C, 5M, 9E, 9G*)

1873 GIBBS, J. W.:
1. "Graphical methods in the thermodynamics of fluids", *Transactions of the Connecticut Academy of Arts and Sciences* **2**, 309–342 = pp. 1–32 of GIBBS' *Collected Works*, Volume 1, 1906.
(*5D, 9E*)
2. "A method of geometrical representation of the thermodynamic properties of substances by means of surfaces", *Transactions of the Connecticut Academy of Arts and Sciences* **2**, 382–404 = pp. 33–54 of GIBBS' *Collected Works*, Volume 1, 1906.
(*9E, 10* App.)

1875 GIBBS, J. W.: "On the equilibrium of heterogeneous substances", *Transactions of the Connecticut Academy of Arts and Sciences* **3** (1875/1878), 108–248, 343–524 = pp. 55–353 of GIBBS' *Collected Works*, Volume 1, 1906.
(*10* App.)

1878 MAXWELL, J. C.: "Tait's 'Thermodynamics'", *Nature* **17** (1877/1878), 257–259, 278–280 = pp. 660–671 of MAXWELL's *Scientific Papers*, Volume 2, 1890.

THOMSON, W. (later Lord KELVIN): "Heat", *Encyclopaedia Brittanica*, 9th ed. = pp. 31–66 of *Elasticity and Heat*, Edinburgh, A. & C. Black,

1880 = (without Mathematical Appendix, with more tables) pp. 113–235 of Volume 3 of W. THOMSON's *Mathematical and Physical Papers*, 1890.
(*2C*, *11H*).

1880 ROWLAND, H. A.: "On the mechanical equivalent of heat ...", *Proceedings of the American Academy of Arts and Sciences* **15** = (n.s.) **7**, 75–200. Reprinted as a separate volume, Cambridge University Press.
(*7I*)

1885 NEUMANN, F.: *Vorlesungen über die Theorie der Elasticität*, Leipzig, Teubner.
(*6B*)

1891 MAXWELL, J. C.: 10th ed. of MAXWELL [1871], with corrections and additions by Lord RAYLEIGH, London, Longmans Green and Co.
(*2C*, *9E*, *9G*)

1896 MACH, E.: *Die Principien der Wärmelehre, Historisch-kritisch entwickelt*, Leipzig, Barth. Second edition, 1900.
(*4A*, *11H*)

1941 BRIDGMAN, P. W.: *The Nature of Thermodynamics*, Harvard University Press. Reprinted "with no essential change", Harper Torchbooks, 1961. (I have used only the latter edition.)
(*5A*, *5H*, *7A*, *7B*, *9B*, *11K*)

1950 [NEUMANN, F.]: "Ein Kapitel aus der Vorlesung von Franz Neumann über mechanische Wärmetheorie, Königsberg 1854/55, ausgearbeitet von CARL NEUMANN", ed. E. R. NEUMANN, *Abhandlungen der Bayerischen Akademie der Wissenschaften* (2) **59**, 3–27.
(*10A*)

1953 BRIDGMAN, P. W.: "Reflections on thermodynamics", *Proceedings of the American Academy of Arts & Sciences* **82**, 301–309 = pp. 226–236 of the 1961 edition of BRIDGMAN [1941].
(*1*)

1960 TRUESDELL, C., & R. TOUPIN: "The classical field theories", pp. 226–793 of *Encyclopedia of Physics*, Volume 3, Part 1, ed. S. FLÜGGE, Berlin *etc.*, Springer.
(*2B*, *4F*, *10A*, *10D*)

1969 TRUESDELL, C.: *Rational Thermodynamics, A Course of Lectures on Selected Topics*, New York *etc.*, McGraw Hill.
(*4F*)

1973 TRUESDELL, C.:

1. "Theoria de effectibus mechanicis caloris pridem ab ill^{mo} Sadi Carnoto verbis physicis promulgata nunc primum mathematice enucleata", *Atti della Accademia delle Scienze dell'Istituto di Bologna*, Classe di Scienze Fisiche (12) **10**, 29–41.
(*3F*, *7H*)

2. "The efficiency of a homogeneous heat engine", *Journal of*

Mathematical and Physical Sciences (Madras) **7**, 349–371; **9** (1975), 193–194.
(*10* Postscr.)

1976 TRUESDELL, C.: "Improved estimates of the efficiencies of irreversible heat engines", *Annali di Matematica Pura ed Applicata* (4) **108**, 305–323.
(*10* Postscr.)

1977 TRUESDELL, C., & S. BHARATHA: *The Concepts and Logic of Classical Thermodynamics as a Theory of Heat Engines, Rigorously Constructed upon the Foundation Laid by S. Carnot and F. Reech*, New York *etc.*, Springer-Verlag.
(*1, 4D, 5C, 5L, 5M, 5O, 7A, 9B, 10A, 10D, 10* App., *11H*)

1979 TRUESDELL, C.: "Absolute temperatures as a consequence of Carnot's General Axiom", *Archive for History of Exact Sciences* **20**, 357–380.
(*9F, 11H*)

Index of Persons Mentioned

Translators and editors are entered only in respect to remarks regarding their work.

Index of Matters Treated

This index refers mainly to first instances of a term or concept and to some of the major developments in the order in which they appear. A conceptual and chronological outline of the book is provided by the Table of Contents, which this index does not duplicate. To find analyses of the works of persons, the reader should consult the List of Sources and the Index of Persons Mentioned.